Particles and Nuclei
An Introduction to the Physical Concepts

Springer
Berlin
Heidelberg
New York
Barcelona
Hong Kong
London
Milan
Paris
Singapore
Tokyo

Bogdan Povh   Klaus Rith
Christoph Scholz   Frank Zetsche

Particles
and Nuclei

An Introduction
to the Physical Concepts

Translated by Martin Lavelle

Second, Revised and Enlarged Edition
With 144 Figures, 12 Tables, and 58 Problems and Solutions

 Springer

Professor Dr. Bogdan Povh
Max-Planck-Institut für Kernphysik
Postfach 10 39 80, D-69029 Heidelberg
Germany

Professor Dr. Klaus Rith
Physikalisches Institut der Universität
Erlangen-Nürnberg
Erwin-Rommel-Strasse 1, D-91058 Erlangen
Germany

Dr. Christoph Scholz
SAP AG
Postfach 1461, D-69185 Walldorf
Germany

Dr. Frank Zetsche
Universität Hamburg und
Deutsches Elektronen-Synchrotron
Notkestrasse 85, D-22603 Hamburg
Germany

Translator:
Dr. Martin Lavelle
Institut de Física d'Altes Energies
Facultat de Ciències
Universitat Autònoma de Barcelona
E-08193 Bellaterra (Barcelona)
Spain

Title of the original German edition:
B. Povh, K. Rith, C. Scholz, F. Zetsche: *Teilchen und Kerne.*
Eine Einführung in die physikalischen Konzepte.
(Fünfte Auflage)
© Springer-Verlag Berlin Heidelberg 1993, 1994, 1995, 1997, 1999

ISBN 3-540-66115-8 2nd Edition Springer-Verlag Berlin Heidelberg New York
ISBN 3-540-59439-6 1st Edition Springer-Verlag Berlin Heidelberg New York

Library of Congress Cataloging-in-Publication Data applied for.

Die Deutsche Bibliothek - CIP-Einheitsaufnahme
Particles and nuclei : an introduction to the physical concepts ; with 58 problems and solutions /
Bogdan Povh ... Transl. by Martin Lavelle. - 2., rev. and enl. ed. - Berlin ; Heidelberg ; New York ;
Barcelona ; Hong Kong ; London ; Milan ; Paris ; Singapore ; Tokyo : Springer, 1999
Dt. Ausg. u.d.T.: Teilchen und Kerne
ISBN 3-540-66115-8

Typesetting: Data conversion by Jürgen Sawinski, Heidelberg
Cover design: Design Concept, Emil Smejkal, Heidelberg
Computer to plate: Mercedes Druck, Berlin

SPIN: 10728252   55/3144/di - 5 4 3 2 1 0 – Printed on acid-free paper

Preface

The second English edition has been updated from the fifth edition of the original German text. The principal addition is a chapter on nuclear thermodynamics. We consider in this chapter the behaviour of nuclear matter at high temperature, how it may be studied in the laboratory, via heavy ion experiments and how it was of great importance in the initial stages of the universe. Such a phase of matter may be described and interpreted using the tools of thermodynamics. In this way a connection between particle and nuclear physics and the currently exciting research areas of cosmology and astrophysics may be constructed.

We would like to thank Martin Lavelle (Plymouth) for the translation of the new chapter and for revising the old text and Jürgen Sawinski (Heidelberg) for the excellent work he has done in reformatting the book.

Heidelberg, May 1999   *The Authors*

Preface to the First Edition

The aim of PARTICLES AND NUCLEI is to give a unified description of nuclear and particle physics because the experiments which have uncovered the substructure of atomic nuclei and nucleons are conceptually similar. With the progress of experimental and theoretical methods, atoms, nuclei, nucleons, and finally quarks have been analysed during the course of this century. The intuitive assumption that our world is composed of a few constituents — an idea which seems attractive, but could not be taken for granted — appears to be confirmed. Moreover, the interactions between these constituents of matter can be formulated elegantly, and are well understood conceptionally, within the so-called "standard model".

Once we have arrived at this underlying theory we are immediately faced with the question of how the complex structures around us are produced by it. On the way from elementary particles to nucleons and nuclei we learn that the "fundamental" laws of the interaction between elementary particles are less and less recognizable in composite systems because many-body interactions cause greater and greater complexity for larger systems.

This book is therefore divided into two parts. In the first part we deal with the reduction of matter in all its complication to a few elementary constituents and interactions, while the second part is devoted to the composition of hadrons and nuclei from their constituents.

We put special emphasis on the description of the experimental concepts but we mostly refrain from explaining technical details. The appendix contains a short description of the principles of accelerators and detectors. The exercises predominantly aim at giving the students a feeling for the sizes of the phenomena of nuclear and particle physics.

Wherever possible, we refer to the similarities between atoms, nuclei, and hadrons, because applying analogies has not only turned out to be a very effective research tool but is also very helpful for understanding the character of the underlying physics.

We have aimed at a concise description but have taken care that all the fundamental concepts are clearly described. Regarding our selection of topics, we were guided by pedagogical considerations. This is why we describe experiments which — from today's point of view — can be interpreted in a straightforward way. Many historically significant experiments, whose results can nowadays be much more simply obtained, were deliberately omitted.

PARTICLES AND NUCLEI (TEILCHEN UND KERNE) is based on lectures on nuclear and particle physics given at the University of Heidelberg to students in their 6th semester and conveys the fundamental knowledge in this area, which is required of a student majoring in physics. On traditional grounds these lectures, and therefore this book, strongly emphasise the physical concepts.

We are particularly grateful to J. Hüfner (Heidelberg) and M. Rosina (Ljubljana) for their valuable contributions to the nuclear physics part of the book. We would like to thank D. Dubbers (Heidelberg), A. Fäßler (Tübingen), G. Garvey (Los Alamos), H. Koch (Bochum), K. Königsmann (Freiburg), U. Lynen (GSI Darmstadt), G. Mairle (Mannheim), O. Nachtmann (Heidelberg), H. J. Pirner (Heidelberg), B. Stech (Heidelberg), and Th. Walcher (Mainz) for their critical reading and helpful comments on some sections. Many students who attended our lecture in the 1991 and 1992 summer semesters helped us through their criticism to correct mistakes and improve unclear passages. We owe special thanks to M. Beck, Ch. Büscher, S. Fabian, Th. Haller, A. Laser, A. Mücklich, W. Wander, and E. Wittmann.

M. Lavelle (Barcelona) has translated the major part of the book and put it in the present linguistic form. We much appreciated his close collaboration with us. The English translation of this book was started by H. Hahn and M. Moinester (Tel Aviv) whom we greatly thank.

Numerous figures from the German text have been adapted for the English edition by J. Bockholt, V. Träumer, and G. Vogt of the Max-Planck-Institut für Kernphysik in Heidelberg.

We would like to extend our thanks to Springer-Verlag, in particular W. Beiglböck for his support and advice during the preparation of the German and, later on, the English editions of this book.

Heidelberg, May 1995

Bogdan Povh
Klaus Rith
Christoph Scholz
Frank Zetsche

Table of Contents

1. Hors d'œuvre

1.1 Fundamental Constituents of Matter

In their search for the fundamental building blocks of matter, physicists have found smaller and smaller constituents which in their turn have proven to themselves be composite systems. By the end of the 19th century, it was known that all matter is composed of atoms. However, the existence of close to 100 elements showing periodically recurring properties was a clear indication that atoms themselves have an internal structure, and are not indivisible.

The modern concept of the atom emerged at the beginning of the 20th century, in particular as a result of Rutherford's experiments. An atom is composed of a dense nucleus surrounded by an electron cloud. The nucleus itself can be decomposed into smaller particles. After the discovery of the neutron in 1932, there was no longer any doubt that the building blocks of nuclei are protons and neutrons (collectively called nucleons). The electron, neutron and proton were later joined by a fourth particle, the neutrino, which was postulated in 1930 in order to reconcile the description of β-decay with the fundamental laws of conservation of energy, momentum and angular momentum.

Thus, by the mid-thirties, these four particles could describe all the then known phenomena of atomic and nuclear physics. Today, these particles are still considered to be the main constituents of matter. But this simple, closed picture turned out in fact to be incapable of describing other phenomena.

Experiments at particle accelerators in the fifties and sixties showed that protons and neutrons are merely representatives of a large family of particles now called *hadrons*. More than 100 hadrons, sometimes called the "hadronic zoo", have thus far been detected. These hadrons, like atoms, can be classified in groups with similar properties. It was therefore assumed that they cannot be understood as fundamental constituents of matter. In the late sixties, the quark model established order in the hadronic zoo. All known hadrons could be described as combinations of two or three quarks.

Figure 1.1 shows different scales in the hierarchy of the structure of matter. As we probe the atom with increasing magnification, smaller and smaller structures become visible: the nucleus, the nucleons, and finally the quarks.

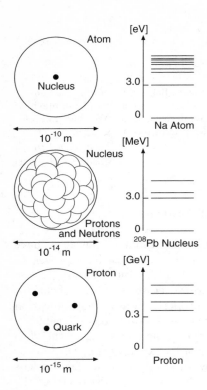

Fig. 1.1. Length scales and structural hierarchy in atomic structure. To the right, typical excitation energies and spectra are shown. Smaller bound systems possess larger excitation energies.

Leptons and quarks. The two fundamental types of building blocks are the *leptons*, which include the electron and the neutrino, and the *quarks*. In scattering experiments, these were found to be smaller than 10^{-18} m. They are possibly point-like particles. For comparison, protons are as large as $\approx 10^{-15}$ m. Leptons and quarks have spin 1/2, i. e. they are fermions. In contrast to atoms, nuclei and hadrons, no excited states of quarks or leptons have so far been observed. Thus, they appear to be elementary particles.

Today, however, we know of 6 leptons and 6 quarks as well as their antiparticles. These can be grouped into so-called "generations" or "families", according to certain characteristics. Thus, the number of leptons and quarks is relatively large; furthermore, their properties recur in each generation. Some physicists believe these two facts are a hint that leptons and quarks are not elementary building blocks of matter. Only experiment will teach us the truth.

1.2 Fundamental Interactions

Together with our changing conception of elementary particles, our understanding of the basic forces of nature and so of the fundamental interactions between elementary particles has evolved. Around the year 1800, four forces were considered to be basic: *gravitation, electricity, magnetism* and the barely comprehended forces between atoms and molecules. By the end of the 19th century, electricity and magnetism were understood to be manifestations of the same force: *electromagnetism*. Later it was shown that atoms have a structure and are composed of a positively charged nucleus and an electron cloud; the whole held together by the electromagnetic interaction. Overall, atoms are electrically neutral. At short distances, however, the electric fields between atoms do not cancel out completely, and neighbouring atoms and molecules influence each other. The different kinds of "chemical forces" (e. g., the Van-der-Waals force) are thus expressions of the electromagnetic force.

When nuclear physics developed, two new short-ranged forces joined the ranks. These are the *nuclear force*, which acts between nucleons, and the *weak force*, which manifests itself in nuclear β-decay. Today, we know that the nuclear force is not fundamental. In analogy to the forces acting between atoms being effects of the electromagnetic interaction, the nuclear force is a result of the *strong force* binding quarks to form protons and neutrons. These strong and weak forces lead to the corresponding fundamental interactions between the elementary particles.

Intermediate bosons. The four fundamental interactions on which all physical phenomena are based are gravitation, the electromagnetic interaction, the strong interaction and the weak interaction.

Gravitation is important for the existence of stars, galaxies, and planetary systems (and for our daily life), it is of no significance in subatomic physics, being far too weak to noticeably influence the interaction between elementary particles. We mention it only for completeness.

According to today's conceptions, interactions are mediated by the exchange of vector bosons, i.e. particles with spin 1. These are *photons* in electromagnetic interactions, *gluons* in strong interactions and the W^+, W^- and Z^0 bosons in weak interactions. The diagrams on the next page show examples of interactions between two particles by the exchange of vector bosons: In our diagrams we depict leptons and quarks by straight lines, photons by wavy lines, gluons by spirals, and $W^\pm$ and Z^0 bosons by dashed lines.

Each of these three interactions is associated with a charge: electric charge, weak charge and strong charge. The strong charge is also called *colour charge* or *colour* for short. A particle is subject to an interaction if and only if it carries the corresponding charge:

– Leptons and quarks carry weak charge.
– Quarks are electrically charged, so are some of the leptons (e. g., electrons).
– Colour charge is only carried by quarks (not by leptons).

The W and Z bosons, masses $M_W \approx 80$ GeV$/c^2$ and $M_Z \approx 91$ GeV$/c^2$, are very heavy particles. According to the Heisenberg uncertainty principle, they can only be produced as virtual, intermediate particles in scattering

e^- e^-
γ
Photon
Mass=0
e^- e^-

q q
g
Gluon
Mass=0
q q

ν_e e^-
W
W-Boson
Mass $\approx$ 80 GeV/c^2
e^- ν_e

e^- ν_μ
Z^0
Z-Boson
Mass $\approx$ 91 GeV/c^2
e^- ν_μ

processes for extremely short times. Therefore, the weak interaction is of very short range. The rest mass of the photon is zero. Therefore, the range of the electromagnetic interaction is infinite.

The gluons, like the photons, have zero rest mass. Whereas photons, however, have no electrical charge, gluons carry colour charge. Hence they can interact with each other. As we will see, this causes the strong interaction to be also very short ranged.

1.3 Symmetries and Conservation Laws

Symmetries are of great importance in physics. The conservation laws of classical physics (energy, momentum, angular momentum) are a consequence of the fact that the interactions are invariant with respect to their canonically conjugate quantities (time, space, angles). In other words, physical laws are independent of the time, the location and the orientation in space under which they take place.

An additional important property in non-relativistic quantum mechanics is reflection symmetry.[1] Depending on whether the sign of the wave function changes under reflection or not, the system is said to have negative or positive *parity* (P), respectively. For example, the spatial wave function of a bound system with angular momentum $\ell\hbar$ has parity $P = (-1)^\ell$. For those laws of nature with left-right symmetry, i.e., invariant under a reflection in space $\mathcal{P}$, the parity quantum number P of the system is conserved. Conservation of parity leads, e. g., in atomic physics to selection rules for electromagnetic transitions.

The concept of parity has been generalised in relativistic quantum mechanics. One has to ascribe an *intrinsic parity* P to particles and antiparticles.

---

[1] As is well known, reflection around a point is equivalent to reflection in a plane with simultaneous rotation about an axis perpendicular to that plane.

Bosons and antibosons have the same intrinsic parity, fermions and antifermions have opposite parities. An additional important symmetry relates particles and antiparticles. An operator $\mathcal{C}$ is introduced which changes particles into antiparticles and vice versa. Since the charge reverses its sign under this operation, it is called *charge conjugation*. Eigenstates of $\mathcal{C}$ have a quantum number *C-parity* which is conserved whenever the interaction is symmetric with respect to $\mathcal{C}$.

Another symmetry derives from the fact that certain groups ("multiplets") of particles behave practically identically with respect to the strong or the weak interaction. Particles belonging to such a multiplet may be described as different states of the same particle. These states are characterised by a quantum number referred to as strong or weak *isospin*. Conservation laws are also applicable to these quantities.

1.4 Experiments

Experiments in nuclear and elementary particle physics have, with very few exceptions, to be carried out using particle accelerators. The development and construction of accelerators with ever greater energies and beam intensities has made it possible to discover more and more elementary particles. A short description of the most important types of accelerators can be found in the appendix. The experiments can be classified as *scattering* or *spectroscopic* experiments.

Scattering. In scattering experiments, a beam of particles with known energy and momentum is directed toward the object to be studied (the *target*). The beam particles then interact with the object. From the changes in the kinematical quantities caused by this process, we may learn about the properties both of the target and of the interaction.

Consider, as an example, elastic electron scattering which has proven to be a reliable method for measuring radii in nuclear physics. The structure of the target becomes visible via diffraction only when the de Broglie wavelength $\lambda = h/p$ of the electron is comparable to the target's size. The resulting diffraction pattern of the scattered particles yields the size of the nucleus rather precisely.

Figure 1.1 shows the geometrical dimensions of various targets. To determine the size of an atom, X-rays with an energy of $\approx 10^4$ eV suffice. Nuclear radii are measured with electron beams of about 10^8 eV, proton radii with electron beams of some 10^8 to 10^9 eV. Even with today's energies, $9 \cdot 10^{10}$ eV for electrons and 10^{12} eV for protons, there is no sign of a substructure in either quarks or leptons.

Spectroscopy. The term "spectroscopy" is used to describe those experiments which determine the decay products of excited states. In this way, one can study the properties of the excited states as well as the interactions between the constituents.

From Fig. 1.1 we see that the excitation energies of a system increase as its size decreases. To produce these excited states high energy particles are needed. Scattering experiments to determine the size of a system and to produce excited states require similar beam energies.

Detectors. Charged particles interact with gases, liquids, amorphous solids, and crystals. These interactions produce electrical or optical signals in these materials which betray the passage of the particles. Neutral particles are detected indirectly through secondary particles: photons produce free electrons or electron-positron pairs, by the photoelectric or Compton effects, and pair production, respectively. Neutrons and neutrinos produce charged particles through reactions with nuclei.

Particle detectors can be divided into the following categories:

- Scintillators provide fast time information, but have only moderate spatial resolution.
- Gaseous counters covering large areas (wire chambers) provide good spatial resolution, and are used in combination with magnetic fields to measure momentum.
- Semiconductor counters have a very good energy and spatial resolution.
- Cherenkov counters and counters based on transition radiation are used for particle identification.
- Calorimeters measure the total energy at very high energies.

The basic types of counters for the detection of charged particles are compiled in Appendix A.2.

1.5 Units

The common units for length and energy in nuclear and elementary particle physics are the *femtometre* (fm, or *Fermi*) and the *electron volt* (eV). The Fermi is a standard SI-unit, defined as 10^{-15} m, and corresponds approximately to the size of a proton. An electron volt is the energy gained by a particle with charge $1e$ by traversing a potential difference of 1 V:

$$1\,\text{eV} = 1.602 \cdot 10^{-19}\,\text{J}\,. \tag{1.1}$$

For the decimal multiples of this unit, the usual prefixes are employed: keV, MeV, GeV, etc. Usually, one uses units of MeV/c^2 or GeV/c^2 for particle masses, according to the mass-energy equivalence $E = mc^2$.

Length and energy scales are connected in subatomic physics by the uncertainty principle. The Planck constant is especially easily remembered in the form

$$\hbar \cdot c \approx 200 \, \text{MeV} \cdot \text{fm} \,. \tag{1.2}$$

Another quantity which will be used frequently is the coupling constant for electromagnetic interactions. It is defined by:

$$\alpha = \frac{e^2}{4\pi\varepsilon_0 \hbar c} \approx \frac{1}{137} \,. \tag{1.3}$$

For historical reasons, it is also called the *fine structure constant*.

A system of physical quantities which is frequently used in elementary particle physics has identical dimensions for mass, momentum, energy, inverse length and inverse time. In this system, the units may be chosen such that $\hbar = c = 1$. In atomic physics, it is common to define $4\pi\varepsilon_0 = 1$ and therefore $\alpha = e^2$ (Gauss system). In particle physics, $\varepsilon_0 = 1$ and $\alpha = e^2/4\pi$ is more commonly used (Heavyside-Lorentz system). However, we will utilise the SI-system [SY78] used in all other fields of physics and so retain the constants everywhere.

Part I

Analysis:
The Building Blocks of Matter

Mens agitat molem.
Vergil
Aeneid 6, 727

2. Global Properties of Nuclei

The discovery of the electron and of radioactivity marked the beginning of a new era in the investigation of matter. At that time, some signs of the atomic structure of matter were already clearly visible: e. g. the integer stoichiometric proportions of chemistry, the thermodynamics of gases, the periodic system of the elements or Brownian motion. But the existence of atoms was not yet generally accepted. The reason was simple: nobody was able to really picture these building blocks of matter, the atoms. The new discoveries showed for the first time "particles" emerging from matter which had to be interpreted as its constituents.

It now became possible to use the particles produced by radioactive decay to bombard other elements in order to study the constituents of the latter. This experimental ansatz is the basis of modern nuclear and particle physics. Systematic studies of nuclei became possible by the late thirties with the availability of modern particle accelerators. But the fundamental building blocks of atoms – the electron, proton and neutron – were detected beforehand. A pre-condition for these discoveries were important technical developments in vacuum techniques and in particle detection. Before we turn to the global properties of nuclei from a modern viewpoint, we will briefly discuss these historical experiments.

2.1 The Atom and its Constituents

The electron. The first building block of the atom to be identified was the electron. In 1897 Thomson was able to produce electrons as beams of free particles in discharge tubes. By deflecting them in electric and magnetic fields, he could determine their velocity and the ratio of their mass and charge. The results turned out to be independent of the kind of cathode and gas used. He had in other words found a universal constituent of matter. He then measured the charge of the electron independently — using a method that was in 1910 significantly refined by Millikan (the drop method) — this of course also fixed the electron mass.

The atomic nucleus. Subsequently, different models of the atom were discussed, one of them being the model of Thomson. In this model, the electrons, and an equivalent number of positively charged particles are uniformly distributed throughout the atom. The resulting atom is electrically neutral. Rutherford, Geiger and Marsden succeeded in disproving this picture. In their famous experiments, where they scattered α-particles off heavy atoms, they were able to show that the positively charged particles are closely packed together. They reached this conclusion from the angular distribution of the scattered α-particles. The angular distribution showed α-particle scattering at large scattering angles which was incompatible with a homogeneous charge distribution. The explanation of the scattering data was a central Coulomb field caused by a massive, positively charged nucleus. The method of extracting the properties of the scattering potential from the angular distribution of the scattered projectiles is still of great importance in nuclear and particle physics, and we will encounter it repeatedly in the following chapters. These experiments established the existence of the atom as a positively charged, small, massive nucleus with negatively charged electrons orbiting it.

The proton. Rutherford also bombarded light nuclei with α-particles which themselves were identified as ionised helium atoms. In these reactions, he was looking for a conversion of elements, i.e., for a sort of inverse reaction to radioactive α-decay, which itself is a conversion of elements. While bombarding nitrogen with α-particles, he observed positively charged particles with an unusually long range, which must have been ejected from the atom as well. From this he concluded that the nitrogen atom had been destroyed in these reactions, and a light constituent of the nucleus had been ejected. He had already discovered similar long-ranged particles when bombarding hydrogen. From this he concluded that these particles were hydrogen nuclei which, therefore, had to be constituents of nitrogen as well. He had indeed observed the reaction

$$^{14}\mathrm{N} + {}^4\mathrm{He} \rightarrow {}^{17}\mathrm{O} + \mathrm{p},$$

in which the nitrogen nucleus is converted into an oxygen nucleus, by the loss of a proton. The hydrogen nucleus could therefore be regarded as an elementary constituent of atomic nuclei. Rutherford also assumed that it would be possible to disintegrate additional atomic nuclei by using α-particles with higher energies than those available to him. He so paved the way for modern nuclear physics.

The neutron. The neutron was also detected by bombarding nuclei with α-particles. Rutherford's method of visually detecting and counting particles by their scintillation on a zinc sulphide screen is not applicable to neutral particles. The development of ionisation and cloud chambers significantly simplified the detection of charged particles, but did not help here. Neutral particles could only be detected indirectly. Chadwick in 1932 found an appropriate experimental approach. He used the irradiation of beryllium with α-particles from a polonium source, and thereby established the neutron as a fundamental constituent of nuclei. Previously, a "neutral radiation" had been observed in similar experiments, but its origin and identity was not understood. Chadwick arranged for this neutral radiation to collide with hydrogen, helium and nitrogen, and measured the recoil energies of these nuclei in a ionisation chamber. He deduced from the laws of collision that the mass of the neutral radiation particle was similar to that of the proton. Chadwick named this particle the "neutron".

Nuclear force and binding. With these discoveries, the building blocks of the atom had been found. The development of ion sources and mass spectrographs now permitted the investigation of the forces binding the nuclear constituents, i.e., the proton and the neutron. These forces were evidently much stronger than the electromagnetic forces holding the atom together, since atomic nuclei could only be broken up by bombarding them with highly energetic α-particles.

The binding energy of a system gives information about its binding and stability. This energy is the difference between the mass of a system and the sum of the masses of its constituents. It turns out that for nuclei this difference is close to 1 % of the nuclear mass. This phenomenon, historically called the mass defect, was one of the first experimental proofs of the mass-energy relation $E = mc^2$. The mass defect is of fundamental importance in the study of strongly interacting bound systems. We will therefore describe nuclear masses and their systematics in this chapter at some length.

2.2 Nuclides

The atomic number. The atomic number Z gives the number of protons in the nucleus. The charge of the nucleus is, therefore, $Q = Ze$, the elementary charge being $e = 1.6 \cdot 10^{-19}$ C. In a neutral atom, there are Z electrons, which balance the charge of the nucleus, in the electron cloud. The atomic number of a given nucleus determines its chemical properties.

The classical method of determining the charge of the nucleus is the measurement of the characteristic X-rays of the atom to be studied. For this purpose the atom is excited by electrons, protons or synchrotron radiation. Moseley's law says that the energy of the K_α-line is proportional to $(Z - 1)^2$. Nowadays, the detection of these characteristic X-rays is used to identify elements in material analysis.

Atoms are electrically neutral, which shows the equality of the absolute values of the positive charge of the proton and the negative charge of the electron. Experiments measuring the deflection of molecular beams in electric fields yield an upper limit for the difference between the proton and electron charges [Dy73]:

$$|e_p + e_e| \le 10^{-18} e . \tag{2.1}$$

Today cosmological estimates give an even smaller upper limit for any difference between these charges.

The mass number. In addition to the Z protons, N neutrons are found in the nucleus. The mass number A gives the number of nucleons in the nucleus, where $A = Z + N$. Different combinations of Z and N (or Z and A) are called *nuclides*.

- Nuclides with the same mass number A are called *isobars*.
- Nuclides with the same atomic number Z are called *isotopes*.
- Nuclides with the same neutron number N are called *isotones*.

The binding energy B is usually determined from atomic masses [AM93], since they can be measured to a considerably higher precision than nuclear masses. We have:

$$B(Z, A) = \left[ZM(^1\mathrm{H}) + (A - Z)M_\mathrm{n} - M(A, Z) \right] \cdot c^2 . \tag{2.2}$$

Here, $M(^1\mathrm{H}) = M_\mathrm{p} + m_\mathrm{e}$ is the mass of the hydrogen atom (the 13.6 eV binding energy of the H-atom is negligible), M_n is the mass of the neutron and $M(A, Z)$ is the mass of an atom with Z electrons whose nucleus contains A nucleons. The rest masses of these particles are:

$$
\begin{aligned}
M_\mathrm{p} &= 938.272 \ \mathrm{MeV}/c^2 &= 1836.149 \ m_\mathrm{e} \\
M_\mathrm{n} &= 939.566 \ \mathrm{MeV}/c^2 &= 1838.679 \ m_\mathrm{e} \\
m_\mathrm{e} &= 0.511 \ \mathrm{MeV}/c^2.
\end{aligned}
$$

The conversion factor into SI units is $1.783 \cdot 10^{-30}$ kg/(MeV/c^2).

In nuclear physics, nuclides are denoted by $^A\mathrm{X}$, X being the chemical symbol of the element. E.g., the stable carbon isotopes are labelled $^{12}\mathrm{C}$ and $^{13}\mathrm{C}$; while the radioactive carbon isotope frequently used for isotopic dating is labelled $^{14}\mathrm{C}$. Sometimes the notations $^A_Z\mathrm{X}$ or $^A_Z\mathrm{X}_N$ are used, whereby the atomic number Z and possibly the neutron number N are explicitly added.

Determining masses from mass spectroscopy. The binding energy of an atomic nucleus can be calculated if the atomic mass is accurately known. At the start of the 20th century, the method of mass spectrometry was developed for precision determinations of atomic masses (and nucleon binding energies). The deflection of an ion with charge Q in an electric and magnetic field allows the simultaneous measurement of its momentum $p = Mv$ and its kinetic energy $E_\mathrm{kin} = Mv^2/2$. From these, its mass can be determined. This is how most mass spectrometers work.

While the radius of curvature r_E of the ionic path in an electrical sector field is proportional to the energy:

$$r_\mathrm{E} = \frac{M}{Q} \cdot \frac{v^2}{E} , \tag{2.3}$$

in a magnetic field B, the radius of curvature r_M of the ion is proportional to its momentum:

$$r_\mathrm{M} = \frac{M}{Q} \cdot \frac{v}{B} . \tag{2.4}$$

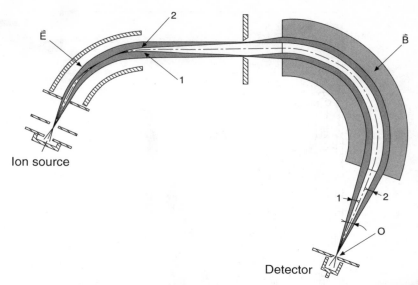

Fig. 2.1. Doubly focusing mass spectrometer [Br64]. The spectrometer focuses ions of a certain specific charge to mass ratio Q/M. For clarity, only the trajectories of particles at the edges of the beam are drawn (*1 and 2*). The electric and magnetic sector fields draw the ions from the ion source into the collector. Ions with a different Q/M ratio are separated from the beam in the magnetic field and do not pass through the slit O.

Figure 2.1 shows a common spectrometer design. After leaving the ion source, the ions are accelerated in an electric field to about 40 keV. In an electric field, they are then separated according to their energy and, in a magnetic field, according to their momentum. By careful design of the magnetic fields, ions with identical Q/M ratios leaving the ion source at various angles are focused at a point at the end of the spectrometer where a detector can be placed.

For technical reasons, it is very convenient to use the ^{12}C nuclide as the reference mass. Carbon and its many compounds are always present in a spectrometer and are well suited for mass calibration. An atomic mass unit u was therefore defined as $1/12$ of the atomic mass of the ^{12}C nuclide. We have:

$$1u = \frac{1}{12}\,M_{^{12}\mathrm{C}} = 931.494\,\mathrm{MeV}/c^2 = 1.660\,54 \cdot 10^{-27}\,\mathrm{kg}\,.$$

Mass spectrometers are still widely used both in research and industry.

Nuclear abundance. A current application of mass spectroscopy in fundamental research is the determination of isotope abundances in the solar system. The relative abundance of the various nuclides as a function of their mass number A is shown in Fig. 2.2. The relative abundances of isotopes in terrestrial, lunar, and meteoritic probes are, with few exceptions, identical

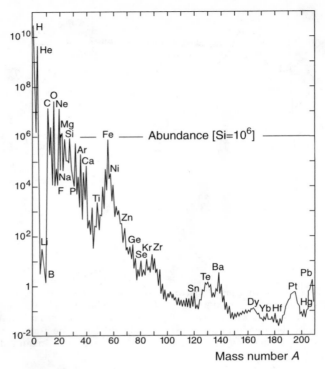

Fig. 2.2. Abundance of the elements in the solar system as a function of their mass number A, normalised to the abundance of silicon ($= 10^6$).

and coincide with the nuclide abundances in cosmic rays from outside the solar system. According to current thinking, the synthesis of the presently existing deuterium and helium from hydrogen fusion mainly took place at the beginning of the universe (minutes after the big bang [Ba80]). Nuclei up to ^{56}Fe, the most stable nucleus, were produced by nuclear fusion in stars. Nuclei heavier than this last were created in the explosion of very heavy stars (supernovae) [Bu57].

Deviations from the universal abundance of isotopes occur locally when nuclides are formed in radioactive decays. Figure 2.3 shows the abundances of various xenon isotopes in a drill core which was found at a depth of 10 km. The isotope distribution strongly deviates from that which is found in the earth's atmosphere. This deviation is a result of the atmospheric xenon being, for the most part, already present when the earth came into existence, while the xenon isotopes from the core come from radioactive decays (spontaneous fission of uranium isotopes).

Determining masses from nuclear reactions. Binding energies may also be determined from systematic studies of nuclear reactions. Consider, as an example, the capture of thermal neutrons ($E_{\mathrm{kin}} \approx 1/40 \, \mathrm{eV}$) by hydrogen,

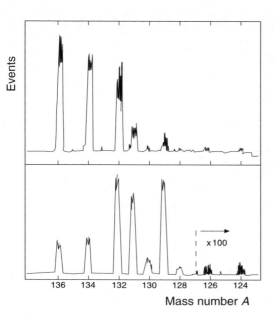

Fig. 2.3. Mass spectrum of xenon isotopes, found in a roughly $2.7 \cdot 10^9$ year old gneiss sample from a drill core produced in the Kola peninsula (*top*) and, for comparison, the spectrum of Xe-isotopes as they occur in the atmosphere (*bottom*). The Xe-isotopes in the gneiss were produced by spontaneous fission of uranium. (*Picture courtesy of Klaus Schäfer, Max-Planck-Institut für Kernphysik.*)

$$\mathrm{n} + {}^1\mathrm{H} \to {}^2\mathrm{H} + \gamma. \tag{2.5}$$

The energy of the emitted photon is directly related to the binding energy B of the deuterium nucleus ${}^2\mathrm{H}$:

$$B = (M_\mathrm{n} + M_{{}^1\mathrm{H}} - M_{{}^2\mathrm{H}}) \cdot c^2 = E_\gamma + \frac{E_\gamma^2}{2 M_{{}^2\mathrm{H}} c^2} = 2.225 \,\mathrm{MeV}, \tag{2.6}$$

where the last term takes into account the recoil energy of the deuteron. As a further example, we consider the reaction

$${}^1\mathrm{H} + {}^6\mathrm{Li} \to {}^3\mathrm{He} + {}^4\mathrm{He}.$$

The energy balance of this reaction is given by

$$E_{{}^1\mathrm{H}} + E_{{}^6\mathrm{Li}} = E_{{}^3\mathrm{He}} + E_{{}^4\mathrm{He}}, \tag{2.7}$$

where the energies E_X each represent the total energy of the nuclide X, i.e., the sum of its rest mass and kinetic energy. If three of these nuclide masses are known, and if all of the kinetic energies have been measured, then the binding energy of the fourth nuclide can be determined.

The measurement of binding energies from nuclear reactions was mainly accomplished using low-energy (van de Graaff, cyclotron, betatron) accelerators. Following two decades of measurements in the fifties and sixties, the systematic errors of both methods, mass spectrometry and the energy balance of nuclear reactions, have been considerably reduced and both now provide

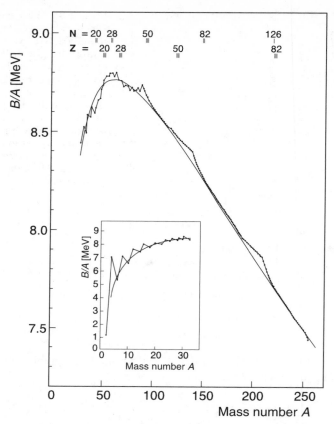

Fig. 2.4. Binding energy per nucleon of nuclei with even mass number A. The solid line corresponds to the Weizsäcker mass formula (2.8). Nuclei with a small number of nucleons display relatively large deviations from the general trend, and should be considered on an individual basis. For heavy nuclei deviations in the form of a somewhat stronger binding per nucleon are also observed for certain proton and neutron numbers. These so-called "magic numbers" will be discussed in Sect. 17.3.

high precision results which are consistent with each other. Figure 2.4 shows schematically the results of the binding energies per nucleon measured for stable nuclei. Nuclear reactions even provide mass determinations for nuclei which are so short-lived that that they cannot be studied by mass spectroscopy.

2.3 Parametrisation of Binding Energies

Apart from the lightest elements, the binding energy per nucleon for most nuclei is about 8-9 MeV. Depending only weakly on the mass number, it can be described with the help of just a few parameters. The parametrisation of nuclear masses as a function of A and Z, which is known as the *Weizsäcker formula* or the *semi-empirical mass formula,* was first introduced in 1935 [We35, Be36]. It allows the calculation of the binding energy according to (2.2). The mass of an atom with Z protons and N neutrons is given by the following phenomenological formula:

$$M(A, Z) = NM_n + ZM_p + Zm_e - a_v A + a_s A^{2/3}$$

$$+ a_c \frac{Z^2}{A^{1/3}} + a_a \frac{(N - Z)^2}{4A} + \frac{\delta}{A^{1/2}} \qquad (2.8)$$

with $N = A - Z$.

The exact values of the parameters a_v, a_s, a_c, a_a and δ depend on the range of masses for which they are optimised. One possible set of parameters is given below:

$$a_v = 15.67 \, \text{MeV}/c^2$$
$$a_s = 17.23 \, \text{MeV}/c^2$$
$$a_c = 0.714 \, \text{MeV}/c^2$$
$$a_a = 93.15 \, \text{MeV}/c^2$$
$$\delta = \begin{cases} -11.2 \, \text{MeV}/c^2 & \text{for even } Z \text{ and } N \text{ (even-even nuclei)} \\ 0 \, \text{MeV}/c^2 & \text{for odd } A \text{ (odd-even nuclei)} \\ +11.2 \, \text{MeV}/c^2 & \text{for odd } Z \text{ and } N \text{ (odd-odd nuclei).} \end{cases}$$

To a great extent the mass of an atom is given by the sum of the masses of its constituents (protons, neutrons and electrons). The nuclear binding responsible for the deviation from this sum is reflected in five additional terms. The physical meaning of these five terms can be understood by recalling that the nuclear radius R and mass number A are connected by the relation

$$R \propto A^{1/3}. \qquad (2.9)$$

The experimental proof of this relation and a quantitative determination of the coefficient of proportionality will be discussed in Sect. 5.4. The individual terms can be interpreted as follows:

Volume term. This term, which dominates the binding energy, is proportional to the number of nucleons. Each nucleon in the interior of a (large) nucleus contributes an energy of about 16 MeV. From this we deduce that the nuclear force has a short range, corresponding approximately to the distance between two nucleons. This phenomenon is called saturation. If each nucleon would interact with each of the other nucleons in the nucleus, the total binding energy would be proportional to $A(A - 1)$ or approximately to A^2. Due to saturation, the central density of nucleons is the same for all nuclei, with few exceptions. The central density is

$$\varrho_0 \approx 0.17 \text{ nucleons/fm}^3 = 3 \cdot 10^{17} \text{ kg/m}^3 \, . \tag{2.10}$$

The average nuclear density, which can be deduced from the mass and radius (see 5.56), is smaller (0.13 nucleons/fm^3). The average inter-nucleon distance in the nucleus is about 1.8 fm.

Surface term. For nucleons at the surface of the nucleus, which are surrounded by fewer nucleons, the above binding energy is reduced. This contribution is proportional to the surface area of the nucleus (R^2 or $A^{2/3}$).

Coulomb term. The electrical repulsive force acting between the protons in the nucleus further reduces the binding energy. This term is calculated to be

$$E_{\text{Coulomb}} = \frac{3}{5} \frac{Z(Z-1)\,\alpha\,\hbar c}{R} \, . \tag{2.11}$$

This is approximately proportional to $Z^2/A^{1/3}$.

Asymmetry term. As long as mass numbers are small, nuclei tend to have the same number of protons and neutrons. Heavier nuclei accumulate more and more neutrons, to partly compensate for the increasing Coulomb repulsion by increasing the nuclear force. This creates an asymmetry in the number of neutrons and protons. For, e.g., ^{208}Pb it amounts to $N-Z = 44$. The dependence of the nuclear force on the surplus of neutrons is described by the asymmetry term $(N-Z)^2/(4A)$. This shows that the symmetry decreases as the nuclear mass increases. We will further discuss this point in Sect. 17.1. The dependence of the above terms on A is shown in Fig. 2.5.

Pairing term. A systematic study of nuclear masses shows that nuclei are more stable when they have an even number of protons and/or neutrons. This observation is interpreted as a coupling of protons and neutrons in pairs. The pairing energy depends on the mass number, as the overlap of the wave functions of these nucleons is smaller, in larger nuclei. Empirically this is described by the term $\delta \cdot A^{-1/2}$ in (2.8).

All in all, the global properties of the nuclear force are rather well described by the mass formula (2.8). However, the details of nuclear structure which we will discuss later (mainly in Chap. 17) are not accounted for by this formula.

The Weizsäcker formula is often mentioned in connection with the *liquid drop model*. In fact, the formula is based on some properties known from liquid drops: constant density, short-range forces, saturation, deformability and surface tension. An essential difference, however, is found in the mean free path of the particles. For molecules in liquid drops, this is far smaller than the size of the drop; but for nucleons in the nucleus, it is large. Therefore, the nucleus has to be treated as a quantum liquid, and not as a classical one. At low excitation energies, the nucleus may be even more simply described as a Fermi gas; i. e., as a system of free particles only weakly interacting with each other. This model will be discussed in more detail in Sect. 17.1.

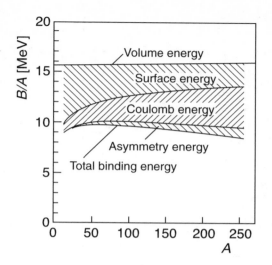

Fig. 2.5. The different contributions to the binding energy per nucleon versus mass number A. The horizontal line at ≈ 16 MeV represents the contribution of the volume energy. This is reduced by the surface energy, the asymmetry energy and the Coulomb energy to the effective binding energy of ≈ 8 MeV (*lower line*). The contributions of the asymmetry and Coulomb terms increase rapidly with A, while the contribution of the surface term decreases.

2.4 Charge Independence of the Nuclear Force and Isospin

Protons and neutrons not only have nearly equal masses, they also have similar nuclear interactions. This is particularly visible in the study of *mirror nuclei*. Mirror nuclei are pairs of isobars, in which the proton number of one of the nuclides equals the neutron number of the other and vice versa.

Figure 2.6 shows the lowest energy levels of the mirror nuclei $^{14}_{6}\text{C}_8$ and $^{14}_{8}\text{O}_6$, together with those of $^{14}_{7}\text{N}_7$. The energy-level diagrams of $^{14}_{6}\text{C}_8$ and $^{14}_{8}\text{O}_6$ are very similar with respect to the quantum numbers J^P of the levels as well as with respect to the distances between them. The small differences and the global shift of the levels as a whole in $^{14}_{6}\text{C}_8$, as compared to $^{14}_{8}\text{O}_6$ can be explained by differences in the Coulomb energy. Further examples of mirror nuclei will be discussed in Sect. 17.3 (Fig. 17.7). The energy levels of $^{14}_{6}\text{C}_8$ and $^{14}_{8}\text{O}_6$ are also found in the isobaric nucleus $^{14}_{7}\text{N}_7$. Other states in $^{14}_{7}\text{N}_7$ have no analogy in the two neighbouring nuclei. We therefore can distinguish between triplet and singlet states.

These multiplets of states are reminiscent of the multiplets known from the coupling of angular momenta (spins). The symmetry between protons and neutrons may therefore be described by a similar formalism, called *isospin I*. The proton and neutron are treated as two states of the nucleon which form a doublet ($I = 1/2$).

$$\text{Nucleon:} \quad I = 1/2 \quad \begin{cases} \text{proton:} & I_3 = +1/2 \\ \text{neutron:} & I_3 = -1/2 \end{cases} \quad (2.12)$$

Formally, isospin is treated as a quantum mechanical angular momentum. For example, a proton-neutron pair can be in a state of total isospin 1 or 0.

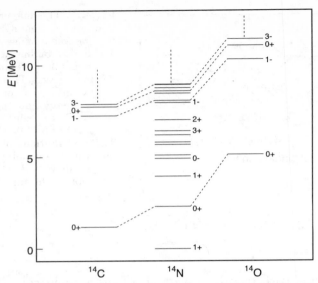

Fig. 2.6. Low-lying energy levels of the three most stable $A = 14$ isobars. Angular momentum J and parity P are shown for the most important levels. The analogous states of the three nuclei are joined by dashed lines. The zero of the energy scale is set to the ground state of $^{14}_{7}N_{7}$.

The third (z-) component of isospin is additive:

$$I_3^{\text{nucleus}} = \sum I_3^{\text{nucleon}} = \frac{Z - N}{2}. \qquad (2.13)$$

This enables us to describe the appearance of similar states in Fig. 2.6: $^{14}_{6}C_{8}$ and $^{14}_{8}O_{6}$, have respectively $I_3 = -1$ and $I_3 = +1$. Therefore, their isospin cannot be less than $I = 1$. The states in these nuclei thus necessarily belong to a triplet of similar states in $^{14}_{6}C_{8}$, $^{14}_{7}N_{7}$ and $^{14}_{8}O_{6}$. The I_3 component of the nuclide $^{14}_{7}N_{7}$, however, is 0. This nuclide can, therefore, have additional states with isospin $I = 0$.

Since $^{14}_{7}N_{7}$ is the most stable $A = 14$ isobar, its ground state is necessarily an isospin singlet since otherwise $^{14}_{6}C_{8}$ would possess an analogous state, which, with less Coulomb repulsion, would be lower in energy and so more stable. $I = 2$ states are not shown in Fig. 2.6. Such states would have analogous states in $^{14}_{5}B_{9}$ and in $^{14}_{9}F_{5}$. These nuclides, however, are very unstable (i. e., highly energetic), and lie above the energy range of the diagram. The $A = 14$ isobars are rather light nuclei in which the Coulomb energy is not strongly felt. In heavier nuclei, the influence of the Coulomb energy grows, which increasingly disturbs the isospin symmetry.

The concept of isospin is of great importance not only in nuclear physics, but also in particle physics. As we will see quarks, and particles composed of quarks, can be classified by isospin into isospin multiplets. In dynamical processes of the strong-interaction type, the isospin of the system is conserved.

_____ **Problems**

1. **Isospin symmetry**

   One could naively imagine the three nucleons in the ^{3}H and ^{3}He nuclei as being rigid spheres. If one solely attributes the difference in the binding energies of these two nuclei to the electrostatic repulsion of the protons in ^{3}He, how large must the separation of the protons be? (The maximal energy of the electron in the β^--decay of ^{3}H is 18.6 keV.)

3. Nuclear Stability

Stable nuclei only occur in a very narrow band in the $Z - N$ plane (Fig. 3.1). All other nuclei are unstable and decay spontaneously in various ways. Isobars with a large surplus of neutrons gain energy by converting a neutron into a proton. In the case of a surplus of protons, the inverse reaction may occur: i.e., the conversion of a proton into a neutron. These transformations are called *β-decays* and they are manifestations of the weak interaction. After dealing with the weak interaction in Chap. 10, we will discuss these decays in more detail in Sects. 15.5 and 17.6. In the present chapter, we will merely survey certain general properties, paying particular attention to the energy balance of β-decays.

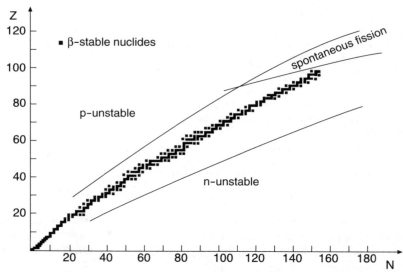

Fig. 3.1. β-stable nuclei in the $Z - N$ plane (from [Bo69]).

Fe- and Ni-isotopes possess the maximum binding energy per nucleon and they are therefore the most stable nuclides. In heavier nuclei the binding energy is smaller because of the larger Coulomb repulsion. For still heavier masses nuclei become unstable to fission and decay spontaneously into two or more lighter nuclei should the mass of the original atom be larger than the sum of the masses of the daughter atoms. For a two-body decay, this condition has the form:

$$M(A,Z) > M(A - A', Z - Z') + M(A',Z'). \tag{3.1}$$

This relation takes into account the conservation of the number of protons and neutrons. However, it does not give any information about the probability of such a decay. An isotope is said to be stable if its lifetime is considerably longer than the age of the solar system. We will not consider many-body decays any further since they are much rarer than two-body decays. It is very often the case that one of the daughter nuclei is a ^{4}He nucleus, i.e., $A' = 4$, $Z' = 2$. This decay mode is called α-decay, and the Helium nucleus is called an α-particle. If a heavy nucleus decays into two similarly massive daughter nuclei we speak of spontaneous fission. The probability of spontaneous fission exceeds that of α-decay only for nuclei with $Z \gtrsim 110$ and is a fairly unimportant process for the naturally occuring heavy elements.

Decay constants. The probability per unit time for a radioactive nucleus to decay is known as the decay constant λ. It is related to the lifetime τ and the half life $t_{1/2}$ by:

$$\tau = \frac{1}{\lambda} \qquad \text{and} \qquad t_{1/2} = \frac{\ln 2}{\lambda}. \tag{3.2}$$

The measurement of the decay constants of radioactive nuclei is based upon finding the activity (the number of decays per unit time):

$$A = -\frac{dN}{dt} = \lambda N \tag{3.3}$$

where N is the number of radioactive nuclei in the sample. The unit of activity is defined to be

$$1 \text{ Bq [Bequerel]} = 1 \text{ decay /s}. \tag{3.4}$$

For short-lived nuclides, the fall-off over time of the activity:

$$A(t) = \lambda N(t) = \lambda N_0 e^{-\lambda t} \qquad \text{where} \quad N_0 = N(t=0) \tag{3.5}$$

may be measured using fast electronic counters. This method of measuring is not suitable for lifetimes larger than about a year. For longer-lived nuclei both the number of nuclei in the sample and the activity must be measured in order to obtain the decay constant from (3.3).

3.1 β-Decay

Let us consider nuclei with equal mass number A (isobars). Equation 2.8 can be transformed into:

$$M(A, Z) = \alpha \cdot A - \beta \cdot Z + \gamma \cdot Z^2 + \frac{\delta}{A^{1/2}}, \qquad (3.6)$$

where
$$\alpha = M_{\mathrm{n}} - a_{\mathrm{v}} + a_{\mathrm{s}} A^{-1/3} + \frac{a_{\mathrm{a}}}{4},$$
$$\beta = a_{\mathrm{a}} + (M_{\mathrm{n}} - M_{\mathrm{p}} - m_{\mathrm{e}}),$$
$$\gamma = \frac{a_{\mathrm{a}}}{A} + \frac{a_{\mathrm{c}}}{A^{1/3}},$$
$$\delta = \text{as in (2.8)}.$$

The nuclear mass is now a quadratic function of Z. A plot of such nuclear masses, for constant mass number A, as a function of Z, the charge number, yields a parabola for odd A. For even A, the masses of the even-even and the odd-odd nuclei are found to lie on two vertically shifted parabolas. The odd-odd parabola lies at twice the pairing energy $(2\delta/\sqrt{A})$ above the even-even one. The minimum of the parabolas is found at $Z = \beta/2\gamma$. The nucleus with the smallest mass in an isobaric spectrum is stable with respect to β-decay.

β-decay in odd mass nuclei. In what follows we wish to discuss the different kinds of β-decay, using the example of the $A = 101$ isobars. For this mass number, the parabola minimum is at the isobar ^{101}Ru which has $Z = 44$. Isobars with more neutrons, such as $^{101}_{42}$Mo and $^{101}_{43}$Tc, decay through the conversion:

$$\mathrm{n} \rightarrow \mathrm{p} + \mathrm{e}^- + \bar{\nu}_{\mathrm{e}}. \qquad (3.7)$$

The charge number of the daughter nucleus is one unit larger than that of the the parent nucleus (Fig. 3.2). An electron and an e-antineutrino are also produced:

$$^{101}_{42}\mathrm{Mo} \rightarrow {}^{101}_{43}\mathrm{Tc} + \mathrm{e}^- + \bar{\nu}_{\mathrm{e}},$$
$$^{101}_{43}\mathrm{Tc} \rightarrow {}^{101}_{44}\mathrm{Ru} + \mathrm{e}^- + \bar{\nu}_{\mathrm{e}}.$$

Historically such decays where a negative electron is emitted are called β^--decays. Energetically, β^--decay is possible whenever the mass of the daughter atom $M(A, Z + 1)$ is smaller than the mass of its isobaric neighbour:

$$M(A, Z) > M(A, Z + 1). \qquad (3.8)$$

We consider here the mass of the whole atom and not just that of the nucleus alone and so the rest mass of the electron created in the decay is automatically taken into account. The tiny mass of the (anti-)neutrino $(< 15\,\mathrm{eV}/c^2)$ [PD98] is negligible in the mass balance.

Isobars with a proton excess, compared to $^{101}_{44}$Ru, decay through proton conversion:

$$\mathrm{p} \rightarrow \mathrm{n} + \mathrm{e}^+ + \nu_{\mathrm{e}}. \qquad (3.9)$$

The stable isobar $^{101}_{44}$Ru is eventually produced via

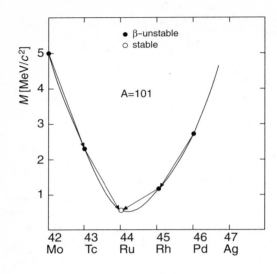

Fig. 3.2. Mass parabola of the $A = 101$ isobars (from [Se77]). Possible β-decays are shown by arrows. The abscissa co-ordinate is the atomic number, Z. The zero point of the mass scale was chosen arbitrarily.

$$^{101}_{46}\mathrm{Pd} \quad \rightarrow \quad ^{101}_{45}\mathrm{Rh} + e^+ + \nu_e\,, \quad \text{and}$$
$$^{101}_{45}\mathrm{Rh} \quad \rightarrow \quad ^{101}_{44}\mathrm{Ru} + e^+ + \nu_e\,.$$

Such decays are called β^+-decays. Since the mass of a free neutron is larger than the proton mass, the process (3.9) is only possible inside a nucleus. By contrast, neutrons outside nuclei can and do decay (3.7). Energetically, β^+-decay is possible whenever the following relationship between the masses $M(A, Z)$ and $M(A, Z - 1)$ (of the parent and daughter atoms respectively) is satisfied:

$$M(A, Z) > M(A, Z - 1) + 2m_e\,. \tag{3.10}$$

This relationship takes into account the creation of a positron and the existence of an excess electron in the parent atom.

β-decay in even nuclei. Even mass number isobars form, as we described above, two separate (one for even-even and one for odd-odd nuclei) parabolas which are split by an amount equal to twice the pairing energy.

Often there is more than one β-stable isobar, especially in the range $A > 70$. Let us consider the example of the nuclides with $A = 106$ (Fig. 3.3). The even-even $^{106}_{46}\mathrm{Pd}$ and $^{106}_{48}\mathrm{Cd}$ isobars are on the lower parabola, and $^{106}_{46}\mathrm{Pd}$ is the stablest. $^{106}_{48}\mathrm{Cd}$ is β-stable, since its two odd-odd neighbours both lie above it. The conversion of $^{106}_{48}\mathrm{Cd}$ is thus only possible through a double β-decay into $^{106}_{46}\mathrm{Pd}$:

$$^{106}_{48}\mathrm{Cd} \quad \rightarrow \quad ^{106}_{46}\mathrm{Pd} + 2e^+ + 2\nu_e\,.$$

The probability for such a process is so small that $^{106}_{48}\mathrm{Cd}$ may be considered to be a stable nuclide.

Odd-odd nuclei always have at least one more strongly bound, even-even neighbour nucleus in the isobaric spectrum. They are therefore unstable. The

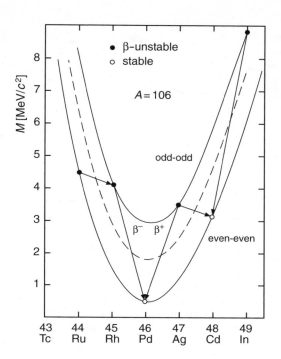

β-unstable

stable

$A = 106$

odd-odd

β^- β^+

even-even

| 43 | 44 | 45 | 46 | 47 | 48 | 49 |
| Tc | Ru | Rh | Pd | Ag | Cd | In |

Fig. 3.3. Mass parabolas of the $A = 106$-isobars (from [Se77]). Possible β-decays are indicated by arrows. The abscissa coordinate is the charge number Z. The zero point of the mass scale was chosen arbitrarily.

only exceptions to this rule are the very light nuclei $^2_1$H, $^6_3$Li, $^{10}_5$B and $^{14}_7$N, which are stable to β-decay, since the increase in the asymmetry energy would exceed the decrease in the pairing energy. Some odd-odd nuclei can undergo both β^--decay and β^+-decay. Well-known examples of this are $^{40}_{19}$K (Fig. 3.4) and $^{64}_{29}$Cu.

Electron capture. Another possible decay process is the capture of an electron from the cloud surrounding the atom. There is a finite probability of finding such an electron inside the nucleus. In such circumstances it can combine with a proton to form a neutron and a neutrino in the following way:

$$p + e^- \to n + \nu_e \,. \tag{3.11}$$

This reaction occurs mainly in heavy nuclei where the nuclear radii are larger and the electronic orbits are more compact. Usually the electrons that are captured are from the innermost (the "K") shell since such K-electrons are closest to the nucleus and their radial wave function has a maximum at the centre of the nucleus. Since an electron is missing from the K-shell after such a *K-capture*, electrons from higher energy levels will successively cascade downwards and in so doing they emit characteristic X-rays.

Electron capture reactions compete with β^+-decay. The following condition is a consequence of energy conservation

$$M(A, Z) > M(A, Z - 1) + \varepsilon \,, \tag{3.12}$$

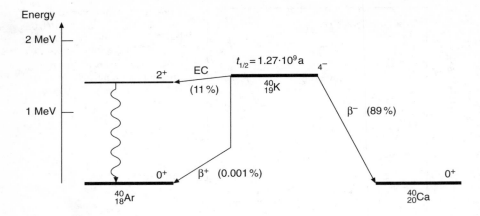

Fig. 3.4. The β-decay of ^{40}K. In this nuclear conversion, β^-- and β^+-decay as well as electron capture (EC) compete with each other. The relative frequency of these decays is given in parentheses. The bent arrow in β^+-decay indicates that the production of an e^+ and the presence of the surplus electron in the ^{40}Ar atom requires 1.022 MeV, and the remainder is carried off as kinetic energy by the positron and the neutrino. The excited state of ^{40}Ar produced in the electron capture reaction decays by photon emission into its ground state.

where ε is the excitation energy of the atomic shell of the daughter nucleus (electron capture always leads to a hole in the electron shell). This process has, compared to β^+-decay, more kinetic energy ($2m_ec^2 - \varepsilon$ more) available to it and so there are some cases where the mass difference between the initial and final atoms is too small for conversion to proceed via β^+-decay and yet K-capture can take place.

Lifetimes. The lifetimes τ of β-unstable nuclei vary between a few ms and 10^{16} years. They strongly depend upon both the energy E which is released ($1/\tau \propto E^5$) and upon the nuclear properties of the mother and daughter nuclei. The decay of a free neutron into a proton, an electron and an antineutrino releases 0.78 MeV and this particle has a lifetime of $\tau = 886.7 \pm 1.9\,\text{s}$ [PD98]. No two neighbouring isobars are known to be β-stable.[1]

A well-known example of a long-lived β-emitter is the nuclide ^{40}K. It transforms into other isobars by both β^-- and β^+-decay. Electron capture in ^{40}K also competes here with β^+-decay. The stable daughter nuclei are ^{40}Ar and ^{40}Ca respectively, which is a case of two stable nuclei having the same mass number A (Fig. 3.4).

---

[1] In some cases, however, one of two neighbouring isobars is stable and the other is extremely long-lived. The most common isotopes of indium (^{115}In, 96 %) and rhenium (^{187}Re, 63 %) β^--decay into stable nuclei (^{115}Sn and ^{187}Os), but they are so long-lived ($\tau = 3\cdot10^{14}$ yrs and $\tau = 3\cdot10^{11}$ yrs respectively) that they may also be considered stable.

The ^{40}K nuclide was chosen here because it contributes considerably to the radiation exposure of human beings and other biological systems. Potassium is an essential element: for example, signal transmission in the nervous system functions by an exchange of potassium ions. The fraction of radioactive ^{40}K in natural potassium is 0.01 %, and the decay of ^{40}K in the human body contributes about 16 % of the total natural radiation which we are exposed to.

3.2 α-Decay

Protons and neutrons have binding energies, even in heavy nuclei, of about 8 MeV (Fig. 2.4) and cannot generally escape from the nucleus. In many cases, however, it is energetically possible for a bound system of a group of nucleons to be emitted, since the binding energy of this system increases the total energy available to the process. The probability for such a system to be formed in a nucleus decreases rapidly with the number of nucleons required. In practice the most significant decay process is the emission of a ^{4}He nucleus; i. e., a system of 2 protons and 2 neutrons. Contrary to systems of 2 or 3 nucleons, this so-called *α-particle* is extraordinarily strongly bound — 7 MeV/nucleon (cf. Fig. 2.4). Such decays are called *α-decays*.

Figure 3.5 shows the potential energy of an α-particle as a function of its separation from the centre of the nucleus. Beyond the nuclear force range, the α-particle feels only the Coulomb potential $V_C(r) = 2(Z-2)\alpha\hbar c/r$, which increases closer to the nucleus. Within the nuclear force range a strongly attractive nuclear potential prevails. Its strength is characterised by the depth of the potential well. Since we are considering α-particles which are energetically allowed to escape from the nuclear potential, the total energy of this α-particle is positive. This energy is released in the decay.

The range of lifetimes for the α-decay of heavy nuclei is extremely large. Experimentally, lifetimes have been measured between 10 ns and 10^{17} years.

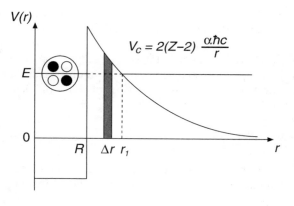

Fig. 3.5. Potential energy of an α-particle as a function of its separation from the centre of the nucleus. The probability that it tunnels through the Coulomb barrier can be calculated as the superposition of tunnelling processes through thin potential walls of thickness Δr (cf. Fig. 3.6).

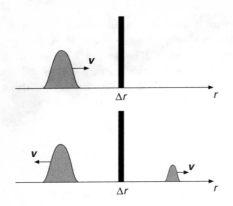

Fig. 3.6. Illustration of the tunnelling probability of a wave packet with energy E and velocity v faced with a potential barrier of height V and thickness Δr.

These lifetimes can be calculated in quantum mechanics by treating the α-particle as a wave packet. The probability for the α-particle to escape from the nucleus is given by the probability for its penetrating the *Coulomb barrier* (the tunnel effect). If we divide the Coulomb barrier into thin potential walls and look at the probability of the α-particle tunnelling through one of these (Fig. 3.6), then the transmission T is given by:

$$T \approx e^{-2\kappa \Delta r} \qquad \text{where} \quad \kappa = \sqrt{2m|E - V|}/\hbar , \qquad (3.13)$$

and Δr is the thickness of the barrier and V is its height. E is the energy of the α-particle. A Coulomb barrier can be thought of as a barrier composed of a large number of thin potential walls of different heights. The transmission can be described accordingly by:

$$T = e^{-2G} . \qquad (3.14)$$

The *Gamow factor* G can be approximated by the integral [Se77]:

$$G = \frac{1}{\hbar} \int_R^{r_1} \sqrt{2m|E - V|} \, dr \approx \frac{\pi \cdot 2 \cdot (Z - 2) \cdot \alpha}{\beta} , \qquad (3.15)$$

where $\beta = v/c$ is the velocity of the outgoing α-particle and R is the nuclear radius.

The probability per unit time λ for an α-particle to escape from the nucleus is therefore proportional to: the probability $w(\alpha)$ of finding such an α-particle in the nucleus, the number of collisions ($\propto v_0/2R$) of the α-particle with the barrier and the transmission probability:

$$\lambda = w(\alpha) \, \frac{v_0}{2R} \, e^{-2G} , \qquad (3.16)$$

where v_0 is the velocity of the α-particle in the nucleus ($v_0 \approx 0.1\,c$). The large variation in the lifetimes is explained by the Gamow factor in the exponent:

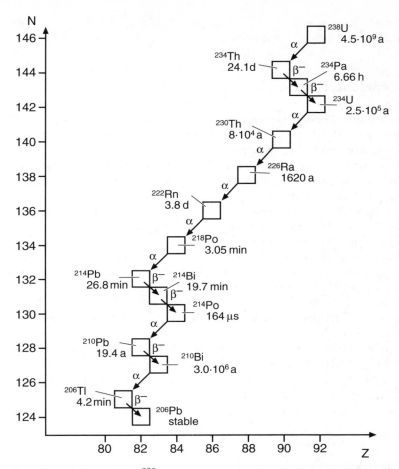

Fig. 3.7. Illustration of the ^{238}U decay chain in the N–Z plane. The half life of each of the nuclides is given together with its decay mode.

since $G \propto Z/\beta \propto Z/\sqrt{E}$, small differences in the energy of the α-particle have a strong effect on the lifetime.

Most α-emitting nuclei are heavier than lead. For lighter nuclei with $A \lesssim 140$, α-decay is energetically possible, but the energy released is extremely small. Therefore, their nuclear lifetimes are so long that decays are usually not observable.

An example of a α-unstable nuclide with a long lifetime, ^{238}U, is shown in Fig. 3.7. Since uranium compounds are common in granite, uranium and its radioactive daughters are a part of the stone walls of buildings. They therefore contribute to the environmental radiation background. This is particularly true of the inert gas ^{222}Rn, which escapes from the walls and is inhaled into the lungs. The α-decay of ^{222}Rn is responsible for about 40 % of the average natural human radiation exposure.

3.3 Nuclear Fission

Spontaneous fission. The largest binding energy per nucleon is found in those nuclei in the region of ^{56}Fe. For heavier nuclei, it decreases as the nuclear mass increases (Fig. 2.4). A nucleus with $Z > 40$ can thus, in principle, split into two lighter nuclei. The potential barrier which must be tunneled through is, however, so large that such spontaneous fission reactions are generally speaking extremely unlikely.

The lightest nuclides where the probability of spontaneous fission is comparable to that of α-decay are certain uranium isotopes. The shape of the fission barrier is shown in Fig. 3.8.

It is interesting to find the charge number Z above which nuclei become fission unstable, i.e., the point from which the mutual Coulombic repulsion of the protons outweighs the attractive nature of the nuclear force. An estimate can be obtained by considering the surface and the Coulomb energies during the fission deformation. As the nucleus is deformed the surface energy increases, while the Coulomb energy decreases. If the deformation leads to an energetically more favourable configuration, the nucleus is unstable. Quantitatively, this can be calculated as follows: keeping the volume of the nucleus constant, we deform its spherical shape into an ellipsoid with axes $a = R(1 + \varepsilon)$ and $b = R(1 - \varepsilon/2)$ (Fig. 3.9).

The surface energy then has the form:

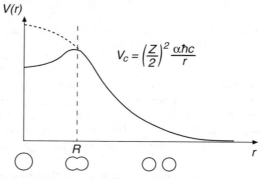

Fig. 3.8. Potential energy during different stages of a fission reaction. A nucleus with charge Z decays spontaneously into two daughter nuclei. The solid line corresponds to the shape of the potential in the parent nucleus. The height of the barrier for fission determines the probability of spontaneous fission. The fission barrier disappears for nuclei with $Z^2/A \gtrsim 48$ and the shape of the potential then corresponds to the dashed line.

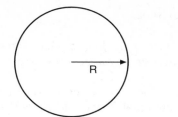
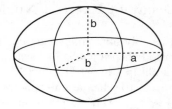

Fig. 3.9. Deformation of a heavy nucleus. For a constant volume V ($V = 4\pi R^3/3 = 4\pi ab^2/3$), the surface energy of the nucleus increases and its Coulomb energy decreases.

$$E_s = a_s A^{2/3} \left(1 + \frac{2}{5}\varepsilon^2 + \cdots\right), \qquad (3.17)$$

while the Coulomb energy is given by:

$$E_c = a_c Z^2 A^{-1/3} \left(1 - \frac{1}{5}\varepsilon^2 + \cdots\right). \qquad (3.18)$$

Hence a deformation ε changes the total energy by:

$$\Delta E = \frac{\varepsilon^2}{5}\left(2a_s A^{2/3} - a_c Z^2 A^{-1/3}\right). \qquad (3.19)$$

If ΔE is negative, a deformation is energetically favoured. The fission barrier disappears for:

$$\frac{Z^2}{A} \geq \frac{2a_s}{a_c} \approx 48. \qquad (3.20)$$

This is the case for nuclei with $Z > 114$ and $A > 270$.

Induced fission. For very heavy nuclei ($Z \approx 92$) the fission barrier is only about 6 MeV. This energy may be supplied if one uses a flow of low energy neutrons to induce neutron capture reactions. These push the nucleus into an excited state above the fission barrier and it splits up. This process is known as *induced nuclear fission.*

Neutron capture by nuclei with an odd neutron number releases not just some binding energy but also a pairing energy. This small extra contribution to the energy balance makes a decisive difference to nuclide fission properties: in neutron capture by ^{238}U, for example, 4.9 MeV binding energy is released, which is below the threshold energy of 5.5 MeV for nuclear fission of ^{239}U. Neutron capture by ^{238}U can therefore only lead to immediate nuclear fission if the neutron possesses a kinetic energy at least as large as this difference ("fast neutrons"). On top of this the reaction probability is proportional to v^{-1}, where v is the velocity of the neutron (4.21), and so it is very small. By contrast neutron capture in ^{235}U releases 6.4 MeV and the fission barrier of ^{236}U is just 5.5MeV. Thus fission may be induced in ^{235}U with the help of low-energy (thermal) neutrons. This is exploited in nuclear reactors and nuclear weapons. Similarly both ^{233}Th and ^{239}Pu are suitable fission materials.

3.4 Decay of Excited Nuclear States

Nuclei usually have many excited states. Most of the lowest-lying states are understood theoretically, at least in a qualitative way as will be discussed in more detail in Chaps. 17 and 18.

Figure 3.10 schematically shows the energy levels of an even-even nucleus with $A \approx 100$. Above the ground state, individual discrete levels with specific J^P quantum numbers can be seen. The excitation of even-even nuclei generally corresponds to the break up of nucleon pairs, which requires about 1–2 MeV. Even-even nuclei with $A \gtrsim 40$, therefore, rarely possess excitations below 2 MeV.[2] In odd-even and odd-odd nuclei, the number of low-energy states (with excitation energies of a few 100 keV) is considerably larger.

---

[2] Collective states in deformed nuclei are an exception to this: they cannot be understood as single particle excitations (Chap. 18).

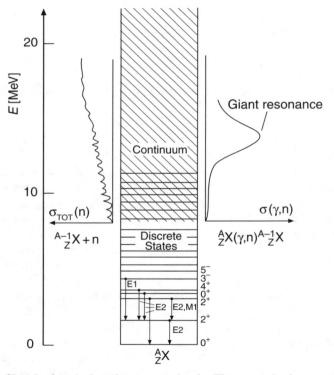

Fig. 3.10. Sketch of typical nuclear energy levels. The example shows an even-even nucleus whose ground state has the quantum numbers 0^+. To the left the total cross-section for the reaction of the nucleus $^{A-1}_{Z}X$ with neutrons (elastic scattering, inelastic scattering, capture) is shown; to the right the total cross-section for γ-induced neutron emission $^A_Z X + \gamma \rightarrow {}^{A-1}_{Z}X + n$.

Electromagnetic decays. Low lying excited nuclear states usually decay by emitting electromagnetic radiation. This can be described in a series expansion as a superposition of different multipolarities each with its characteristic angular distribution. Electric dipole, quadrupole, octupole radiation etc. are denoted by E1, E2, E3, etc. Similarly, the corresponding magnetic multipoles are denoted by M1, M2, M3 etc. Conservation of angular momentum and parity determine which multipolarities are possible in a transition. A photon of multipolarity $E\ell$ has angular momentum ℓ and parity $(-1)^{\ell}$, an $M\ell$ photon has angular momentum ℓ and parity $(-1)^{(\ell+1)}$. In a transition $J_i \rightarrow J_f$, conservation of angular momentum means that the triangle inequality $|J_i - J_f| \leq \ell \leq J_i + J_f$ must be satisfied.

Table 3.1. Selection rules for some electromagnetic transitions.

Multi-polarity	Electric			Magnetic		
	$E\ell$	$|\Delta \boldsymbol{J}|$	ΔP	$M\ell$	$|\Delta \boldsymbol{J}|$	ΔP
Dipole	E1	1	$-$	M1	1	$+$
Quadrupole	E2	2	$+$	M2	2	$-$
Octupole	E3	3	$-$	M3	3	$+$

The lifetime of a state strongly depends upon the multipolarity of the γ-transitions by which it can decay. The lower the multipolarity, the larger the transition probability. A magnetic transition $M\ell$ has approximately the same probability as an electric $E(\ell + 1)$ transition. A transition $3^+ \rightarrow 1^+$, for example, is in principle a mixture of E2, M3, and E4, but will be easily dominated by the E2 contribution. A $3^+ \rightarrow 2^+$ transition will usually consist of an M1/E2 mixture, even though M3, E4, and M5 transitions are also possible. In a series of excited states 0^+, 2^+, 4^+, the most probable decay is by a cascade of E2-transitions $4^+ \rightarrow 2^+ \rightarrow 0^+$, and not by a single $4^+ \rightarrow 0^+$ E4-transition. The lifetime of a state and the angular distribution of the electromagnetic radiation which it emits are signatures for the multipolarity of the transitions, which in turn betray the spin and parity of the nuclear levels. The decay probability also strongly depends upon the energy. For radiation of multipolarity ℓ it is proportional to $E_\gamma^{2\ell+1}$ (cf. Sect. 18.1).

The excitation energy of a nucleus may also be transferred to an electron in the atomic shell. This process is called *internal conversion*. It is most important in transitions for which γ-emission is suppressed (high multipolarity, low energy) and the nucleus is heavy (high probability of the electron being inside the nucleus).

$0^+ \rightarrow 0^+$ transitions cannot proceed through photon emission. If a nucleus is in an excited 0^+-state, and all its lower lying levels also have 0^+ quantum numbers (e. g. in ^{16}O or ^{40}Ca – cf. Fig. 18.6), then this state can only decay in a different way: by internal conversion, by emission of 2 photons or by the emission of an e^+e^--pair, if this last is energetically possible. Parity conservation does not permit internal conversion transitions between two levels with $J = 0$ and opposite parity.

The lifetime of excited nuclear states typically varies between 10^{-9} s and 10^{-15} s, which corresponds to a state width of less than 1 eV. States which can only decay by low energy and high multipolarity transitions have considerably longer lifetimes. They are called *isomers* and are designated by an "mßuperscript on the symbol of the element. An extreme example is the second excited state of ^{110}Ag, whose quantum numbers are $J^P = 6^+$ and excitation energy is 117.7 keV. It relaxes via an M4-transition into the first excited state (1.3 keV; 2^-) since a decay directly into the ground state (1^+) is even more improbable. The half life of ^{110}Ag^m is extremely long ($t_{1/2} = 235$ d) [Le78].

Continuum states. Most nuclei have a binding energy per nucleon of about 8 MeV (Fig. 2.4). This is approximately the energy required to separate a single nucleon from the nucleus *(separation energy)*. States with excitation energies above this value can therefore emit single nucleons. The emitted nucleons are primarily neutrons since they are not hindered by the Coulomb threshold. Such a strong interaction process is clearly preferred to γ-emission.

The excitation spectrum above the threshold for particle emission is called the *continuum*, just as in atomic physics. Within this continuum there are also discrete, quasi-bound states. States below this threshold decay only by (relatively slow) γ-emission and are, therefore, very narrow. But for excitation energies above the particle threshold, the lifetimes of the states decrease dramatically, and their widths increase. The density of states increases approximately exponentially with the excitation energy. At higher excitation energies, the states therefore start to overlap, and states with the same quantum numbers can begin to mix.

The continuum can be especially effectively investigated by measuring the cross-sections of neutron capture and neutron scattering. Even at high excitation energies, some narrow states can be identified. These are states with exotic quantum numbers (high spin) which therefore cannot mix with neighbouring states.

Figure 3.10 shows schematically the cross-sections for neutron capture and γ-induced neutron emission *(nuclear photoelectric effect)*. A broad resonance is observed, the *giant dipole resonance,* which will be interpreted in Sect. 18.2.

_____ **Problems**

1. **α-decay**
   The α-decay of a ^{238}Pu (τ=127 yrs) nuclide into a long lived ^{234}U ($\tau = 3.5 \cdot 10^5$ yrs) daughter nucleus releases 5.49 MeV kinetic energy. The heat so produced can be converted into useful electricity by radio-thermal generators (RTG's). The *Voyager 2* space probe, which was launched on the 20.8.1977, flew past four planets, including Saturn which it reached on the 26.8.1981. Saturn's separation from the sun is 9.5 AU; 1 AU = separation of the earth from the sun.
   a) How much plutonium would an RTG on Voyager 2 with 5.5 % efficiency have to carry so as to deliver at least 395 W electric power when the probe flies past Saturn?
   b) How much electric power would then be available at Neptune (24.8.1989; 30.1 AU separation)?
   c) To compare: the largest ever "solar paddles" used in space were those of the space laboratory *Skylab* which would have produced 10.5 kW from an area of 730 m^2 if they had not been damaged at launch. What area of solar cells would *Voyager 2* have needed?

2. **Radioactivity**
   Naturally occuring uranium is a mixture of the ^{238}U (99.28 %) and ^{235}U (0.72 %) isotopes.
   a) How old must the material of the solar system be if one assumes that at its creation both isotopes were present in equal quantities? How do you interpret this result? The lifetime of ^{235}U is $\tau = 1.015 \cdot 10^9$ yrs. For the lifetime of ^{238}U use the data in Fig. 3.7.
   b) How much of the ^{238}U has decayed since the formation of the earth's crust $2.5 \cdot 10^9$ years ago?
   c) How much energy per uranium nucleus is set free in the decay chain ^{238}U $\rightarrow$ ^{206}Pb? A small proportion of ^{238}U spontaneously splits into, e. g., $^{142}_{54}$Xe und $^{96}_{38}$Sr.

3. **Radon activity**
   After a lecture theatre whose walls, floor and ceiling are made of concrete ($10 \times 10 \times 4$ m^3) has not been aired for several days, a specific activity A from ^{222}Rn of 100 Bq/m^3 is measured.
   a) Calculate the activity of ^{222}Rn as a function of the lifetimes of the parent and daughter nuclei.
   b) How high is the concentration of ^{238}U in the concrete if the effective thickness from which the ^{222}Rn decay product can diffuse is 1.5 cm?

4. **Mass formula**
   Isaac Asimov in his novel *The Gods Themselves* describes a universe where the stablest nuclide with $A = 186$ is not $^{186}_{74}$W but rather $^{186}_{94}$Pu. This is claimed to be a consequence of the ratio of the strengths of the strong and electromagnetic interactions being different to that in our universe. Assume that only the electromagnetic coupling constant α differs and that both the strong interaction and the nucleon masses are unchanged. How large must α be in order that $^{186}_{82}$Pb, $^{186}_{88}$Ra and $^{186}_{94}$Pu are stable?

5. **α-decay**
   The binding energy of an α particle is 28.3 MeV. Estimate, using the mass formula (2.8), from which mass number A onwards α-decay is energetically allowed for all nuclei.

6. Quantum numbers

An even-even nucleus in the ground state decays by α emission. Which J^P states are available to the daughter nucleus?

4. Scattering

4.1 General Observations About Scattering Processes

Scattering experiments are an important tool of nuclear and particle physics. They are used both to study details of the interactions between different particles and to obtain information about the internal structure of atomic nuclei and their constituents. These experiments will therefore be discussed at length in the following.

In a typical scattering experiment, the object to be studied (the *target*) is bombarded with a beam of particles with (mostly) well-defined energy. Occasionally, a reaction of the form

$$a + b \rightarrow c + d$$

between the projectile and the target occurs. Here, a and b denote the beam- and target particles, and c and d denote the products of the reaction. In inelastic reactions, the number of the reaction products may be larger than two. The rate, the energies and masses of the reaction products and their angles relative to the beam direction may be determined with suitable systems of detectors.

It is nowadays possible to produce beams of a broad variety of particles (electrons, protons, neutrons, heavy ions, ...). The beam energies available vary between 10^{-3} eV for "cold" neutrons up to 10^{12} eV for protons. It is even possible to produce beams of secondary particles which themselves have been produced in high energy reactions. Some such beams are very short-lived, such as muons, $\pi-$ or K-mesons, or hyperons ($\Sigma^{\pm}$, Ξ^-, Ω^-).

Solid, liquid or gaseous targets may be used as scattering material or, in storage ring experiments, another beam of particles may serve as the target. Examples of this last are the electron-positron storage ring LEP (Large Electron Positron collider) at CERN[1] in Geneva (maximum beam energy at present: $E_{e^+,e^-} = 86$ GeV), the "Tevatron" proton-antiproton storage ring at the Fermi National Accelerator Laboratory (FNAL) in the USA ($E_{p,\bar{p}} = 900$ GeV) and HERA (Hadron-Elektron-Ringanlage), the electron-proton storage ring at DESY[2] in Hamburg ($E_e = 30$ GeV, $E_p = 920$ GeV), which last was brought on-line in 1992.

Figure 4.1 shows some scattering processes. We distinguish between elastic and inelastic scattering reactions.

---

[1] Conseil Européen pour la Recherche Nucléaire
[2] Deutsches Elektronen-Synchrotron

Elastic scattering. In an elastic process (Fig. 4.1a):

$$a + b \rightarrow a' + b',$$

the same particles are presented both before and after the scattering. The target b remains in its ground state, absorbing merely the recoil momentum and hence changing its kinetic energy. The apostrophe indicates that the particles in the initial and in the final state are identical up to momenta and energy. The scattering angle and the energy of the a' particle and the production angle and energy of b' are unambigously correlated. As in optics, conclusions about the spatial shape of the scattering object can be drawn from the dependence of the scattering rate upon the beam energy and scattering angle.

It is easily seen that in order to resolve small target structures, larger beam energies are required. The reduced de-Broglie wave-length $\lambdabar = \lambda/2\pi$ of a particle with momentum p is given by

$$\lambdabar = \frac{\hbar}{p} = \frac{\hbar c}{\sqrt{2mc^2 E_{\text{kin}} + E_{\text{kin}}^2}} \approx \begin{cases} \hbar/\sqrt{2mE_{\text{kin}}} & \text{for } E_{\text{kin}} \ll mc^2 \\ \hbar c/E_{\text{kin}} \approx \hbar c/E & \text{for } E_{\text{kin}} \gg mc^2 . \end{cases} \quad (4.1)$$

The largest wavelength that can resolve structures of linear extension Δx, is of the same order: $\lambdabar \lesssim \Delta x$.

From Heisenberg's uncertainty principle the corresponding particle momentum is:

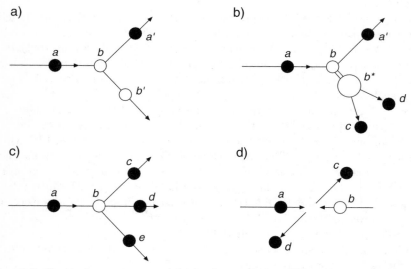

Fig. 4.1. Scattering processes: (**a**) elastic scattering; (**b**) inelastic scattering – production of an excited state which then decays into two particles; (**c**) inelastic production of new particles; (**d**) reaction of colliding beams.

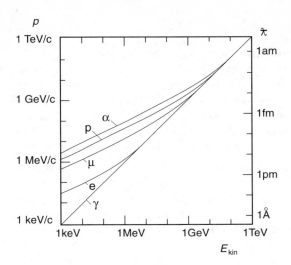

Fig. 4.2. The connection between kinetic energy, momentum and reduced wavelength of photons (γ), electrons (e), muons (μ), protons (p), and ^{4}He nuclei (α). Atomic diameters are typically a few Å (10^{-10} m), nuclear diameters a few fm (10^{-15} m).

$$p \gtrsim \frac{\hbar}{\Delta x} \; , \qquad pc \gtrsim \frac{\hbar c}{\Delta x} \approx \frac{200 \, \mathrm{MeV \, fm}}{\Delta x} \; . \qquad (4.2)$$

Thus to study nuclei, whose radii are of a few fm, beam momenta of the order of $10-100\,\mathrm{MeV}/c$ are necessary. Individual nucleons have radii of about 0.8 fm; and may be resolved if the momenta are above $\approx 100\,\mathrm{MeV}/c$. To resolve the constituents of a nucleon, the quarks, one has to penetrate deeply into the interior of the nucleon. For this purpose, beam momenta of many GeV/c are necessary (see Fig. 1.1).

Inelastic scattering. In inelastic reactions (Fig. 4.1b):

$$\mathrm{a+b} \;\; \rightarrow \;\; \mathrm{a'+b^*}$$
$$\llcorner_{\!\!\rightarrow} \mathrm{c+d} \, ,$$

part of the kinetic energy transferred from a to the target b excites it into a higher energy state b*. The excited state will afterwards return to the ground state by emitting a light particle (e. g. a photon or a π-meson) or it may decay into two or more different particles.

A measurement of a reaction in which only the scattered particle a′ is observed (and the other reaction products are not), is called an *inclusive* measurement. If all reaction products are detected, we speak of an *exclusive* measurement.

When allowed by the laws of conservation of lepton and baryon number (see Sect.s 8.2 and 10.1), the beam particle may completely disappear in the reaction (Fig. 4.1c,d). Its total energy then goes into the excitation of the target or into the production of new particles. Such inelastic reactions represent the basis of nuclear and particle *spectroscopy*, which will be discussed in more detail in the second part of this book.

4.2 Cross Sections

The reaction rates measured in scattering experiments, and the energy spectra and angular distributions of the reaction products yield, as we have already mentioned, information about the dynamics of the interaction between the projectile and the target, i. e., about the shape of the interaction potential and the coupling strength. The most important quantity for the description and interpretation of these reactions is the so-called *cross-section* σ, which is a yardstick of the probability of a reaction between the two colliding particles.

Geometric reaction cross-section. We consider an idealised experiment, in order to elucidate this concept. Imagine a thin scattering target of thickness d with N_b scattering centres b and with a particle density n_b. Each target particle has a cross-sectional area σ_b, to be determined by experiment. We bombard the target with a monoenergetic beam of point-like particles a. A reaction occurs whenever a beam particle hits a target particle, and we assume that the beam particle is then removed from the beam. We do not distinguish between the final target states, i. e., whether the reaction is elastic or inelastic. The total reaction rate $\dot{N}$, i. e. the total number of reactions per unit time, is given by the difference in the beam particle rate $\dot{N}_a$ upstream and downstream of the target. This is a direct measure for the cross-sectional area σ_b (Fig. 4.3).

We further assume that the beam has cross-sectional area A and particle density n_a. The number of projectiles hitting the target per unit area and per unit time is called the *flux* Φ_a. This is just the product of the particle density and the particle velocity v_a:

$$\Phi_a = \frac{\dot{N}_a}{A} = n_a \cdot v_a , \qquad (4.3)$$

and has dimensions $[(\text{area} \times \text{time})^{-1}]$.

The total number of target particles within the beam area is $N_b = n_b \cdot A \cdot d$. Hence the reaction rate $\dot{N}$ is given by the product of the incoming flux and the total cross-sectional area seen by the particles:

$$\dot{N} = \Phi_a \cdot N_b \cdot \sigma_b . \qquad (4.4)$$

This formula is valid as long as the scattering centres do not overlap and particles are only scattered off individual scattering centres. The area presented by a single scattering centre to the incoming projectile a, will be called the *geometric reaction cross-section*: in what follows:

$$
\begin{aligned}
\sigma_b &= \frac{\dot{N}}{\Phi_a \cdot N_b} \\[2mm]
&= \frac{\text{number of reactions per unit time}}{\text{beam particles per unit time per unit area} \times \text{scattering centres}} .
\end{aligned}
\qquad (4.5)
$$

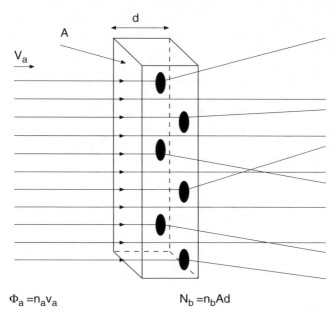

$$\Phi_a = n_a v_a \qquad\qquad N_b = n_b A d$$

Fig. 4.3. Measurement of the geometric reaction cross-section. The particle beam, a, coming from the left with velocity v_a and density n_a, corresponds to a particle flux $\Phi_a = n_a v_a$. It hits a (macroscopic) target of thickness d and cross-sectional area A. Some beam particles are scattered by the scattering centres of the target, i.e., they are deflected from their original trajectory. The frequency of this process is a measure of the cross-sectional area of the scattering particles.

This definition assumes a homogeneous, constant beam (e.g., neutrons from a reactor). In experiments with particle accelerators, the formula used is:

$$\sigma_b = \frac{\text{number of reactions per unit time}}{\text{beam particles per unit time} \times \text{scattering centres per unit area}},$$

since the beam is then generally not homogeneous but the area density of the scattering centres is.

Cross sections. This naive description of the geometric reaction cross-section as the effective cross-sectional area of the target particles, (if necessary convoluted with the cross-sectional area of the beam particles) is in many cases a good approximation to the true reaction cross-section. An example is high-energy proton-proton scattering where the geometric extent of the particles is comparable to their interaction range.

The reaction probability for two particles is, however, generally very different to what these geometric considerations would imply. Furthermore a strong energy dependence is also observed. The reaction rate for the capture of thermal neutrons by uranium, for example, varies by several orders of magnitude within a small energy range. The reaction rate for scattering of (point-like) neutrinos, which only feel the weak interaction, is much smaller than that for the scattering of (also point-like) electrons which feel the electromagnetic interaction.

The shape, strength and range of the interaction potential, and not the geometric forms involved in the scattering process, primarily determine the effective cross-sectional area. The interaction can be determined from the reaction rate if the flux of the incoming beam particles, and the area density of the scattering centres are known, just as in the model above. The *total cross-section* is defined analogously to the geometric one:

$$\sigma_{\text{tot}} = \frac{\text{number of reactions per unit time}}{\text{beam particles per unit time} \times \text{scattering centres per unit area}} .$$

In analogy to the *total* cross-section, cross-sections for *elastic* reactions σ_{el} and for *inelastic* reactions σ_{inel} may also be defined. The inelastic part can be further divided into different reaction channels. The *total cross-section* is the sum of these parts:

$$\sigma_{\text{tot}} = \sigma_{\text{el}} + \sigma_{\text{inel}} . \tag{4.6}$$

The cross-section is a physical quantity with dimensions of [area], and is independent of the specific experimental design. A commonly used unit is the *barn,* which is defined as:

$$1 \text{ barn } = 1 \text{ b } = 10^{-28} \text{ m}^2$$
$$1 \text{ millibarn } = 1 \text{ mb } = 10^{-31} \text{ m}^2$$
$$\text{etc.}$$

Typical total cross-sections at a beam energy of $10\,\text{GeV}$, for example, are

$$\sigma_{\text{pp}}(10 \text{ GeV}) \approx 40 \text{ mb} \tag{4.7}$$

for proton-proton scattering; and

$$\sigma_{\nu\text{p}}(10 \text{ GeV}) \approx 7 \cdot 10^{-14} \text{ b} = 70 \text{ fb} \tag{4.8}$$

for neutrino-proton scattering.

Luminosity. The quantity

$$\mathcal{L} = \Phi_{\text{a}} \cdot N_{\text{b}} \tag{4.9}$$

is called the *luminosity.* Like the flux, it has dimensions of $[(\text{area} \times \text{time})^{-1}]$. From (4.3) and $N_{\text{b}} = n_{\text{b}} \cdot d \cdot A$ we have

$$\mathcal{L} = \Phi_{\text{a}} \cdot N_{\text{b}} = \dot{N}_{\text{a}} \cdot n_{\text{b}} \cdot d = n_{\text{a}} \cdot v_{\text{a}} \cdot N_{\text{b}} . \tag{4.10}$$

Hence the luminosity is the product of the number of incoming beam particles per unit time $\dot{N}_{\text{a}}$, the target particle density in the scattering material n_{b},

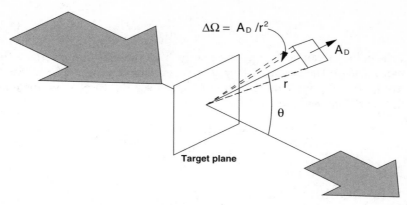

Fig. 4.4. Description of the differential cross-section. Only particles scattered into the small solid angle $\Delta\Omega$ are recorded by the detector of cross-sectional area A_D.

and the target's thickness d; or the beam particle density n_a, their velocity v_a and the number of target particles N_b exposed to the beam.

There is an analogous equation for the case of two particle beams colliding in a storage ring. Assume that j particle packets, each of N_a or N_b particles, have been injected into a ring of circumference U. The two particle types circulate with velocity v in opposite directions. Steered by magnetic fields, they collide at an interaction point $j \cdot v/U$ times per unit time. The luminosity is then:

$$\mathcal{L} = \frac{N_\mathrm{a} \cdot N_\mathrm{b} \cdot j \cdot v/U}{A} \qquad (4.11)$$

where A is the beam cross-section at the collision point. For a Gaussian distribution of the beam particles around the beam centre (with horizontal and vertical standard deviations σ_x and σ_y respectively), A is given by:

$$A = 4\pi\sigma_x\sigma_y . \qquad (4.12)$$

To achieve a high luminosity, the beams must be focused at the interaction point into the smallest possible cross-sectional area possible. Typical beam diameters are of the order of tenths of millimeters or less.

An often used quantity in storage ring experiments is the integrated luminosity $\int \mathcal{L}\,\mathrm{d}t$. The number of reactions which can be observed in a given reaction time is just the product of the integrated luminosity and the cross-section. With a 1 nb cross-section and a 100 pb^{-1} integrated luminosity, for example, 10^5 reactions would be expected.

Differential cross sections. In practice, only a fraction of all the reactions are measured. A detector of area A_D is placed at a distance r and at an angle θ with respect to the beam direction, covering a solid angle $\Delta\Omega = A_\mathrm{D}/r^2$ (Fig. 4.4). The rate of reactions seen by this detector is then proportional to the *differential cross-section* $\mathrm{d}\sigma(E,\theta)/\mathrm{d}\Omega$:

$$\dot{N}(E,\theta,\Delta\Omega) = \mathcal{L} \cdot \frac{d\sigma(E,\theta)}{d\Omega} \Delta\Omega \ . \tag{4.13}$$

If the detector can determine the energy E' of the scattered particles then one can measure the *doubly differential* cross-section $d^2\sigma(E,E',\theta)/d\Omega\,dE'$. The total cross-section σ is then the integral over the total solid angle and over all scattering energies:

$$\sigma_{\text{tot}}(E) = \int_0^{E'_{\text{max}}} \int_{4\pi} \frac{d^2\sigma(E,E',\theta)}{d\Omega\,dE'}\ d\Omega\,dE' \ . \tag{4.14}$$

4.3 The "Golden Rule"

The cross-section can be experimentally determined from the reaction rate $\dot{N}$, as we saw above. We now outline how it may be found from theory.

First, the reaction rate is dependent upon the properties of the interaction potential described by the Hamilton operator $\mathcal{H}_{\text{int}}$. In a reaction, this potential transforms the initial-state wave function ψ_i into the final-state wave function ψ_f. The *transition matrix element* is given by:

$$\mathcal{M}_{fi} = \langle \psi_f | \mathcal{H}_{\text{int}} | \psi_i \rangle = \int \psi_f^* \, \mathcal{H}_{\text{int}} \, \psi_i \ dV \ . \tag{4.15}$$

This matrix element is also called the *probability amplitude* for the transition.

Furthermore, the reaction rate will depend upon the number of final states available to the reaction. According to the uncertainty principle, each particle occupies a volume $h^3 = (2\pi\hbar)^3$ in *phase space*, the six-dimensional space of momentum and position. Consider a particle scattered into a volume V and into a momentum interval between p' and $p' + dp'$. In momentum space, the interval corresponds to a spherical shell with inner radius p' and thickness dp' which has a volume $4\pi p'^2 dp'$. Excluding processes where the spin changes, the number of final states available is:

$$dn(p') = \frac{V \cdot 4\pi p'^2}{(2\pi\hbar)^3} dp' \ . \tag{4.16}$$

The energy and momentum of a particle are connected by:

$$dE' = v'dp'. \tag{4.17}$$

Hence the density of final states in the energy interval dE' is given by:

$$\varrho(E') = \frac{dn(E')}{dE'} = \frac{V \cdot 4\pi p'^2}{v' \cdot (2\pi\hbar)^3} \ . \tag{4.18}$$

The connection between the reaction rate, the transition matrix element and the density of final states is expressed by Fermi's *second golden rule*.

Its derivation can be found in quantum mechanics textbooks (e. g. [Sc95]). It expresses the reaction rate W per target particle and per beam particle in the form:

$$W = \frac{2\pi}{\hbar} \, |\mathcal{M}_{fi}|^2 \cdot \varrho(E') \, . \tag{4.19}$$

We also know, however, from (4.3) and (4.4) that:

$$W = \frac{\dot{N}(E)}{N_{\mathrm{b}} \cdot N_{\mathrm{a}}} = \frac{\sigma \cdot v_{\mathrm{a}}}{V}, \tag{4.20}$$

where $V = N_{\mathrm{a}}/n_{\mathrm{a}}$ is the spatial volume occupied by the beam particles. Hence, the cross-section is:

$$\sigma = \frac{2\pi}{\hbar \cdot v_{\mathrm{a}}} \, |\mathcal{M}_{fi}|^2 \cdot \varrho\,(E') \cdot V \, . \tag{4.21}$$

If the interaction potential is known, the cross-section can be calculated from (4.21). Otherwise, the cross-section data and equation (4.21) can be used to determine the transition matrix element.

The golden rule applies to both scattering and spectroscopic processes. Examples of the latter are the decay of unstable particles, excitation of particle resonances and transitions between different atomic or nuclear energy states. In these cases we have

$$W = \frac{1}{\tau} \, , \tag{4.22}$$

and the transition probability per unit time can be either directly determined by measuring the lifetime τ or indirectly read off from the energy width of the state $\Delta E = \hbar/\tau$.

4.4 Feynman Diagrams

> In QED, as in other quantum field theories, we can use the little pictures invented by my colleague Richard Feynman, which are supposed to give the illusion of understanding what is going on in quantum field theory.
>
> *M. Gell-Mann* [Ge80]

Elementary processes such as the scattering of two particles off each other or the decay of a single particle are nowadays commonly depicted by *Feynman diagrams*. Originally, these diagrams were introduced by Feynman as a sort of shorthand for the individual terms in his calculations of transition matrix elements $\mathcal{M}_{fi}$ in electromagnetic processes in the framework of *quantum electrodynamics* (QED). Each symbol in such a space-time diagram corresponds to a term in the matrix element. The meaning of the individual terms and the links between them are fixed by the *Feynman rules*. Similarly to the QED rules, corresponding prescriptions exist for the calculation of weak and strong processes as well, in *quantum chromodynamics* (QCD). We will not use such diagrams for quantitative calculations, since this requires knowledge of relativistic field theory. Instead, they will serve as pictorial illustrations of the processes that occur. We will therefore merely treat a few examples below and explain some of the definitions and rules.

Figure 4.5 shows some typical diagrams. We use the convention that the time axis runs upwards and the space axis from left to right. The straight lines in the graphs correspond to the wave functions of the initial and final fermions. Antiparticles (in our examples: the positron e^+, the positive muon μ^+ and the electron antineutrino $\overline{\nu}_e$) are symbolized by arrows pointing backwards in time; photons by wavy lines; heavy vector bosons by dashed lines; and gluons by corkscrew-like lines.

As we mentioned in Chap. 1, the electromagnetic interaction between charged particles proceeds via photon exchange. Figure 4.5a depicts schematically the elastic scattering of an electron off a positron. The interaction process corresponds to a photon being emitted by the electron and absorbed by the positron. Particles appearing neither in the initial nor in the final

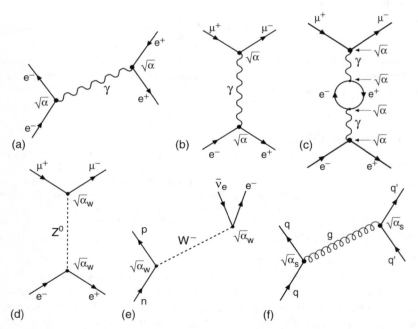

Fig. 4.5. Feynman diagrams for the electromagnetic (**a, b, c**), weak (**d, e**) and strong interactions (**f**).

state, such as this exchanged photon, are called *virtual particles*. Because of the uncertainty principle, virtual particles do not have to satisfy the energy-momentum relation $E^2 = \boldsymbol{p}^2 c^2 + m^2 c^4$. This may be interpreted as meaning that the exchanged particle has a mass different from that of a free (real) particle, or that energy conservation is violated for a brief period of time.

Points at which three or more particles meet are called *vertices*. Each vertex corresponds to a term in the transition matrix element which includes the structure and strength of the interaction. In (a), the exchanged photon couples to the charge of the electron at the left vertex and to that of the positron at the right vertex. For each vertex the transition amplitude contains a factor which is proportional to e, i.e., $\sqrt{\alpha}$.

Figure 4.5b represents the annihilation of an electron-positron pair. A photon is created as an intermediate state which then decays into a negatively charged μ^- and its positively charged antiparticle, a μ^+. Figure 4.5c shows a slightly more complicated version of the same process. Here, the photon, by vacuum polarisation, is briefly transformed into an intermediate state made up of an $e^+ e^-$ pair. This and additional, more complicated, diagrams contributing to the same process are called *higher-order diagrams*.

The transition matrix element includes the superposition of amplitudes of all diagrams leading to the same final state. Because the number of vertices is greater in higher-order diagrams these graphs include higher powers of α. The amplitude of diagram (b) is proportional to α, while diagram (c)'s is proportional to α^2. The cross-section for conversion of an electron-positron pair into a $\mu^+ \mu^-$ pair is therefore given to a good approximation by diagram (b). Diagram (c) and other diagrams of even higher order produce only small corrections to (b).

Figure 4.5d shows electron-positron annihilation followed by muon pair production in a weak interaction proceeding through exchange of the neutral, heavy vector boson Z^0. In Fig. 4.5e, we see a neutron transform into a proton via β-decay in which it emits a negatively charged heavy vector boson W^- which subsequently decays into an electron and antineutrino $\overline{\nu}_e$. Figure 4.5f depicts a strong interaction process between two quarks q and q' which exchange a gluon, the field quantum of the strong interaction.

In weak interactions, a heavy vector boson is exchanged which couples to the "weak charge" g and not to the electric charge e. Accordingly, $\mathcal{M}_{fi} \propto g^2 \propto \alpha_w$. In strong interactions the gluons which are exchanged between the quarks couple to the "colour charge" of the quarks, $\mathcal{M}_{fi} \propto \sqrt{\alpha_s} \cdot \sqrt{\alpha_s} = \alpha_s$.

The exchange particles contribute a *propagator* term to the transition matrix element. This contribution has the general form

$$\frac{1}{Q^2 + M^2 c^2} \, . \tag{4.23}$$

Here Q^2 is the square of the four-momentum (cf. 5.3 and 6.3) which is transferred in the interaction and M is the mass of the exchange particle. In the case of a virtual photon, this results in a factor $1/Q^2$ in the amplitude and $1/Q^4$ in the cross-section. In the weak interaction, the large mass of the exchanged vector boson causes the cross-section to be much smaller than that of the electromagnetic interaction — although at very high momentum transfers, of the order of the masses of the vector bosons, the two cross-sections become comparable in size.

Problems

1. **Cross-section**
   Deuterons with an energy $E_{kin} = 5\,\mathrm{MeV}$ are perpendicularly incident upon a tritium target, which has a mass occupation density $\mu_t = 0.2\,\mathrm{mg/cm^2}$, so as to investigate the reaction $^3\mathrm{H(d,n)^4He}$.
   a) How many neutrons per second pass through a detector with a reception area of $A = 20\,\mathrm{cm^2}$ which is at a distance $R = 3\,\mathrm{m}$ from the target and an angle $\theta = 30°$ to the deuteron beam direction, if the differential cross-section $d\sigma/d\Omega$ at this angle is $13\,\mathrm{mb/sr}$ and the deuteron current applied to the target is $I_d = 2\,\mu\mathrm{A}$?
   b) How many neutrons per second does the detector receive if the target is tilted so that the same deuteron current now approaches it at $80°$ instead of $90°$?

2. **Absorption length**
   A particle beam is incident upon a thick layer of an absorbing material (with n absorbing particles per unit volume). How large is the absorption length, i.e., the distance over which the intensity of the beam is reduced by a factor of $1/e$ for the following examples?
   a) Thermal neutrons ($E \approx 25\,\mathrm{meV}$) in cadmium ($\varrho = 8.6\,\mathrm{g/cm^3}$, $\sigma = 24\,506$ barn).
   b) $E_\gamma = 2\,\mathrm{MeV}$ photons in lead ($\varrho = 11.3\,\mathrm{g/cm^3}$, $\sigma = 15.7$ barn/atom).
   c) Antineutrinos from a reactor in earth ($\varrho = 5\,\mathrm{g/cm^3}$, $\sigma \approx 10^{-19}$ barn/electron; interactions with nuclei may be neglected; $Z/A \approx 0.5$).

5. Geometric Shapes of Nuclei

In this chapter we shall study nuclear sizes and shapes. In principle, this information may be obtained from scattering experiments (e.g., scattering of protons or α particles) and when Rutherford discovered that nuclei have a radial extent of less than 10^{-14} m, he employed α scattering. In practice, however, there are difficulties in extracting detailed information from such experiments. Firstly, these projectiles are themselves extended objects. Therefore, the cross-section reflects not only the structure of the target, but also that of the projectile. Secondly, the nuclear forces between the projectile and the target are complex and not well understood.

Electron scattering is particularly valuable for investigating small objects. As far as we know electrons are point-like objects without any internal structure. The interactions between an electron and a nucleus, nucleon or quark take place via the exchange of a virtual photon — this may be very accurately calculated inside quantum electrodynamics (QED). These processes are in fact manifestations of the well known electromagnetic interaction, whose coupling constant $\alpha \approx 1/137$ is much less than one. This last means that higher order corrections play only a tiny role.

5.1 Kinematics of Electron Scattering

In electron scattering experiments one employs highly relativistic particles. Hence it is advisable to use four-vectors in kinematical calculations. The zero component of space–time four-vectors is time, the zero component of four-momentum vectors is energy:

$$
\begin{aligned}
x &= (x_0, x_1, x_2, x_3) &&= (ct, \boldsymbol{x}) \,, \\
p &= (p_0, p_1, p_2, p_3) &&= (E/c, \boldsymbol{p}) \,.
\end{aligned}
\tag{5.1}
$$

Three-vectors are designated by bold-faced type to distinguish them from four-vectors. The Lorentz-invariant scalar product of two four-vectors a and b is defined by

$$
a \cdot b = a_0 b_0 - a_1 b_1 - a_2 b_2 - a_3 b_3 = a_0 b_0 - \boldsymbol{a} \cdot \boldsymbol{b} \,.
\tag{5.2}
$$

In particular, this applies to the four-momentum squared:

$$p^2 = \frac{E^2}{c^2} - \boldsymbol{p}^2 \ . \tag{5.3}$$

This squared product is equal to the square of the rest mass m (multiplied by c^2). This is so since a reference frame in which the particle is at rest can always be found and there $\boldsymbol{p} = 0$, and $E = mc^2$. The quantity

$$m = \sqrt{p^2} \, /c \tag{5.4}$$

is called the *invariant mass*. From (5.3) and (5.4) we obtain the relativistic energy-momentum relation:

$$E^2 - \boldsymbol{p}^2 c^2 = m^2 c^4 \tag{5.5}$$

and thus

$$E \approx |\boldsymbol{p}| \, c \quad \text{if} \quad E \gg mc^2 \ . \tag{5.6}$$

For electrons, this approximation is already valid at energies of a few MeV.

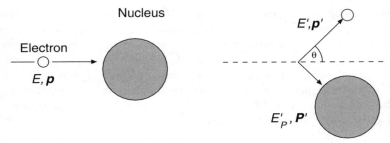

Fig. 5.1. Kinematics of elastic electron-nucleus scattering.

■ Consider the scattering of an electron with four-momentum p off a particle with four-momentum P (Fig. 5.1). Energy and momentum conservation imply that the sums of the four-momenta before and after the reaction are identical:

$$p + P = p' + P' \ , \tag{5.7}$$

or squared:

$$p^2 + 2pP + P^2 = p'^2 + 2p'P' + P'^2 \ . \tag{5.8}$$

In elastic scattering the invariant masses m_e and M of the colliding particles are unchanged. Hence from:

$$p^2 = p'^2 = m_e^2 c^2 \quad \text{and} \quad P^2 = P'^2 = M^2 c^2 \tag{5.9}$$

it follows that:

$$p \cdot P = p' \cdot P' \ . \tag{5.10}$$

Usually only the scattered electron is detected and not the recoiling particle. In this case the relation:

$$p \cdot P = p' \cdot (p + P - p') = p'p + p'P - m_e^2 c^2 \qquad (5.11)$$

is used. Consider the laboratory frame where the particle with four-momentum P is at rest before the collision. Then the four-momenta can be written as:

$$p = (E/c, \boldsymbol{p}) \quad p' = (E'/c, \boldsymbol{p}') \quad P = (Mc, \boldsymbol{0}) \quad P' = (E_P'/c, \boldsymbol{P}') \ . \qquad (5.12)$$

Hence (5.11) yields:

$$E \cdot Mc^2 = E'E - \boldsymbol{p}\boldsymbol{p}'c^2 + E'Mc^2 - m_e^2 c^4 \ . \qquad (5.13)$$

At high energies, $m_e^2 c^4$ may be neglected and $E \approx |\boldsymbol{p}| \cdot c$ (Eq. 5.6) can be safely used. One thus obtains a relation between between the angle and the energy:

$$E \cdot Mc^2 = E'E \cdot (1 - \cos\theta) + E' \cdot Mc^2 \ . \qquad (5.14)$$

In the laboratory system, the energy E' of the scattered electron is:

$$E' = \frac{E}{1 + E/Mc^2 \cdot (1 - \cos\theta)} \ . \qquad (5.15)$$

The angle θ through which the electron is deflected is called the *scattering angle*. The recoil which is transferred to the target is given by the difference $E - E'$. In elastic scattering, a one to one relationship (5.15) exists between the scattering angle θ and the energy E' of the scattered electron; (5.15) does not hold for inelastic scattering.

The angular dependence of the scattering energy E' is described by the term $(1 - \cos\theta)$ multiplied by E/Mc^2. Hence the recoil energy of the target increases with the ratio of the relativistic electron mass E/c^2 to the target mass M. This is in accordance with the classical laws of collision.

In electron scattering at the relatively low energy of 0.5 GeV off a nucleus with mass number $A = 50$ the scattering energy varies by only 2 % between forward and backward scattering. The situation is very different for 10 GeV-electrons scattering off protons. The scattering energy E' then varies between 10 GeV ($\theta \approx 0°$) and 445 MeV ($\theta = 180°$) (cf. Fig. 5.2).

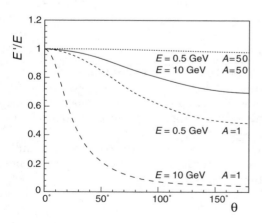

Fig. 5.2. Angular dependence of the scattering energy of electrons normalised to beam energy, E'/E, in elastic electron-nucleus scattering. The curves show this dependence for two different beam energies (0.5 GeV and 10 GeV) and for two nuclei with different masses ($A = 1$ and $A = 50$).

5.2 The Rutherford Cross-Section

We will now consider the cross-section for an electron with energy E scattering off an atomic nucleus with charge Ze. For the calculation of the reaction kinematics to be sufficiently precise, it must be both relativistic and quantum mechanical. We will approach this goal step by step. Firstly, we introduce the Rutherford scattering formula. By definition, this formula yields the cross-section up to spin effects. For heavy nuclei and low energy electrons, the recoil can, from (5.15), be neglected. In this case, the energy E and the modulus of the momentum $\boldsymbol{p}$ are the same before and after the scattering. The kinematics can be calculated in the same way as, for example, the hyperbolic trajectory of a comet which is deflected by the sun as it traverses the solar system. As long as the radius of the scattering centre (nucleus, sun) is smaller than the closest approach of the projectile (electron, comet) then the spatial extension of the scattering centre does not affect this purely classical calculation. This leads to the Rutherford formula for the scattering of a particle with charge ze and kinetical energy E_{kin} on a target nucleus with charge Ze:[1]

$$\left(\frac{\mathrm{d}\sigma}{\mathrm{d}\Omega}\right)_{\text{Rutherford}} = \frac{(zZe^2)^2}{(4\pi\varepsilon_0)^2 \cdot (4E_{\text{kin}})^2 \sin^4 \frac{\theta}{2}} \, . \tag{5.16}$$

Exactly the same equation is obtained by a calculation of this cross-section in non-relativistic quantum mechanics using Fermi's golden rule. This we will now demonstrate. To avoid unnecessary repetitions we will consider the case of a central charge with finite spatial distribution.

Scattering off an extended charge distribution. Consider the case of a target so heavy that the recoil is negligible. We can then use three-momenta. If Ze is small, i. e. if:

$$Z\alpha \ll 1 \, , \tag{5.17}$$

the *Born approximation* can be applied, and the wave functions ψ_i and ψ_f of the incoming and of the outgoing electron can be described by plane waves:

$$\psi_i = \frac{1}{\sqrt{V}} \, \mathrm{e}^{i\boldsymbol{p}\boldsymbol{x}/\hbar} \qquad \psi_f = \frac{1}{\sqrt{V}} \, \mathrm{e}^{i\boldsymbol{p}'\boldsymbol{x}/\hbar} \, . \tag{5.18}$$

We can sidestep any difficulties related to the normalisation of the wave functions by considering only a finite volume V. We need this volume to be large compared to the scattering centre, and also large enough that the discrete energy states in this volume can be approximated by a continuum. The physical results have, of course, to be independent of V.

We consider an electron beam with a density of n_a particles per unit volume. With the volume of integration chosen to be sufficiently large, the normalisation condition is given by:

---

[1] The derivation may be found in, e.g., [Ge92].

$$\int_V |\psi_i|^2 \, \mathrm{d}V = n_a \cdot V \qquad \text{where} \quad V = \frac{N_a}{n_a} \, , \tag{5.19}$$

i.e. V is the normalisation volume that must be chosen for a single beam particle.

According to (4.20), the reaction rate W is given by the product of the cross-section σ and the beam particle velocity v_a divided by the above volume. When applying the golden rule (4.19), we get:

$$\frac{\sigma v_a}{V} = W = \frac{2\pi}{\hbar} \, |\langle \psi_f | \mathcal{H}_{\mathrm{int}} | \psi_i \rangle|^2 \, \frac{\mathrm{d}n}{\mathrm{d}E_f} \, . \tag{5.20}$$

Here, E_f is the total energy (kinetic energy and rest mass) of the final state. Since we neglect the recoil and since the rest mass is a constant, $\mathrm{d}E_f = \mathrm{d}E' = \mathrm{d}E$.

The density n of possible final states in phase space (cf. 4.16) is:

$$\mathrm{d}n(|\boldsymbol{p}'|) = \frac{4\pi |\boldsymbol{p}'|^2 \mathrm{d}|\boldsymbol{p}'| \cdot V}{(2\pi\hbar)^3} \, . \tag{5.21}$$

Therefore the cross-section for the scattering of an electron into a solid angle element $\mathrm{d}\Omega$ is:

$$\mathrm{d}\sigma \cdot v_a \cdot \frac{1}{V} = \frac{2\pi}{\hbar} \, |\langle \psi_f | \mathcal{H}_{\mathrm{int}} | \psi_i \rangle|^2 \, \frac{V |\boldsymbol{p}'|^2 \mathrm{d}|\boldsymbol{p}'|}{(2\pi\hbar)^3 \mathrm{d}E_f} \, \mathrm{d}\Omega \, . \tag{5.22}$$

The velocity v_a can be replaced, to a good approximation, by the velocity of light c. For large electron energies, $|\boldsymbol{p}'| \approx E'/c$ applies, and we obtain:

$$\frac{\mathrm{d}\sigma}{\mathrm{d}\Omega} = \frac{V^2 E'^2}{(2\pi)^2 (\hbar c)^4} \, |\langle \psi_f | \mathcal{H}_{\mathrm{int}} | \psi_i \rangle|^2 \, . \tag{5.23}$$

The interaction operator for a charge e in an electric potential ϕ is $\mathcal{H}_{\mathrm{int}} = e\phi$. Hence, the matrix element is:

$$\langle \psi_f | \mathcal{H}_{\mathrm{int}} | \psi_i \rangle = \frac{e}{V} \int \mathrm{e}^{-i\boldsymbol{p}'\boldsymbol{x}/\hbar} \, \phi(\boldsymbol{x}) \, \mathrm{e}^{i\boldsymbol{p}\boldsymbol{x}/\hbar} \mathrm{d}^3 x \, . \tag{5.24}$$

Defining the *momentum transfer* $\boldsymbol{q}$ by:

$$\boldsymbol{q} = \boldsymbol{p} - \boldsymbol{p}', \tag{5.25}$$

we may re-write the matrix element as:

$$\langle \psi_f | \mathcal{H}_{\mathrm{int}} | \psi_i \rangle = \frac{e}{V} \int \phi(\boldsymbol{x}) \, \mathrm{e}^{i\boldsymbol{q}\boldsymbol{x}/\hbar} \mathrm{d}^3 x \, . \tag{5.26}$$

■ Green's theorem permits us to use a clever trick here: for two arbitrarily chosen scalar fields u and v, which fall off fast enough at large distances, the following equation holds for a sufficiently large integration volume:

$$\int (u\triangle v - v\triangle u)\,\mathrm{d}^3 x = 0\,, \qquad \text{with} \quad \triangle = \nabla^2\,. \tag{5.27}$$

Inserting:

$$\mathrm{e}^{i\boldsymbol{qx}/\hbar} = \frac{-\hbar^2}{|\boldsymbol{q}|^2}\cdot\triangle\mathrm{e}^{i\boldsymbol{qx}/\hbar} \tag{5.28}$$

into (5.26), we may rewrite the matrix element as:

$$\langle\psi_f|\mathcal{H}_{\mathrm{int}}|\psi_i\rangle \;=\; \frac{-e\hbar^2}{V|\boldsymbol{q}|^2}\int \triangle\phi(\boldsymbol{x})\,\mathrm{e}^{i\boldsymbol{qx}/\hbar}\,\mathrm{d}^3 x\,. \tag{5.29}$$

The potential $\phi(\boldsymbol{x})$ and the charge density $\varrho(\boldsymbol{x})$ are related by Poisson's equation:

$$\triangle\phi(\boldsymbol{x}) = \frac{-\varrho(\boldsymbol{x})}{\varepsilon_0}\,. \tag{5.30}$$

In the following, we will assume the charge density $\varrho(\boldsymbol{x})$ to be static, i. e. independent of time.

We now define a charge distribution function f by $\varrho(\boldsymbol{x}) = Zef(\boldsymbol{x})$ which satisfies the normalisation condition $\int f(\boldsymbol{x})\,\mathrm{d}^3 x = 1$, and re-write the matrix element as:

$$\begin{aligned}\langle\psi_f|\mathcal{H}_{\mathrm{int}}|\psi_i\rangle \;&=\; \frac{e\hbar^2}{\varepsilon_0\cdot V|\boldsymbol{q}|^2}\int \varrho(\boldsymbol{x})\,\mathrm{e}^{i\boldsymbol{qx}/\hbar}\mathrm{d}^3 x \\[2mm] &=\; \frac{Z\cdot 4\pi\alpha\hbar^3 c}{|\boldsymbol{q}|^2\cdot V}\int f(\boldsymbol{x})\,\mathrm{e}^{i\boldsymbol{qx}/\hbar}\mathrm{d}^3 x\,.\end{aligned} \tag{5.31}$$

The integral

$$F(\boldsymbol{q}) = \int \mathrm{e}^{i\boldsymbol{qx}/\hbar}f(\boldsymbol{x})\mathrm{d}^3 x \tag{5.32}$$

is the Fourier transform of the charge function $f(\boldsymbol{x})$, normalised to the total charge. It is called the *form factor* of the charge distribution. The form factor contains all the information about the spatial distribution of the charge of the object being studied. We will discuss form factors and their meaning in the following chapters in some detail.

To calculate the Rutherford cross section we, by definition, neglect the spatial extension — i. e., we replace the charge distribution by a δ-function. Hence, the form factor is fixed to unity. By inserting the matrix element into (5.23) we obtain:

$$\left(\frac{\mathrm{d}\sigma}{\mathrm{d}\Omega}\right)_{\mathrm{Rutherford}} = \frac{4Z^2\alpha^2(\hbar c)^2 E'^2}{|\boldsymbol{q}c|^4}\,. \tag{5.33}$$

The $1/\boldsymbol{q}^4$-dependence of the electromagnetic cross-section implies very low event rates for electron scattering with large momentum transfers. The event rates drop off so sharply that small measurement errors in $\boldsymbol{q}$ can significantly falsify the results.

■ Since recoil is neglected in Rutherford scattering, the electron energy and the magnitude of its momentum do not change in the interaction:

$$E = E' , \qquad |\boldsymbol{p}| = |\boldsymbol{p'}| . \qquad (5.34)$$

The magnitude of the momentum transfer $\boldsymbol{q}$ is therefore:

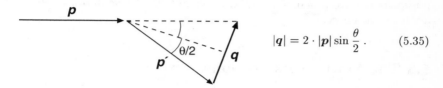

$$|\boldsymbol{q}| = 2 \cdot |\boldsymbol{p}| \sin \frac{\theta}{2} . \qquad (5.35)$$

If we recall that $E = |\boldsymbol{p}| \cdot c$ is a good approximation we obtain the relativistic Rutherford scattering formula:

$$\left(\frac{d\sigma}{d\Omega} \right)_{\text{Rutherford}} = \frac{Z^2 \alpha^2 (\hbar c)^2}{4 E^2 \sin^4 \frac{\theta}{2}} . \qquad (5.36)$$

The classical Rutherford formula (5.16) may be obtained from (5.33) by applying nonrelativistic kinematics: $\boldsymbol{p} = m\boldsymbol{v}$, $E_{\text{kin}} = m\boldsymbol{v}^2/2$ and $E' \approx mc^2$.

Field-theoretical considerations. The sketch on the right is a pictorial representation of a scattering process. In the language of field theory, the electromagnetic interaction of an electron with the charge distribution is mediated by the exchange of a photon, the field quantum of this interaction. The photon which does not itself carry any charge, couples to the charges of the two interacting particles. In the transition matrix element, this yields a factor $Ze \cdot e$ and in the cross-section we have a term $(Ze^2)^2$. The three-momentum transfer $\boldsymbol{q}$ defined in (5.25)

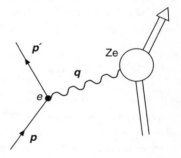

is the momentum transferred by the exchanged photon. Hence the reduced de-Broglie wavelength of the photon is:

$$\lambdabar = \frac{\hbar}{|\boldsymbol{q}|} = \frac{\hbar}{|\boldsymbol{p}|} \cdot \frac{1}{2 \sin \frac{\theta}{2}} . \qquad (5.37)$$

If $\lambdabar$ is considerably larger than the spatial extent of the target particle, internal structures cannot be resolved, and the target particle may be considered to be pointlike. The Rutherford cross-section from (5.33) was obtained for this case.

In the form (5.33), the dependence of the cross-section on the momentum transfer is clearly expressed. To lowest order the interaction is mediated by the exchange of a photon. Since the photon is massless, the propagator

(4.23) in the matrix element is $1/Q^2$, or $1/|\boldsymbol{q}|^2$ in a non-relativistic approxi-
mation. The propagator enters the cross-section squared which leads to the
characteristic fast $1/|\boldsymbol{q}|^4$ fall-off of the cross-section.

If the Born approximation condition (5.17) no longer holds, then our simple picture must be modified. Higher order corrections (exchange of several
photons) must be included and more complicated calculations (phase shift
analyses) are necessary.

5.3 The Mott Cross-Section

Up to now we have neglected the spins of the electron and of the target.
At relativistic energies, however, the Rutherford cross-section is modified by
spin effects. The *Mott cross-section*, which describes electron scattering and
includes effects due to the electron spin, may be written as:

$$\left(\frac{\mathrm{d}\sigma}{\mathrm{d}\Omega}\right)^*_{\mathrm{Mott}} = \left(\frac{\mathrm{d}\sigma}{\mathrm{d}\Omega}\right)_{\mathrm{Rutherford}} \cdot \left(1 - \beta^2 \sin^2\frac{\theta}{2}\right), \quad \text{with } \beta = \frac{v}{c}. \quad (5.38)$$

The asterisk indicates that the recoil of the nucleus has been neglected in
deriving this equation. The expression shows that, at relativistic energies,
the Mott cross-section drops off more rapidly at large scattering angles than
does the Rutherford cross-section. In the limiting case of $\beta \to 1$, and using
$\sin^2 x + \cos^2 x = 1$, the Mott cross-section can be written in a simpler form:

$$\left(\frac{\mathrm{d}\sigma}{\mathrm{d}\Omega}\right)^*_{\mathrm{Mott}} = \left(\frac{\mathrm{d}\sigma}{\mathrm{d}\Omega}\right)_{\mathrm{Rutherford}} \cdot \cos^2\frac{\theta}{2} = \frac{4Z^2\alpha^2(\hbar c)^2 E'^2}{|\boldsymbol{q}c|^4}\cos^2\frac{\theta}{2}. \quad (5.39)$$

The additional factor in (5.39) can be understood by considering the
extreme case of scattering through 180°. For relativistic particles in the limit
$\beta \to 1$, the projection of their spin $\boldsymbol{s}$ on the direction of their motion $\boldsymbol{p}/|\boldsymbol{p}|$ is
a conserved quantity. This conservation law follows from the solution of the
Dirac equation in relativistic quantum mechanics [Go86]. It is usually called
conservation of *helicity* rather than conservation of the projection of the spin.
Helicity is defined by:

$$h = \frac{\boldsymbol{s} \cdot \boldsymbol{p}}{|\boldsymbol{s}| \cdot |\boldsymbol{p}|}. \quad (5.40)$$

Particles with spin pointing in the direction of their motion have helicity $+1$,
particles with spin pointing in the opposite direction have helicity -1.

Figure 5.3 shows the kinematics of scattering through 180°. We here choose the momentum direction of the incoming electron as the axis of quantisation $\hat{z}$. Because of conservation of helicity, the projection of the spin on
the $\hat{z}$-axis would have to turn over (spin-flip). This, however, is impossible
with a spinless target, because of conservation of total angular momentum.
The orbital angular momentum $\boldsymbol{\ell}$ is perpendicular to the direction of motion

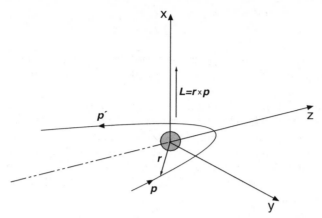

Fig. 5.3. Helicity, $h = \boldsymbol{s} \cdot \boldsymbol{p}/(|\boldsymbol{s}| \cdot |\boldsymbol{p}|)$, is conserved in the $\beta \to 1$ limit. This means that the spin projection on the z-axis would have to change its sign in scattering through 180°. This is impossible if the target is spinless, because of conservation of angular momentum.

$\hat{z}$. It therefore cannot cause any change in the z-component of the angular momentum. Hence in the limiting case $\beta \to 1$, scattering through 180° must be completely suppressed.

If the target has spin, the spin projection of the electron can be changed, as conservation of angular momentum can be compensated by a change in the spin direction of the target. In this case, the above reasoning is not valid, and scattering through 180° is possible.

5.4 Nuclear Form Factors

In actual scattering experiments with nuclei or nucleons, we see that the Mott cross-sections agree with the experimental cross-sections only in the limit $|\boldsymbol{q}| \to 0$. At larger values of $|\boldsymbol{q}|$, the experimental cross-sections are systematically smaller. The reason for this lies in the spatial extension of nuclei and nucleons. At larger values of $|\boldsymbol{q}|$, the reduced wavelength of the virtual photon decreases (5.37), and the resolution increases. The scattered electron no longer sees the total charge, but only parts of it. Therefore, the cross-section decreases.

As we have seen, the spatial extension of a nucleus is described by a form factor (5.32). In the following, we will restrict the discussion to the form factors of spherically symmetric systems which have no preferred orientation in space. In this case, the form factor only depends on the momentum transfer $\boldsymbol{q}$. We symbolize this fact by writing the form factor as $F(\boldsymbol{q}^2)$.

Experimentally, the magnitude of the form factor is determined by the ratio of the measured cross-section to the Mott cross-section:

$$\left(\frac{d\sigma}{d\Omega}\right)_{\text{exp.}} = \left(\frac{d\sigma}{d\Omega}\right)^{*}_{\text{Mott}} \cdot \left|F(\boldsymbol{q}^2)\right|^2 . \tag{5.41}$$

One therefore measures the cross-section for a fixed beam energy at various angles (and thus different values of $|\boldsymbol{q}|$) and divides by the calculated Mott cross-section.

In Fig. 5.4, a typical experimental set-up for the measurement of form factors is depicted. The electron beam is provided by a linear accelerator and is directed at a thin target. The scattered electrons are measured in a magnetic spectrometer. In an analysing magnet the electrons are deflected according to their momentum, and are then detected in wire chambers. The spectrometer can be rotated around the target in order to allow measurements at different angles θ.

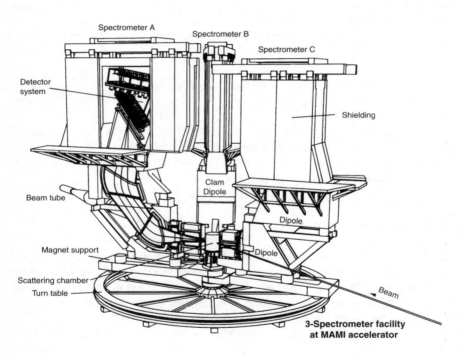

Fig. 5.4. Experimental set-up for the measurement of electron scattering off protons and nuclei at the electron accelerator MAMI-B (Mainzer Microtron). The maximum energy available is 820 MeV. The figure shows three magnetic spectrometers. They can be used individually to detect elastic scattering or in coincidence for a detailed study of inelastic channels. Spectrometer A is shown in cutaway view. The scattered electrons are analysed according to their momentum by two dipole magnets supplemented by a system of detectors made up of wire chambers and scintillation counters. The diameter of the rotating ring is approximately 12 m. (*Courtesy of Arnd P. Liesenfeld (Mainz), who produced this picture*)

Examples of form factors. The first measurements of nuclear form factors were carried out in the early fifties at a linear accelerator at Stanford University, California. Cross-sections were measured for a large variety of nuclei at electron energies of about 500 MeV.

An example of one of the first measurements of form factors can be seen in Fig. 5.5. It shows the ^{12}C cross-section measured as a function of the scattering angle θ. The fast fall-off of the cross-section at large angles corresponds to the $1/|\boldsymbol{q}|^4$-dependence. Superimposed is a typical diffraction pattern associated with the form factor. It has a minimum at $\theta \approx 51°$ or $|\boldsymbol{q}|/\hbar \approx 1.8\,\mathrm{fm}^{-1}$. We want to now discuss this figure and describe what information about the nucleus can be extracted from it.

As we have seen, the form factor $F(\boldsymbol{q}^2)$ is under certain conditions (negligible recoil, Born approximation) the Fourier transform of the charge distribution $f(\boldsymbol{x})$:

$$F(\boldsymbol{q}^2) = \int e^{i\boldsymbol{q}\boldsymbol{x}/\hbar} f(\boldsymbol{x})\,\mathrm{d}^3 x\;. \tag{5.42}$$

For spherically symmetric cases f only depends upon the radius $r = |\boldsymbol{x}|$. Integration over the total solid angle then yields:

$$F(\boldsymbol{q}^2) = 4\pi \int f(r)\,\frac{\sin|\boldsymbol{q}|r/\hbar}{|\boldsymbol{q}|r/\hbar}\,r^2\,\mathrm{d}r, \tag{5.43}$$

with the normalisation:

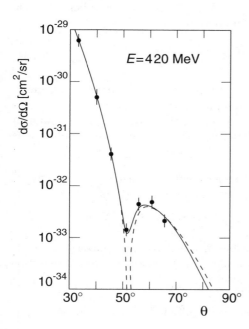

Fig. 5.5. Measurement of the form factor of ^{12}C by electron scattering (from [Ho57]). The figure shows the differential cross-section measured at a fixed beam energy of 420 MeV, at 7 different scattering angles. The dashed line corresponds to scattering of a plane wave off an homogeneous sphere with a diffuse surface (Born approximation). The solid line corresponds to an exact phase shift analysis which was fitted to the experimental data.

$$1 = \int f(\boldsymbol{x})\,\mathrm{d}^3 x = \int_0^\infty \int_{-1}^{+1} \int_0^{2\pi} f(r)\,r^2\,\mathrm{d}\phi\,\mathrm{d}\cos\vartheta\,\mathrm{d}r = 4\pi \int_0^\infty f(r)\,r^2\,\mathrm{d}r\,.$$
$$(5.44)$$

In principle, the radial charge distribution could be determined from the inverse Fourier transform, using the $\boldsymbol{q}^2$–dependence of the experimental form factor:

$$f(r) = \frac{1}{(2\pi)^3} \int F(\boldsymbol{q}^2)\,\mathrm{e}^{-i\boldsymbol{q}\boldsymbol{x}/\hbar}\,\mathrm{d}^3 q\,. \qquad (5.45)$$

In practice, however, the form factor can be measured only over a limited range of momentum transfer $|\boldsymbol{q}|$. The limitation is due to the finite beam energy available and the sharp drop in the cross-section for large momentum transfer. One therefore chooses various parameterisations of $f(r)$, determines the theoretical prediction for $F(\boldsymbol{q}^2)$ and varies the parameters to obtain a best fit between theory and the measured value of $F(\boldsymbol{q}^2)$.

The form factor can be calculated analytically for certain charge distributions described by some simple radial functions $f(r)$. The form factors for some special cases of $f(r)$ are listed in Table 5.1, and are depicted in Fig. 5.6. A charge distribution which drops off gently corresponds to a smooth form factor. The more extended the charge distribution, the stronger the fall-off of the form factor with $\boldsymbol{q}^2$. On the other hand if the object is small, the form factor falls off slowly. In the limit of a pointlike target, the form factor approaches unity.

Scattering off an object with a sharp surface generally results in well-defined diffraction maxima and minima. For a homogeneous sphere with radius R, for example, a minimum is found at

$$\frac{|\boldsymbol{q}|\cdot R}{\hbar} \approx 4.5\,. \qquad (5.46)$$

The location of the minima thus tells us the size of the scattering nucleus.

In Fig. 5.5 we saw that the minimum in the cross-section of electron scattering off ^{12}C (and thus the minimum in the form factor) is found at $|\boldsymbol{q}|/\hbar \approx 1.8$ fm^{-1}. One concludes that the carbon nucleus has a radius $R = 4.5\,\hbar/|\boldsymbol{q}| \approx 2.5$ fm.

Figure 5.7 shows the result of an experiment comparing the two isotopes ^{40}Ca and ^{48}Ca. This picture is interesting in several respects:

– The cross-section was measured over a large range of $|\boldsymbol{q}|$. Within this range, it changes by seven orders of magnitude.[2]
– Not *one* but three minima are visible in the diffraction pattern. This behaviour of the cross-section means that $F(\boldsymbol{q}^2)$ and the charge distribution $\varrho(r)$ can be determined very accurately.

---

[2] Even measurements over 12 (!) orders of magnitude have been carried out (cf., e. g., [Si79]).

Table 5.1. Connection between charge distributions and form factors for some spherically symmetric charge distributions in Born approximation.

Charge distribution $f(r)$		Form Factor $F(q^2)$	
point	$\delta(r)/4\pi$	1	constant
exponential	$(a^3/8\pi)\cdot\exp(-ar)$	$\left(1+q^2/a^2\hbar^2\right)^{-2}$	dipole
Gaussian	$\left(a^2/2\pi\right)^{3/2}\cdot\exp\left(-a^2r^2/2\right)$	$\exp\left(-q^2/2a^2\hbar^2\right)$	Gaussian
homogeneous sphere	$\begin{cases} 3/4\pi R^3 & \text{for } r \le R \\ 0 & \text{for } r > R \end{cases}$	$3\,\alpha^{-3}\left(\sin\alpha-\alpha\cos\alpha\right)$ with $\alpha=|q|R/\hbar$	oscillating

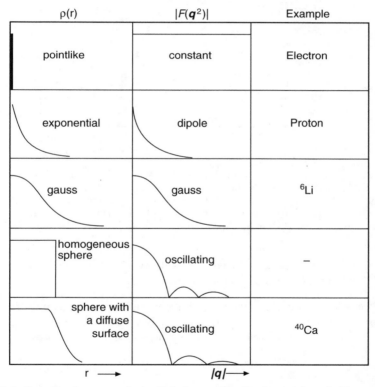

Fig. 5.6. Relation between the radial charge distribution $\varrho(r)$ and the corresponding form factor in Born approximation. A constant form factor corresponds to a pointlike charge (e. g., an electron); a dipole form factor to a charge distribution which falls off exponentially (e. g., a proton); a Gaussian form factor to a Gaussian charge distribution (e. g., ^{6}Li nucleus); and an oscillating form factor corresponds to a homogeneous sphere with a more or less sharp edge. All nuclei except for the lightest ones, display an oscillating form factor.

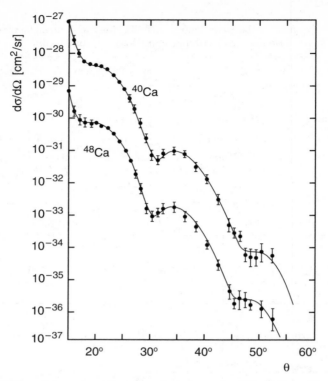

Fig. 5.7. Differential cross-sections for electron scattering off the calcium isotopes ^{40}Ca and ^{48}Ca [Be67]. For clarity, the cross-sections of ^{40}Ca and ^{48}Ca have been multiplied by factors of 10 and 10^{-1}, respectively. The solid lines are the charge distributions obtained from a fit to the data. The location of the minima shows that the radius of ^{48}Ca is larger than that of ^{40}Ca.

– The minima of ^{48}Ca are shifted to slightly lower values of $|\boldsymbol{q}|$ than those of ^{40}Ca. This shows that ^{48}Ca is larger.

Information about the nuclear radius can be obtained not only from the location of the minima of the form factor, but also from its behaviour for $\boldsymbol{q}^2 \to 0$. If the wavelength is considerably larger than the nuclear radius R, then:

$$\frac{|\boldsymbol{q}| \cdot R}{\hbar} \ll 1 \,, \tag{5.47}$$

and $F(\boldsymbol{q}^2)$ can from (5.42) be expanded in powers of $|\boldsymbol{q}|$:

$$F(\boldsymbol{q}^2) \quad = \quad \int f(\boldsymbol{x}) \sum_{n=0}^{\infty} \frac{1}{n!} \left(\frac{i|\boldsymbol{q}||\boldsymbol{x}|\cos\vartheta}{\hbar} \right)^n \mathrm{d}^3 x \qquad \text{with } \vartheta = \sphericalangle(\boldsymbol{x}, \boldsymbol{q})$$

$$= \int_0^\infty \int_{-1}^{+1} \int_0^{2\pi} f(r) \left[1 - \frac{1}{2} \left(\frac{|\boldsymbol{q}|r}{\hbar} \right)^2 \cos^2 \vartheta + \cdots \right] d\phi \, d\cos\vartheta \, r^2 dr$$

$$= 4\pi \int_0^\infty f(r) \, r^2 \mathrm{d}r - \frac{1}{6} \frac{q^2}{\hbar^2} 4\pi \int_0^\infty f(r) \, r^4 \mathrm{d}r + \cdots . \tag{5.48}$$

Defining the *mean square* charge radius according to the normalisation condition (5.44) by:

$$\langle r^2 \rangle = 4\pi \int_0^\infty r^2 \cdot f(r) \, r^2 \mathrm{d}r , \tag{5.49}$$

then

$$F(\boldsymbol{q}^2) = 1 - \frac{1}{6} \frac{\boldsymbol{q}^2 \langle r^2 \rangle}{\hbar^2} + \cdots . \tag{5.50}$$

Hence it is necessary to measure the form factor $F(\boldsymbol{q}^2)$ down to very small values of $\boldsymbol{q}^2$ in order to determine $\langle r^2 \rangle$. The following equation holds:

$$\langle r^2 \rangle = -6 \, \hbar^2 \frac{\mathrm{d}F(\boldsymbol{q}^2)}{\mathrm{d}\boldsymbol{q}^2} \bigg|_{\boldsymbol{q}^2 = 0} . \tag{5.51}$$

Charge distributions of nuclei. Many high-precision measurements of this kind have been carried out at different accelerators since the middle of the 1950's. Radial charge distributions $\varrho(r)$ have been determined from the results. The following has been understood:

– Nuclei are not spheres with a sharply defined surface. In their interior, the charge density is nearly constant. At the surface the charge density

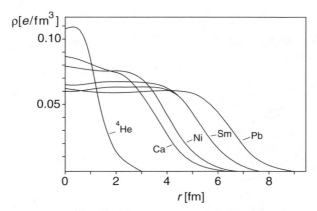

Fig. 5.8. Radial charge distributions of various nuclei. These charge distributions can be approximately described by the Fermi distribution (5.52), i. e., as spheres with diffuse surfaces.

falls off over a relatively large range. The radial charge distribution can be described to good approximation by a Fermi function with two parameters

$$\varrho(r) = \frac{\varrho(0)}{1 + e^{(r-c)/a}} .$$ (5.52)

This is shown in Fig. 5.8 for different nuclei.

– The constant c is the radius at which $\varrho(r)$ has decreased by one half. Empirically, for larger nuclei, c and a are measured to be:

$$c = 1.07 \text{ fm} \cdot A^{1/3} , \qquad a = 0.54 \text{ fm} .$$ (5.53)

– From this charge density, the mean square radius can be calculated. Approximately, for medium and heavy nuclei:

$$\langle r^2 \rangle^{1/2} = r_0 \cdot A^{1/3} \qquad \text{where} \quad r_0 = 0.94 \text{ fm} .$$ (5.54)

The nucleus is often approximated by a homogeneously charged sphere. The radius R of this sphere is then quoted as the nuclear radius. The following connection exists between this radius and the mean square radius:

$$R^2 = \frac{5}{3} \langle r^2 \rangle .$$ (5.55)

Quantitively we have:

$$R = 1.21 \cdot A^{1/3} \text{ fm} .$$ (5.56)

This definition of the radius is used in the mass formula (2.8).

– The surface thickness t is defined as the thickness of the layer over which the charge density drops from 90 % to 10 % of its maximal value:

$$t = r_{(\varrho/\varrho_0 = 0.1)} - r_{(\varrho/\varrho_0 = 0.9)} .$$ (5.57)

Its value is roughly the same for all heavy nuclei, namely:

$$t = 2a \cdot \ln 9 \approx 2.40 \text{ fm} .$$ (5.58)

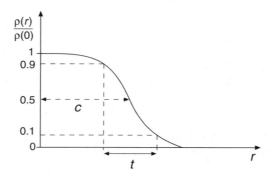

– The charge density $\varrho(0)$ at the centre of the nucleus decreases slightly with increasing mass number. If one takes the presence of the neutrons into account by multiplying by A/Z one finds an almost identical nuclear density in the nuclear interior for nearly all nuclei. For "infinitely largennuclear matter, it would amount to[3]

$$\varrho_n \approx 0.17 \text{ nucleons/fm}^3 \ . \tag{5.59}$$

This corresponds to a value of $c = 1.12\,\text{fm} \cdot A^{1/3}$ in (5.53).

– Some nuclei deviate from a spherical shape and possess ellipsoidal deformations. In particular, this is found in the lanthanides (the "rare earthëlements). Their exact shape cannot be determined by elastic electron scattering. Only a rather diffuse surface can be observed.

– Light nuclei such as ^{6,7}Li, ^{9}Be, and in particular ^{4}He, are special cases. Here, no constant density plateau is formed in the nuclear interior, and the charge density is approximately Gaussian.

This summary describes only the global shape of nuclear charge distributions. Many details specific to individual nuclei are known, but will not be treated further here [Fr82].

5.5 Inelastic Nuclear Excitations

In the above we have mainly discussed elastic scattering off nuclei. In this case the initial and final state particles are identical. The only energy transferred is recoil energy and the target is not excited to a higher energy level. For fixed scattering angles, the incoming and scattering energies are then uniquely connected by (5.15).

The measured energy spectrum of the scattered electrons, at a fixed scattering angle θ, contains events where the energy transfer is larger than we would expect from recoil. These events correspond to inelastic reactions.

Figure 5.9 shows a high-resolution spectrum of electrons with an initial energy of 495 MeV, scattered off ^{12}C and detected at a scattering angle of 65.4°. The sharp peak at $E' \approx 482$ MeV is due to elastic scattering off the ^{12}C nucleus. Below this energy, excitations of individual nuclear energy levels are clearly seen. The prominent maximum at $E' \approx 463$ MeV is caused by the giant dipole resonance (Sect. 18.2). At even lower scattering energies a broad distribution from quasi-elastic scattering off the nucleons bound in the nucleus (Sect. 6.2) is seen.

---

[3] This quantity is usually denoted by ϱ_0 in the literature. To avoid any confusion with the charge density we have used the symbol ϱ_n here.

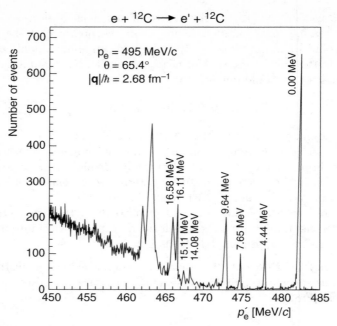

Fig. 5.9. Spectrum of electron scattering off ^{12}C. The sharp peaks correspond to elastic scattering and to the excitation of discrete energy levels in the ^{12}C nucleus by inelastic scattering. The excitation energy of the nucleus is given for each peak. The 495 MeV electrons were accelerated with the linear accelerator MAMI-B in Mainz and were detected using a high-resolution magnetic spectrometer (cf. Fig. 5.4) at a scattering angle of 65.4°. (*Courtesy of Th. Walcher and G. Rosner, Mainz*)

Problems

1. **Kinematics of electromagnetic scattering**
   An electron beam with energy E_e is elastically scattered off a heavy nucleus.
   a) Calculate the maximal momentum transfer.
   b) Calculate the momentum and energy of the backwardly scattered nucleus in this case.
   c) Obtain the same quantities for the elastic scattering of photons with the same energy (nuclear Compton effect).

2. **Wavelength**
   Fraunhofer diffraction upon a circular disc with diameter D produces a ring shaped diffraction pattern. The first minimum appears at $\theta = 1.22\,\lambda/D$.
   Calculate the angular separation of the diffraction minima of α particles with energy $E_{kin} = 100\,\mathrm{MeV}$ scattered off a $^{56}\mathrm{Fe}$ nucleus. The nucleus should be considered as an impenetrable disc.

3. **Rutherford scattering**
   α particles with $E_{kin} = 6\,\mathrm{MeV}$ from a radioactive source are scattered off $^{197}\mathrm{Au}$ nuclei. At which scattering angle are deviations from the cross-section (5.16) to be expected?

4. **Form factor**
   Instead of α particles with $E_{kin} = 6\,\mathrm{MeV}$ we now consider the scattering of electrons with the same de Broglie wavelength off gold. How large must the kinetic energy of the electrons be? How many maxima and minima will be visible in the angular distribution (cf. Fig. 5.7)?
   Since the recoil is small in this case, we may assume that the kinematical quantities are the same in both the centre of mass and laboratory frames.

5. **Elastic scattering of X-rays**
   X-rays are scattered off liquid helium. Which charge carriers in the helium atom are responsible for the scattering? Which of the form factors of Fig. 5.6 corresponds to this scattering off helium?

6. **Compton scattering**
   Compton scattering off bound electrons can be understood in analogy to quasi-elastic and deep inelastic scattering. Gamma rays from positronium annihilation are scattered off helium atoms (binding energy of the "first" electron: 24 eV). Calculate the angular spread of the Compton electrons that are measured in coincidence with photons that are scattered by $\theta_\gamma = 30°$.

6. Elastic Scattering off Nucleons

6.1 Form Factors of the Nucleons

Elastic electron scattering off the lightest nuclei, hydrogen and deuterium, yields information about the nuclear building blocks, the proton and the neutron. Certain subtleties have, however, to be taken into account in any discussion of these experiments.

Recoil. As we will soon see, nucleons have a radius of about 0.8 fm. Their study therefore requires energies from some hundred MeV up to several GeV. Comparing these energies with the mass of the nucleon $M \approx 938$ MeV$/c^2$, we see that they are of the same order of magnitude. Hence the target recoil can no longer be neglected. In the derivation of the cross-sections (5.33) and (5.39) we "prepared" for this by using E' rather than E. On top of this, however, the phase space density $\mathrm{d}n/\mathrm{d}E_f$ in (5.20) must be modified. We so eventually find an additional factor of E'/E in the Mott cross-section [Pe87]:

$$\left(\frac{\mathrm{d}\sigma}{\mathrm{d}\Omega}\right)_{\mathrm{Mott}} = \left(\frac{\mathrm{d}\sigma}{\mathrm{d}\Omega}\right)^{*}_{\mathrm{Mott}} \cdot \frac{E'}{E} \ . \tag{6.1}$$

Since the energy loss of the electron due to the recoil is now significant, it is no longer possible to describe the scattering in terms of a three-momentum transfer. Instead the four-momentum transfer, whose square is Lorentz-invariant:

$$\begin{aligned} q^2 &= (p - p')^2 = 2m_{\mathrm{e}}^2 c^2 - 2\left(EE'/c^2 - |\boldsymbol{p}||\boldsymbol{p}'|\cos\theta\right) \\ &\approx \frac{-4EE'}{c^2}\sin^2\frac{\theta}{2} \ , \end{aligned} \tag{6.2}$$

must be used. In order to only work with positive quantities we define:

$$Q^2 = -q^2 \ . \tag{6.3}$$

In the Mott cross-section, $\boldsymbol{q}^2$ must be replaced by q^2 or Q^2.

Magnetic moment. We must now not only take the interaction of the electron with the nuclear charge into account, but also we have to consider the interaction between the current of the electron and the nucleon's magnetic moment.

The magnetic moment of a charged, spin-1/2 particle which does not possess any internal structure (a Dirac particle) is given by

$$\mu = g \cdot \frac{e}{2M} \cdot \frac{\hbar}{2} \tag{6.4}$$

where M is the mass of the particle and the $g = 2$ factor is a result of relativistic quantum mechanics (the Dirac equation). The magnetic interaction is associated with a flip of the spin of the nucleon. An argument analogous to that of Sect. 5.3 is applicable here: scattering through $0°$ is not consistent with conservation of both angular momentum and helicity and scattering through $180°$ is preferred. The magnetic interaction thus introduces a factor into the interaction which, analogously to (5.39), contains a factor of $\sin^2 \frac{\theta}{2}$. With $\sin^2 \frac{\theta}{2} = \cos^2 \frac{\theta}{2} \cdot \tan^2 \frac{\theta}{2}$ we obtain for the cross-section:

$$\left(\frac{\mathrm{d}\sigma}{\mathrm{d}\Omega} \right)_{\substack{\text{point} \\ \text{spin } 1/2}} = \left(\frac{\mathrm{d}\sigma}{\mathrm{d}\Omega} \right)_{\text{Mott}} \cdot \left[1 + 2\tau \tan^2 \frac{\theta}{2} \right] , \tag{6.5}$$

where

$$\tau = \frac{Q^2}{4M^2c^2} . \tag{6.6}$$

The 2τ factor can be fairly easily made plausible: the matrix element of the interaction is proportional to the magnetic moment of the nucleon (and thus to $1/M$) and to the magnetic field which is produced at the target in the scattering process. Integrated over time, this is then proportional to the deflection of the electron (i.e., to the momentum transfer Q). These quantities then enter the cross-section quadratically.

The magnetic term in (6.5) is large at high four-momentum transfers Q^2 and if the scattering angle θ is large. This additional term causes the cross-section to fall off less strongly at larger scattering angles and a more isotropic distribution is found then the electric interaction alone would produce.

Anomalous magnetic moment. For charged Dirac-particles the g-factor in (6.4) should be exactly 2, while for neutral Dirac particles the magnetic moment should vanish. Indeed measurements of the magnetic moments of electrons and muons yield the value $g = 2$, up to small deviations. These last are caused by quantum electrodynamical processes of higher order, which are theoretically well understood.

Nucleons, however, are not Dirac particles since they are made up of quarks. Therefore their g-factors are determined by their sub-structure. The values measured for protons and neutrons are:

$$\mu_{\mathrm{p}} = \frac{g_{\mathrm{p}}}{2}\mu_{\mathrm{N}} = +2.79 \cdot \mu_{\mathrm{N}} , \tag{6.7}$$

$$\mu_n = \frac{g_n}{2}\mu_N = -1.91 \cdot \mu_N \; , \tag{6.8}$$

where the nuclear magneton μ_N is:

$$\mu_N = \frac{e\hbar}{2M_p} = 3.1525 \cdot 10^{-14} \, \text{MeV} \, \text{T}^{-1} \; . \tag{6.9}$$

Charge and current distributions can be described by form factors, just as in the case of nuclei. For nucleons, two form factors are necessary to characterise both the electric and magnetic distributions. The cross-section for the scattering of an electron off a nucleon is described by the *Rosenbluth formula* [Ro50]:

$$\left(\frac{d\sigma}{d\Omega}\right) = \left(\frac{d\sigma}{d\Omega}\right)_{\text{Mott}} \cdot \left[\frac{G_E^2(Q^2) + \tau G_M^2(Q^2)}{1+\tau} + 2\tau G_M^2(Q^2)\tan^2\frac{\theta}{2}\right] \; . \tag{6.10}$$

Here $G_E(Q^2)$ and $G_M(Q^2)$ are the *electric and magnetic form factors* both of which depend upon Q^2. The measured Q^2-dependence of the form factors gives us information about the radial charge distributions and the magnetic moments. The limiting case $Q^2 \to 0$ is particularly important. In this case G_E coincides with the electric charge of the target, normalised to the elementary charge e; and G_M is equal to the magnetic moment μ of the target, normalised to the nuclear magneton. The limiting values are:

$$\begin{matrix} G_E^p(Q^2 = 0) = 1 & G_E^n(Q^2 = 0) = 0 \\ G_M^p(Q^2 = 0) = 2.79 & G_M^n(Q^2 = 0) = -1.91 \; . \end{matrix} \tag{6.11}$$

In order to independently determine $G_E(Q^2)$ and $G_M(Q^2)$ the cross-sections must be measured at fixed values of Q^2, for various scattering angles θ (i. e., at different beam energies E). The measured cross-sections are then divided by the Mott cross-sections. If we display the results as a function of $\tan^2(\theta/2)$, then the measured points form a straight line (Fig. 6.1), in accordance with the Rosenbluth formula. $G_M(Q^2)$ is then determined by the slope of the line, and the intercept $(G_E^2 + \tau G_M^2)/(1+\tau)$ at $\theta = 0$ then yields $G_E(Q^2)$. If we perform this analysis for various values of Q^2 we can obtain the Q^2 dependence of the form factors.

Measurements of the electromagnetic form factors right up to very high values of Q^2 were carried out mainly in the late sixties and early seventies at accelerators such as the linear accelerator SLAC in Stanford. Fig. 6.2 shows the Q^2 dependence of the two form factors for both protons and neutrons.

It turned out that the proton electric form factor and the magnetic form factors of both the proton and the neutron fall off similarly with Q^2. They can be described to a good approximation by a so-called *dipole fit*:

$$G_E^p(Q^2) = \frac{G_M^p(Q^2)}{2.79} = \frac{G_M^n(Q^2)}{-1.91} = G^{\text{dipole}}(Q^2)$$

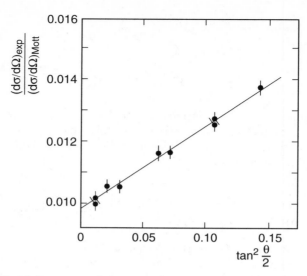

Fig. 6.1. Ratio of the measured cross-section and the Mott cross-section $\sigma_{\mathrm{exp}}/\sigma_{\mathrm{Mott}}$ as a function of $\tan^2\theta/2$ at a four-momentum transfer of $Q^2 = 2.5\ \mathrm{GeV}^2/c^2$ [Ta67].

$$\text{where} \qquad G^{\mathrm{dipole}}(Q^2) = \left(1 + \frac{Q^2}{0.71\,(\mathrm{GeV}/c)^2}\right)^{-2} . \qquad (6.12)$$

The neutron appears from the outside to be electrically neutral and it therefore has a very small electric form factor.

We may obtain the nucleons' charge distributions and magnetic moments from the Q^2 dependence of the form factors, just as we saw could be done for nuclei. The interpretation of the form factors as the Fourier transform of the static charge distribution is, however, only correct for small values of Q^2, since only then are the three- and four-momentum transfers approximately equal. The observed dipole form factor (6.12) corresponds to a charge distribution which falls off exponentially (cf. Fig. 5.6):

$$\varrho(r) = \varrho(0)\,\mathrm{e}^{-ar} \quad \text{with } a = 4.27\,\mathrm{fm}^{-1} . \qquad (6.13)$$

Nucleons are, we see, neither pointlike particles nor homogeneously charged spheres, but rather quite diffuse systems.

The mean square radii of the charge distribution in the proton and of the magnetic moment distributions in the proton and the neutron are similarly large. They may be found from the slope of $G_{\mathrm{E,M}}(Q^2)$ at $Q^2 = 0$. The dipole fit yields:

$$\langle r^2 \rangle_{\mathrm{dipole}} = -6\hbar^2 \left. \frac{\mathrm{d}G^{\mathrm{dipole}}(Q^2)}{\mathrm{d}Q^2} \right|_{Q^2=0} = \frac{12}{a^2} = 0.66\,\mathrm{fm}^2 ,$$

$$\sqrt{\langle r^2 \rangle_{\mathrm{dipole}}} = 0.81\,\mathrm{fm} . \qquad (6.14)$$

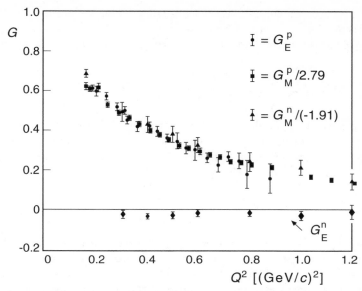

Fig. 6.2. Proton and neutron electric and magnetic form factors as functions of Q^2. The data points are scaled by the factors noted in the diagram so that they coincide and thus more clearly display the global dipole-like behaviour [Hu65].

Precise measurements of the form factors at small values of Q^2 show slight deviations from the dipole parametrisation. The slope at $Q^2 \to 0$ determined from these data yields the present best value [Bo75] of the charge radius of the proton:

$$\sqrt{\langle r^2 \rangle_{\mathrm{p}}} = 0.862 \ \text{fm} \ . \tag{6.15}$$

Determining the neutron electric form factor is rather difficult: targets with free neutrons are not available and so information about $G_{\mathrm{E}}^{\mathrm{n}}(Q^2)$ must be extracted from electron scattering off deuterons. In this case it is necessary to correct the measured data for the effects of the nuclear force between the proton and the neutron. However, an alternative, elegant approach has been developed to determine the charge radius of the free neutron. Low-energy neutrons from a nuclear reactor are scattered off electrons in an atomic shell and the so-ejected electrons are then measured. This reaction corresponds to electron-neutron scattering at small Q^2. The result of these measurements is [Ko95]:

$$-6\hbar^2 \left. \frac{\mathrm{d} G_{\mathrm{E}}^{\mathrm{n}}(Q^2)}{\mathrm{d} Q^2} \right|_{Q^2=0} = -0.113 \pm 0.005 \ \text{fm}^2 \ . \tag{6.16}$$

The neutron, therefore, only appears electrically neutral from the outside. Its interior contains electrically charged constituents which also possess magnetic moments. Since both the charges and their magnetic moments contribute to the electric form factor, we cannot separate their contributions in a Lorentz invariant fashion. Comparisons with model calculations show that, locally inside the neutron, the charges of the constituents almost completely cancel, which also follows naturally from the measured value (6.16).

6.2 Quasi-elastic Scattering

In Sect. 6.1 we considered the elastic scattering of electrons off free protons (neutrons) at rest. In this reaction for a given beam energy E and at a fixed scattering angle θ scattered electrons always have a definite scattering energy E' which is given by (5.15):

$$E' = \frac{E}{1 + \frac{E}{Mc^2}(1 - \cos\theta)} \ . \tag{6.17}$$

Repeating the scattering experiment at the same beam energy and at the same detector angle, but now off a nucleus containing several nucleons, a more complicated energy spectrum is observed. Figure 6.3 shows a spectrum of electrons which were scattered off a thin H_2O target, i. e., some were scattered off free protons, some off oxygen nuclei.

The narrow peak observed at $E' \approx 160$ MeV stems from elastic scattering off the free protons in hydrogen. Superimposed is a broad distribution with a maximum shifted a few MeV towards smaller scattering energies. This part of the spectrum may be identified with the scattering of electrons off

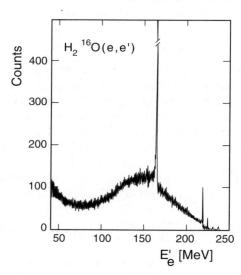

Fig. 6.3. Energy spectrum of electrons scattered off a thin H_2O target. The data were taken at the linear accelerator MAMI-A in Mainz at a beam energy of 246 MeV and at a scattering angle of 148.5°. (*Courtesy of J. Friedrich, Mainz*)

individual nucleons within the ^{16}O nucleus. This process is called *quasi-elastic scattering*. The sharp peaks at high energies are caused by scattering off the ^{16}O nucleus as a whole (cf. Fig. 5.9). At the left side of the picture, the tail of the Δ-resonance can be recognised, this will be discussed in Sect. 7.1.

Both the shift and the broadening of the quasi-elastic spectrum contain information about the internal structure of atomic nuclei. In the *impulse approximation* we assume that the electron interacts with a single nucleon. The nucleon is knocked out of the nuclear system by the scattering process without any further interactions with the remaining nucleons in the nucleus. The shift of the maximum in the energy of the scattered electrons towards lower energies is due to the energy needed to remove the nucleon from the nucleus. From the broadening of the maximum, compared to elastic scattering off free protons in the hydrogen atom, we conclude that the nucleus is not a static object with locally fixed nucleons. The nucleons rather move around "quasi-freely" within the nucleus. This motion causes a change in the kinematics compared to scattering off a nucleon at rest.

Let us consider a bound nucleon moving with momentum P in an effective average nuclear potential of strength S. This nucleon's binding energy is then $S - P^2/2M$. We neglect residual interactions with other nucleons, and the kinetic energy of the remaining nucleus and consider the scattering of an electron off this nucleon.

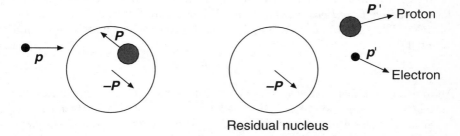

Residual nucleus

In this case, the following kinematic connections apply:

$$p + P = p' + P' \qquad \text{momentum conservation in the e-p system}$$
$$P' = q + P \qquad \text{momentum conservation in the } \gamma\text{-p system}$$
$$E + E_\mathrm{p} = E' + E'_\mathrm{p} \qquad \text{energy conservation in the e-p system}$$

The energy transfer ν from the electron to the proton for $E, E' \gg m_e c^2$ and $|P|, |P'| \ll Mc$ is given by:

$$\nu = E - E' = E'_\mathrm{p} - E_\mathrm{p} = \left(Mc^2 + \frac{P'^2}{2M}\right) - \left(Mc^2 + \frac{P^2}{2M} - S\right)$$

$$= \frac{(P+q)^2}{2M} - \frac{P^2}{2M} + S = \frac{q^2}{2M} + S + \frac{2|q||P|\cos\alpha}{2M}, \qquad (6.18)$$

Table 6.1. Fermi momentum P_F and effective average potential S for various nuclei. These values were obtained from an analysis of quasi-elastic electron scattering at beam energies between 320 MeV and 500 MeV and at a fixed scattering angle of $60°$ [Mo71, Wh74]. The errors are approximately 5 MeV/c (P_F) and 3 MeV (S).

Nucleus	^{6}Li	^{12}C	^{24}Mg	^{40}Ca	^{59}Ni	^{89}Y	^{119}Sn	^{181}Ta	^{208}Pb
P_F [MeV/c]	169	221	235	249	260	254	260	265	265
S [MeV]	17	25	32	33	36	39	42	42	44

where α is the angle between $\boldsymbol{q}$ and $\boldsymbol{P}$. We now assume that the motion of the nucleons within the nucleus is isotropic (i.e. a spherically symmetric distribution). This leads to a symmetric distribution for ν around an average value:

$$\nu_0 = \frac{q^2}{2M} + S \qquad (6.19)$$

with a width of

$$\sigma_\nu = \sqrt{\langle(\nu - \nu_0)^2\rangle} = \frac{|\boldsymbol{q}|}{M}\sqrt{\langle \boldsymbol{P}^2 \cos^2\alpha\rangle} = \frac{|\boldsymbol{q}|}{M}\sqrt{\frac{1}{3}\langle \boldsymbol{P}^2\rangle}\,. \qquad (6.20)$$

Fermi momentum. As we will discuss in Sect. 17.1, the nucleus can be described as a *Fermi gas* in which the nucleons move around like quasi-free particles. The *Fermi momentum* P_F is related to the mean square momentum by (cf. 17.9):

$$P_F^2 = \frac{5}{3}\langle \boldsymbol{P}^2\rangle\,. \qquad (6.21)$$

An analysis of quasi-elastic scattering off different nuclei can thus determine the effective average potential S and the Fermi momentum P_F of the nucleons.

Studies of the A-dependence of S and P_F were first carried out in the early seventies. The results of the first systematic analysis are shown in Table 6.1 and can be summarized as follows:

- The effective average nuclear potential S increases continuously with the mass number A, varying between 17 MeV in Li to 44 MeV in Pb.
- Apart from in the lightest nuclei, the Fermi momentum is nearly independent of A and is:

$$P_F \approx 250 \text{ MeV}/c\,. \qquad (6.22)$$

This behaviour is consistent with the Fermi gas model. The density of nuclear matter is independent of the mass number except for in the lightest nuclei.

6.3 Charge Radii of Pions and Kaons

The charge radii of various other particles can also be measured by the same method that was used for the neutron. For example those of the π-meson [Am84] and the K-meson [Am86], particles which we will introduce in Sect. 8.2. High-energy mesons are scattered off electrons in the hydrogen atom. The form factor is then determined by analysing the angular distribution of the ejected electrons. Since the pion and the kaon are spin-0 particles, they have an electric but not a magnetic form factor.

The Q^2-dependence of these form factors is shown in Fig. 6.4. Both can be described by a *monopole form factor*:

$$G_{\mathrm{E}}(Q^2) = \left(1 + Q^2/a^2\hbar^2\right)^{-1} \qquad \text{with} \quad a^2 = \frac{6}{\langle r^2 \rangle} \,. \tag{6.23}$$

The slopes near the origin yield the mean square charge radii:

$$\langle r^2 \rangle_\pi = 0.44 \pm 0.02 \,\mathrm{fm}^2 \quad ; \quad \sqrt{\langle r^2 \rangle_\pi} = 0.67 \pm 0.02 \,\mathrm{fm}$$

$$\langle r^2 \rangle_{\mathrm{K}} = 0.34 \pm 0.05 \,\mathrm{fm}^2 \quad ; \quad \sqrt{\langle r^2 \rangle_{\mathrm{K}}} = 0.58 \pm 0.04 \,\mathrm{fm} \,.$$

We see that the pion and the kaon have a different charge distribution than the proton, in particular it is less spread out. This may be understood as a result of the different internal structures of these particles. We will see in Chap. 8 that the proton is composed of three quarks, while the pion and kaon are both composed of a quark and an antiquark.

The kaon has a smaller radius than that of the pion. This can be traced back to the fact that the kaon, in contrast to the pion, contains a heavy quark (an s-quark). In Sect. 13.5 we will demonstrate in a heavy quark–antiquark system that the radius of a system of quarks decreases if the mass of its constituents increases.

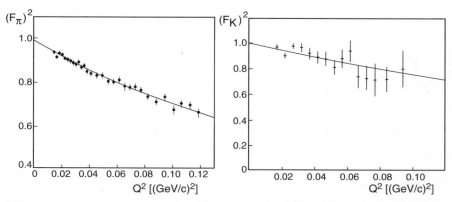

Fig. 6.4. Pion and kaon form factors as functions of Q^2 (from [Am84] and [Am86]). The solid lines correspond to a monopole form factor, $(1 + Q^2/a^2\hbar^2)^{-1}$.

Problems

1. **Electron radius**
   Suppose one wants to obtain an upper bound for the electron's radius by looking for a deviation from the Mott cross-section in electron-electron scattering. What centre of mass energy would be necessary to set an upper limit on the radius of 10^{-3} fm?

2. **Electron-pion scattering**
   State the differential cross-section, $d\sigma/d\Omega$, for elastic electron-pion scattering. Write out explicitly the Q^2 dependence of the form factor part of the cross-section in the limit $Q^2 \to 0$ asuming that $\langle r^2 \rangle_\pi = 0.44\,\text{fm}^2$.

7. Deep Inelastic Scattering

7.1 Excited States of the Nucleons

In Sect. 5.5 we discussed the spectra observed in electron scattering off nuclei. As well as the elastic scattering peak some additional peaks, which we associated with nuclear excitations, were observed. Similar spectra are observed for electron-nucleon scattering.

Figure 7.1 shows a spectrum from electron-proton scattering. It was obtained at an electron energy $E = 4.9$ GeV and at a scattering angle of $\theta = 10°$ by varying the accepted scattering energy of a magnetic spectrometer in small steps. Besides the sharp elastic scattering peak (scaled down by a factor of 15 for clarity) peaks at lower scattering energies are observed associated with in-

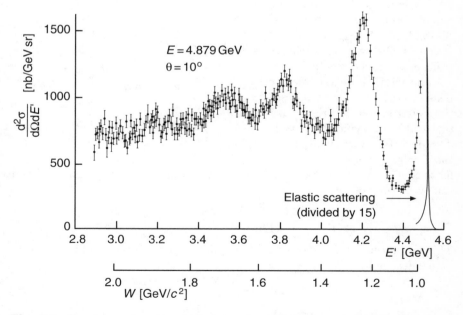

Fig. 7.1. Spectrum of scattered electrons from electron–proton scattering at an electron energy of $E = 4.9$ GeV and a scattering angle of $\theta = 10°$ (from [Ba68]).

elastic excitations of the proton. These peaks correspond to excited states of the nucleon, which we call *nucleon resonances*. The existence of these excited states of the proton demonstrates that the proton is a composite system. In Chap. 15 we will explain the structure of these resonances in the framework of the quark model.

The invariant mass of these states is denoted by W. It is calculated from the four-momenta of the exchanged photon (q) and of the incoming proton (P) according to

$$W^2c^2 = P'^2 = (P+q)^2 = M^2c^2 + 2Pq + q^2 = M^2c^2 + 2M\nu - Q^2 \ . \quad (7.1)$$

Here the Lorentz-invariant quantity ν is defined by

$$\nu = \frac{Pq}{M} \ . \quad (7.2)$$

The target proton is at rest in the laboratory system, which corresponds to $P = (Mc, \mathbf{0})$ and $q = ((E-E')/c, \mathbf{q})$. Therefore the energy transferred by the virtual photon from the electron to the proton in the laboratory frame is:

$$\nu = E - E' \ . \quad (7.3)$$

The $\Delta(1232)$ resonance. The nucleon resonance $\Delta(1232)$, which appears in Fig. 7.1 at about $E' = 4.2$ GeV, has a mass $W = 1232\,\mathrm{MeV}/c^2$. As we will see in Chap. 15, this resonance exists in four different charge states: Δ^{++}, Δ^+, Δ^0, and Δ^-. In Fig. 7.1, the Δ^+ excitation is observed since charge is not transferred in the reaction.

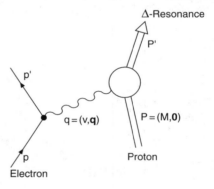

The width observed for the elastic peak is a result of the finite resolution of the spectrometer, but resonances have a real width[1] of typically $\Gamma \approx 100\,\mathrm{MeV}$. The uncertainty principle then implies that such resonances have very short lifetimes. The $\Delta(1232)$ resonance has a width of approximately 120 MeV and thus a lifetime of

$$\tau = \frac{\hbar}{\Gamma} = \frac{6.6 \cdot 10^{-22}\,\mathrm{MeV\,s}}{120\,\mathrm{MeV}} = 5.5 \cdot 10^{-24}\,\mathrm{s} \ . \quad (7.4)$$

This is the typical time scale for strong interaction processes. The Δ^+ resonance decays by:

$$\Delta^+ \ \rightarrow \ \mathrm{p} + \pi^0$$
$$\Delta^+ \ \rightarrow \ \mathrm{n} + \pi^+ \ .$$

---

[1] The exact meaning of "width" will be discussed in Sect. 9.2.

A light particle, the π-meson (or pion) is produced in such decays in addition to the nucleon.

7.2 Structure Functions

Individual resonances cannot be distinguished in the excitation spectrum for invariant masses $W \gtrsim 2.5$ GeV/c^2. Instead, one observes that many further strongly interacting particles (hadrons) are produced.

The dynamics of such production processes may be, similarly to the case of elastic scattering, described in terms of form factors. In the inelastic case they are known as the W_1 and W_2 structure functions.

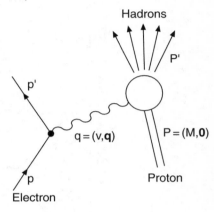

In *elastic* scattering, at a given beam energy E, there is only *one* free parameter. For example, if the scattering angle θ is fixed, kinematics requires that the squared four-momentum transfer Q^2, the energy transfer ν, the energy of the scattered electron E' etc. are also fixed. Since $W = M$, (7.1) yields the relationship:

$$2M\nu - Q^2 = 0 \,. \qquad (7.5)$$

In *inelastic* scattering, however, the excitation energy of the proton adds a further degree of freedom. Hence these structure functions and cross-sections are functions of *two* independent, free parameters, e.g., (E', θ) or (Q^2, ν). Since $W > M$ in this case, we obtain

$$2M\nu - Q^2 > 0 \,. \qquad (7.6)$$

The Rosenbluth formula (6.10) is now replaced by the cross-section:

$$\frac{\mathrm{d}^2\sigma}{\mathrm{d}\Omega\,\mathrm{d}E'} = \left(\frac{\mathrm{d}\sigma}{\mathrm{d}\Omega}\right)^*_{\mathrm{Mott}} \left[W_2(Q^2, \nu) + 2W_1(Q^2, \nu)\tan^2\frac{\theta}{2}\right] \,. \qquad (7.7)$$

The second term again contains the magnetic interaction.

The first *deep inelastic* scattering experiments were carried out in the late sixties at SLAC, using a linear electron accelerator with a maximum energy of 25 GeV. Figure 7.2 shows spectra of scattering off hydrogen obtained at a fixed scattering angle $\theta = 4°$. In the graphs, $\mathrm{d}^2\sigma/\mathrm{d}\Omega\,\mathrm{d}E'$ is plotted as a function of W at several fixed beam energies E between 4.5 GeV and 20 GeV. Each spectrum covers a different range of Q^2 from $0.06 < Q^2 < 0.09$ (GeV/c)^2 to $1.45 < Q^2 < 1.84$ (GeV/c)^2. It can be clearly seen that the cross-sections

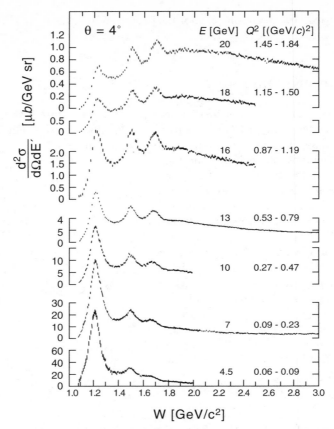

Fig. 7.2. Electron-proton scattering: excitation spectra measured in deep inelastic electron–nucleon scattering as functions of the invariant mass W [St75]. Note the different scales of the y-axis. The measurements were taken at a fixed scattering angle, $\theta = 4°$. The average Q^2-range of the data increases with increasing beam energy E. The resonances, in particular the first one ($W = 1.232$ GeV/c^2), become less and less pronounced, but the continuum ($W \gtrsim 2.5$ GeV/c^2) decreases only slightly.

drop off rapidly with Q^2 in the range of the nucleon resonances. For increasing W, however, the falloff is less pronounced.

The behaviour above the resonance region was, however, surprising. Here, the counting rates were much larger than was expected in view of the results from elastic scattering, or from the excitation of the Δ resonance. Figure 7.3 shows the ratio:

$$\frac{\mathrm{d}^2\sigma}{\mathrm{d}\Omega\,\mathrm{d}E'} \bigg/ \left(\frac{\mathrm{d}\sigma}{\mathrm{d}\Omega}\right)^*_{\mathrm{Mott}} \tag{7.8}$$

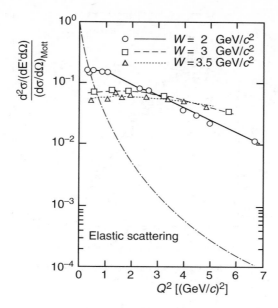

Fig. 7.3. Electron–proton scattering: measured cross-sections normalised to the Mott cross-sections as functions of Q^2 at different values of the invariant mass W [Br69].

measured in these experiments as a function of Q^2 at different values of W. It can be seen that this ratio only depends weakly on Q^2 for $W > 2$ GeV/c^2, in clear contrast to the rapid drop with $|G^{\text{dipole}}|^2 \approx 1/Q^8$ for elastic scattering (6.12). Hence, in deep inelastic scattering, the structure functions W_1 and W_2 are nearly independent of Q^2 for fixed values of the invariant mass W.

To better discuss this result, we introduce a new Lorentz-invariant quantity, the *Bjorken scaling variable*.

$$x := \frac{Q^2}{2Pq} = \frac{Q^2}{2M\nu} \ . \tag{7.9}$$

This dimensionless quantity is a measure of the inelasticity of the process. For elastic scattering, in which $W = M$, (7.5) yields:

$$2M\nu - Q^2 = 0 \quad \Longrightarrow \quad x = 1 \ . \tag{7.10}$$

However, for inelastic processes, in which $W > M$, we have:

$$2M\nu - Q^2 > 0 \quad \Longrightarrow \quad 0 < x < 1 \ . \tag{7.11}$$

The two dimensionful structure functions $W_1(Q^2, \nu)$ and $W_2(Q^2, \nu)$ are usually replaced by two dimensionless structure functions:

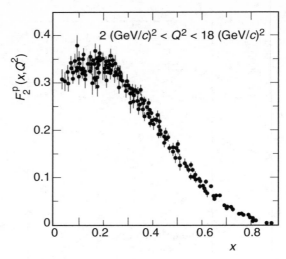

Fig. 7.4. The structure function F_2 of the proton as a function of x, for Q^2 between 2 $(\mathrm{GeV}/c)^2$ and 18 $(\mathrm{GeV}/c)^2$ [At82].

$$
\begin{aligned}
F_1(x, Q^2) &= Mc^2\, W_1(Q^2, \nu) \\
F_2(x, Q^2) &= \nu\, W_2(Q^2, \nu)\,.
\end{aligned}
\tag{7.12}
$$

If one extracts $F_1(x, Q^2)$ and $F_2(x, Q^2)$ from the cross-sections, one observes at fixed values of x that they depend only weakly, or not at all, on Q^2. This is shown in Fig. 7.4 where $F_2(x, Q^2)$ is displayed as a function of x, for data covering a range of Q^2 between 2 $(\mathrm{GeV}/c)^2$ and 18 $(\mathrm{GeV}/c)^2$.

The fact that the structure functions are independent of Q^2 means, according to our previous discussion, that the electrons are scattered off a point charge (cf. Fig. 5.6). Since nucleons are extended objects, it follows from the above result that:

nucleons have a sub-structure made up of point-like constituents.

The F_1 structure function results from the magnetic interaction. It vanishes for scattering off spin zero particles. For spin 1/2 Dirac particles (6.5) and (7.7) imply the relation:

$$
2xF_1(x) = F_2(x)\,.
\tag{7.13}
$$

This is called the *Callan-Gross relation* [Ca69] (see the exercises).

The ratio $2xF_1/F_2$ is shown in Fig. 7.5 as a function of x. It can be seen that the ratio is, within experimental error, consistent with unity. Hence we can further conclude that:

the point-like constituents of the nucleon have spin 1/2.

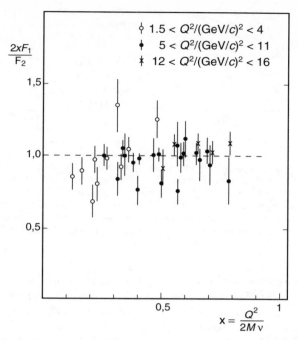

Fig. 7.5. Ratio of the structure functions $2xF_1(x)$ and $F_2(x)$. The data are from experiments at SLAC (from [Pe87]). It can be seen that the ratio is approximately constant (≈ 1).

7.3 The Parton Model

The interpretation of deep inelastic scattering off protons may be considerably simplified if the reference frame is chosen judiciously. The physics of the process is, of course, independent of this choice. If one looks at the proton in a fast moving system, then the transverse momenta and the rest masses of the proton constituents can be neglected. The structure of the proton is then given to a first approximation by the longitudinal momenta of its constituents. This is the basis of the *parton model* of Feynman and Bjorken. In this model the constituents of the proton are called *partons*. Today the charged partons are identified with the quarks and the electrically neutral ones with the gluons, the field quanta of the strong interaction.

Decomposing the proton into free moving partons, the interaction of the electron with the proton can be viewed as the incoherent sum of its interactions with the individual partons. These interactions in turn can be regarded as elastic scattering. This approximation is valid as long as the duration of the photon-parton interaction is so short that the interaction between the partons themselves can be safely neglected (Fig. 7.6). This is the *impulse approximation* which we have already met in quasi-elastic scattering (p. 79). In

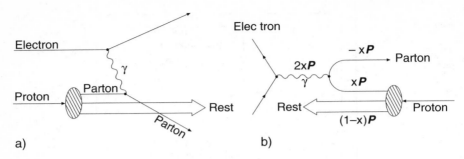

Fig. 7.6. Schematic representation of deep inelastic electron-proton scattering according to the parton model, in the laboratory system (**a**) and in a fast moving system (**b**). This diagram shows the process in two spatial dimensions. The arrows indicate the directions of the momenta. Diagram (**b**) depicts the scattering process in the Breit frame in which the momentum transferred by the virtual photon is zero. Hence the momentum of the struck parton is turned around but its magnitude is unchanged.

deep inelastic scattering this approximation is valid because the interaction between partons at short distances is weak, as we will see Sect. 8.3.

If we make this approximation and assuming both that the parton masses can be safely neglected and that $Q^2 \gg M^2 c^2$, we obtain a direct interpretation of the Bjorken scaling variable $x = Q^2/2M\nu$, which we defined in (7.9). It is that fraction of the four-momentum of the proton which is carried by the struck parton. A photon which, in the laboratory system, has four-momentum $q = (\nu/c, \boldsymbol{q})$ interacts with a parton carrying the four-momentum $x\boldsymbol{P}$. We emphasize that this interpretation of x is only valid in the impulse approximation, and then only if we neglect transverse momenta and the rest mass of the parton; i.e., in a very fast moving system.

A popular reference frame satisfying these conditions is the so-called *Breit frame* (Fig. 7.6b), where the photon does not transfer any energy ($q_0 = 0$). In this system x is the three-momentum fraction of the parton.

The spatial resolution of deep inelastic scattering is given by the reduced wave-length $\lambdabar$ of the virtual photon. This quantity is not Lorentz-invariant, but depends upon the reference frame. In the laboratory system ($q_0 = \nu/c$) it is:

$$\lambdabar = \frac{\hbar}{|\boldsymbol{q}|} = \frac{\hbar c}{\sqrt{\nu^2 + Q^2 c^2}} \approx \frac{\hbar c}{\nu} = \frac{2Mx\hbar c}{Q^2}. \tag{7.14}$$

For example, if $x = 0.1$ and $Q^2 = 4\,(\text{GeV}/c)^2$ one finds $\lambdabar \simeq 10^{-17}\,\text{m}$ in the laboratory system. In the Breit frame, the equation simplifies to

$$\lambdabar = \frac{\hbar}{|\boldsymbol{q}|} = \frac{\hbar}{\sqrt{Q^2}}. \tag{7.15}$$

Q^2 therefore has an obvious interpretation in the Breit frame: it fixes the spatial resolution with which structures can be studied.

7.4 Interpretation of Structure Functions in the Parton Model

Structure functions describe the internal composition of the nucleon. We now assume the nucleon to be built from different types of quarks f carrying an electrical charge $z_f \cdot e$. The cross-section for electromagnetic scattering from a quark is proportional to the square of its charge, and hence to z_f^2.

We denote the distribution function of the quark momenta by $q_f(x)$, i.e. $q_f(x)\mathrm{d}x$ is the expectation value of the number of quarks of type f in the hadron whose momentum fraction lies within the interval $[x, x + \mathrm{d}x]$. The quarks responsible for the quantum numbers of the nucleon are called *valence quarks*. Additionally quark–antiquark pairs are found in the interior of nucleons. They are produced and annihilated as virtual particles from the gluons in the field of the strong interaction. This process is analogous to the production of virtual electron–positron pairs in the Coulomb field. These quarks and antiquarks are called *sea quarks*.

The momentum distribution of the antiquarks is denoted by $\bar{q}_f(x)$, and accordingly that of the gluons by $g(x)$. The structure function F_2 is then the sum of the momentum distributions weighted by x and z_f^2. Here the sum is over all types of quarks and antiquarks:

$$F_2(x) = x \cdot \sum_f z_f^2 \left(q_f(x) + \bar{q}_f(x) \right) . \tag{7.16}$$

The structure functions were determined by scattering experiments on hydrogen, deuterium and heavier nuclei. By convention in scattering off nuclei the structure function is always given per nucleon. Except for small corrections due to the Fermi motion of the nucleons in the deuteron, the structure function of the deuteron F_2^{d} is equal to the average structure function of the nucleons F_2^{N}:

$$F_2^{\mathrm{d}} \approx \frac{F_2^{\mathrm{p}} + F_2^{\mathrm{n}}}{2} =: F_2^{\mathrm{N}} . \tag{7.17}$$

Hence the structure function of the neutron may be determined by subtracting the structure function of the proton from that of the deuteron.

In addition to electrons, muons and neutrinos can also used as beam particles. Like electrons, muons are point-like, charged particles. There is an advantage in using them, as they can be produced with higher energies than electrons. The scattering processes are completely analogous, and the cross-sections are identical. Neutrino scattering yields complementary information about the quark distribution. Neutrinos couple to the weak charge of the quarks via the weak interaction. In neutrino scattering, it is possible to distinguish between the different types of quarks, and also between quarks and antiquarks. Details will be given in Sect. 10.6.

x-dependence of the structure functions. Combining the results of neutrino and antineutrino scattering yields the momentum distribution of the sea quarks and of the valence quarks separately. The shape of the curves in Fig. 7.7 shows that sea quarks contribute to the structure function only at small values of x. Their momentum distribution drops off rapidly with x and is negligible above $x \approx 0.35$. The distribution of the valence quarks has a maximum at about $x \approx 0.2$ and approaches zero for $x \to 1$ and $x \to 0$. The distribution is smeared out by the Fermi motion of the quarks in the nucleon.

For large x, F_2 becomes extremely small. Thus it is very unlikely that *one* quark alone carries the major part of the momentum of the nucleon.

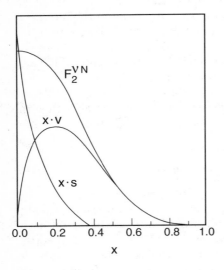

Fig. 7.7. Sketch of the structure function F_2 of the nucleon as measured in (anti)neutrino scattering. Also shown are the momentum distributions, weighted by x, of the valence quarks (v) and sea quarks (s).

Nuclear effects. Typical energies in nuclear physics (e. g., binding energies) are of the order of several MeV and typical momenta (e. g., Fermi momenta) are of the order of 250 MeV/c. These are orders of magnitude less than the Q^2 values of scattering experiments used to determine the structure functions. Therefore one would expect the structure functions to be the same for scattering off free nucleons or scattering off nucleons bound in nuclei, except, of course, for kinematic effects due to the Fermi motion of the nucleons in the nucleus. In practice, however, a definite influence of the surrounding nuclear medium on the momentum distribution of the quarks is observed [Ar94]. This phenomenon is called the *EMC Effect* after the collaboration which detected it in 1983.

For illustration, Fig. 7.8 shows the ratio of the structure functions of calcium and deuterium. The fraction of the isotope ^{40}Ca in natural calcium is 97 %. It is the heaviest stable nuclide which is an isoscalar; i. e., which has equal numbers of protons and neutrons. Deuterium, on the other hand, is

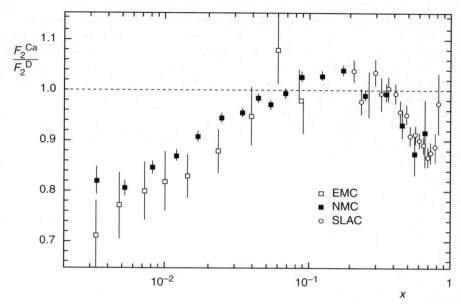

Fig. 7.8. Ratio of the structure functions F_2 of calcium and deuterium as a function of x [Ar88, Go94b, Am95].

only weakly bound, and its proton and neutron can be roughly considered to be free nucleons. The advantage of comparing isoscalar nuclides is that we can study the influence of nuclear binding on the structure function F_2, without having to worry about the differences between F_2^p and F_2^n.

A distinct deviation of the ratio from unity is visible throughout the entire x-range. For $x \lesssim 0.06$, the ratio is smaller than unity, and decreases with decreasing values of x. In this range the structure functions are dominated by the sea quarks. For $0.06 \lesssim x \lesssim 0.3$, the ratio is slightly larger than unity. In the range $0.3 \lesssim x \lesssim 0.8$ where the valence quarks prevail, the ratio is again smaller than unity, with a minimum at $x \approx 0.65$. For large values of x, it increases rapidly with x. The rapid change of the ratio in this region disguises the fact that the absolute changes in F_2 are very small since the structure functions themselves are tiny. Measurements with different nuclei show a strong increase of the EMC effect with increasing mass number A at small values of x and a weak increase in the range of intermediate values of x.

Notwithstanding the small size of the observed effects, they have generated great theoretical interest. They could contain key information for understanding the nuclear force on the basis of the fundamental interaction between quarks and gluons. We will expand upon this point in Sect. 16.3.

Theoretical models for the explanation of the EMC effect are abundant [Ar94]. So far, none of the models is able to convincingly describe the phenomena over the entire range of x. This suggests that the effect is probably due to several factors. Some possible explanations for the EMC effect are: interactions between quarks in different nucleons; the "swelling" of the radius of the nucleon within the nucleus; coalescence of nucleons to form "multiquark clusters" of 6, 9, ... valence quarks; kinematical effects caused by the reduction in the effective nucleon mass due to nuclear binding; correlation between nucleons; Fermi motion — and many other reasons.

In general, despite great theoretical effort, there is no single commonly accepted picture of the physics underlying the dependence of the structure functions on the nuclear environment.

Problems

1. **Compton scattering**

   At the HERA collider ring the spins of the electrons going around the ring align themselves over time antiparallel to the magnetic guide fields (Sokolov-Ternov effect [So64]). This spin polarisation may be measured with the help of the spin dependence of Compton scattering. We solely consider the kinematics below.

   a) Circularly polarised photons from an argon laser (514 nm) hit the electrons (26.67 GeV, straight flight path) head on. What energy does the incoming photon have in the rest frame of the electron?

   b) Consider photon scattering through 90° and 180° in the electron rest frame. What energy does the scattered photon possess in each case? How large are the energies and scattering angles in the lab frame?

   c) How good does the spatial resolution of a calorimeter have to be if it is 64 m away from the interaction vertex and should spatially distinguish between these photons?

2. **Deep inelastic scattering**

   Derive the Callan-Gross relation (7.13). Which value for the mass of the target must be used?

3. **Deep inelastic scattering**

   Deep inelastic electron-proton scattering is studied at the HERA collider. Electrons with 30 GeV are collided head on with 820 GeV protons.

   a) Calculate the centre of mass energy of this reaction. What energy does an electron beam which hits a stationary proton target have to have to reproduce this centre of mass energy?

   b) The relevant kinematical quantities in deep inelastic scattering are the square of the four momentum transfer Q^2 and the Bjorken scaling variable x. Q^2 may, e.g., be found from (6.2). Only the electron's kinematical variables (the beam energy E_e, the energy of the scattered electron E'_e and the scattering angle θ) appear here. In certain kinematical regions it is better to extract Q^2 from other variables since their experimental values give Q^2 with smaller errors. Find a formula for Q^2 where the scattering angles of the electron θ and of the scattered quark γ appear. The latter may be determined experimentally from measurements of the final state hadron energies and momenta. How?

   c) What is the largest possible four momentum transfer Q^2 at HERA? What Q^2 values are attainable in experiments with stationary targets and 300 GeV beam energies? What spatial resolution of the proton does this value correspond to?

   d) Find the kinematical region in Q^2 and x that can be reached with the ZEUS calorimeter which covers the angular region 7° to 178°. The scattered electron needs to have at least 5 Gev energy to be resolved.

   e) The electron-quark interaction can occur through neutral currents (γ, Z^0) or through charged ones ($W^\pm$). Estimate at which value of Q^2 the electromagnetic and weak interaction cross-sections are of the same size.

4. **Spin polarisation**

   Muons are used to carry out deep inelastic scattering experiments at high beam energies. First a static target is bombarded with a proton beam. This produces charged pions which decay in flight into muons and neutrinos.

   a) What is the energy range of the muons in the laboratory frame if magnetic fields are used to select a 350 GeV pion beam?

b) Why are the spins of such a monoenergetic muon beam polarised? How does the polarisation vary as a function of the muon energy?

5. **Parton momentum fractions and x**
Show that in the parton model of deep inelastic scattering, if we do **not** neglect the masses of the nucleon M and of the parton m, the momentum fraction ξ of the scattered parton in a nucleon with momentum P is given by

$$\xi = x \left[1 + \frac{m^2 c^2 - M^2 c^2 x^2}{Q^2} \right].$$

In the deep inelastic domain $\frac{x^2 M^2 c^2}{Q^2} \ll 1$ and $\frac{m^2 c^2}{Q^2} \ll 1$. (Hint: for small ε, ε' we can approximate $\sqrt{1 + \varepsilon(1 + \varepsilon')} \approx 1 + \frac{\varepsilon}{2}(1 + \varepsilon' - \frac{\varepsilon}{4})$.)

8. Quarks, Gluons, and the Strong Interaction

Quark [aus dem Slaw.], aus Milch durch Säue-
rung oder Labfällung und Abtrennen der Molke
gewonnenes Frischkäseprodukt, das vor allem aus
geronnenem, weiß ausgeflocktem (noch stark was-
serhaltigem) Kasein besteht.

Brockhaus-Encyclopaedia, 19th edition

In the previous chapter we learned how deep inelastic scattering may be used as a tool to study the structure and composition of the nucleons. Complementary information about the structure of the nucleons and of other strongly interacting objects (the hadrons) can be obtained from the spectroscopy of these particles. This gives us information about the strong interaction and its field quanta which describe the internal dynamics of the hadrons and the forces acting between them.

The quark model was conceived in the mid-sixties in order to systematise the great diversity of hadrons which had been discovered up to then. In this chapter we will use information from both deep inelastic scattering and spectroscopy to extract the properties of the quarks.

8.1 The Quark Structure of Nucleons

Quarks. By means of deep inelastic scattering, we found that nucleons consist of electrically charged, point-like particles, the *quarks*. It should be possible to reconstruct and to explain the properties of the nucleons (charge, mass, magnetic moment, isospin, etc.) from the quantum numbers of these constituents. For this purpose, we need at least two different types of quarks, which are designated by u (*up*) and d (*down*). The quarks have spin 1/2 and, in the naive quark model, their spins must combine to give the total spin 1/2 of the nucleon. Hence nucleons are built up out of at least 3 quarks. The proton has two u-quarks and one d-quark, while the neutron has two d-quarks and one u-quark.

		u	d	p (uud)	n (udd)
Charge	z	+2/3	−1/3	1	0
Isospin	I	1/2		1/2	
	I_3	+1/2	−1/2	+1/2	−1/2
Spin	s	1/2	1/2	1/2	1/2

The proton and the neutron form an isospin doublet ($I = 1/2$). This is attributed to the fact that u- and d-quarks form an isospin doublet as well. The fact that the charges of the quarks are multiples of $1/3$ is not unequivocally fixed by the charges of the proton and the neutron. It is rather related to other clues; such as the fact that the maximum positive charge found in hadrons is two (e.g., Δ^{++}), and the maximum negative charge is one (e.g., Δ^-). Hence the charges of these hadrons are attributed to 3 u-quarks (charge: $3 \cdot (2e/3) = 2e$) and 3 d-quarks (charge: $3 \cdot (-1e/3) = -1e$) respectively.

Valence quarks and sea quarks. The three quarks that determine the quantum numbers of the nucleons are called *valence quarks*. As well as these, so called *sea quarks*, virtual quark-antiquark pairs, also exist in the nucleon. Their effective quantum numbers average out to zero and do not alter those of the nucleon. Because of their electrical charge, they too are "visible" in deep inelastic scattering. However, they carry only very small fractions x of the nucleon's momentum.

As well as u- and d-quarks, further types of quark–antiquark pairs are found in the "sea"; they will be discussed in more detail in Chap. 9. The different types of quarks are called "flavours". The additional quarks are called s (*strange*), c (*charm*), b (*bottom*) and t (*top*). As we will see later, the six quark types can be arranged in doublets (called *families* or *generations*), according to their increasing mass :

$$\begin{pmatrix} u \\ d \end{pmatrix} \quad \begin{pmatrix} c \\ s \end{pmatrix} \quad \begin{pmatrix} t \\ b \end{pmatrix} .$$

The quarks of the top row have charge number $z_f = +2/3$, those of the bottom row $z_f = -1/3$. The c, b and t quarks are so heavy that they play a very minor role in most experiments at currently attainable Q^2-values. We will therefore neglect them in what follows.

Quark charges. The charge numbers $z_f = +2/3$ and $-1/3$ of the u- and d-quarks are confirmed by comparing the structure functions of the nucleon as measured in deep inelastic neutrino scattering or electron or muon scattering. The structure functions of the proton and the neutron in deep inelastic electron or muon scattering are given, according to (7.16), by:

$$F_2^{e,p}(x) = x \cdot \left[\frac{1}{9} \left(d_v^p + d_s + \bar{d}_s \right) + \frac{4}{9} \left(u_v^p + u_s + \bar{u}_s \right) + \frac{1}{9} \left(s_s + \bar{s}_s \right) \right]$$

$$F_2^{e,n}(x) = x \cdot \left[\frac{1}{9} \left(d_v^n + d_s + \bar{d}_s \right) + \frac{4}{9} \left(u_v^n + u_s + \bar{u}_s \right) + \frac{1}{9} \left(s_s + \bar{s}_s \right) \right] . \quad (8.1)$$

Here $u_v^{p,n}(x)$ is the distribution of the u valence quarks in the proton or neutron, $u_s(x)$ that of the u sea quarks, etc. We proceed on the assumption that the sea quark distributions in the proton and neutron are identical and drop the upper index. Formally, the neutron and the proton may be transformed into each other by interchanging the u- and d-quarks (*isospin symmetry*). Their quark distributions are therefore related by:

$$
\begin{aligned}
u_v^p(x) &= d_v^n(x)\,,\\
d_v^p(x) &= u_v^n(x)\,,\\
u_s^p(x) = d_s^p(x) &= d_s^n(x) = u_s^n(x)\,.
\end{aligned}
\tag{8.2}
$$

The structure function of an "average" nucleon is then given by:

$$
\begin{aligned}
F_2^{e,N}(x) &= \frac{F_2^{e,p}(x) + F_2^{e,n}(x)}{2}\\
&= \frac{5}{18}\,x \cdot \sum_{q=d,u}(q(x) + \bar{q}(x)) + \frac{1}{9}\,x \cdot \left[s_s(x) + \bar{s}_s(x) \right]\,.
\end{aligned}
\tag{8.3}
$$

The second term is small, as s-quarks are only present as sea quarks. Thus the factor 5/18 is approximately the mean square charge of the u- and d-quarks (in units of e^2).

In deep inelastic neutrino scattering, the factors z_f^2 are not present, as the weak charge is the same for all quarks. Because of charge conservation and helicity, neutrinos and antineutrinos couple differently to the different types of quarks and antiquarks. These differences, however, cancel out when the structure function of an average nucleon (7.17) is considered. One then obtains:

$$
F_2^{\nu,N}(x) = x \cdot \sum_f (q_f(x) + \bar{q}_f(x))\,.
\tag{8.4}
$$

Experiments show indeed that except for the factor 5/18, $F_2^{e,N}$ and $F_2^{\nu,N}$ are identical (Fig. 8.1). Hence one can conclude that the charge numbers $+2/3$ and $-1/3$ have been correctly attributed to the u- and d-quarks.

Quark momentum distributions. Combining the results from the scattering of charged leptons and neutrinos, one obtains information about the momentum distribution of sea quarks and valence quarks (see Sect. 10.6). The distribution of the valence quarks has a maximum at $x \approx 0.17$ and a mean value of $\langle x_v \rangle \approx 0.12$. The sea quarks are only relevant at small x; their mean value lies at $\langle x_s \rangle \approx 0.04$.

Further important information is obtained by considering the integral over the structure function $F_2^{\nu,N}$. The integration is carried out over all quark momenta weighted by their distribution functions; hence, the integral yields that fraction of the momentum of the nucleon which is carried by the quarks. Experimentally, one finds

$$
\int_0^1 F_2^{\nu,N}(x)\,dx \approx \frac{18}{5} \int_0^1 F_2^{e,N}(x)\,dx \approx 0.5\,.
\tag{8.5}
$$

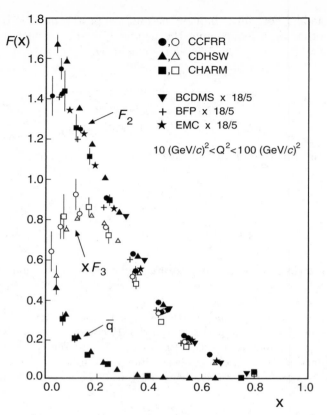

Fig. 8.1. Comparison of the structure functions observed in deep inelastic scatte-ring with charged leptons, and with neutrinos [PD94] (see also Sect. 10.6). As well as the F_2 structure function, the distributions of the antiquarks $\bar{q}(x)$ which yield the sea quark distribution and the distributions of the valence quarks (denoted by $xF_3(x)$) are given (cf. Fig. 7.7).

Thus roughly half of the momentum is carried by particles interacting neither electromagnetically nor weakly. They are identified with the *gluons*.

Figure 8.2 shows the ratio F_2^n/F_2^p. For $x \to 0$, the ratio approaches uni-ty. In this region the sea quarks are absolutely predominant and the small difference in the distribution of the valence quarks has no significant effect on the ratio. As $x \to 1$, it is the other way round and the sea quarks no longer play a part. Hence, one would expect F_2^n/F_2^p to approach the value $2/3$ in this region. This value would correspond to $(2z_d^2 + z_u^2)/(2z_u^2 + z_d^2)$, the ratio of the mean square charges of the valence quarks of the neutron and proton. The measured value, however, is $1/4$, i. e., z_d^2/z_u^2. This implies that large momentum fractions in the proton are carried by u-quarks, and, in the neutron, by d-quarks.

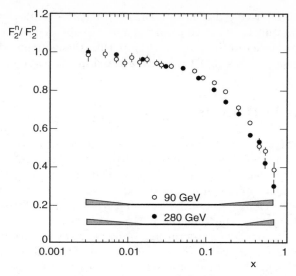

Fig. 8.2. The structure function ratio F_2^n/F_2^p [Am92b]. The data were obtained from muon scattering with beam energies of 90 and 280 GeV, and averaged over Q^2. The error bars denote statistical errors; the horizontal bands denote systematic errors.

Constituent quarks. In (8.5) we saw that only about half of the momentum of a nucleon is carried by valence and sea quarks. In dealing with the spectroscopic properties of nucleons, sea quarks and gluons need not be explicitly dealt with. We can combine them with the valence quarks. One then acts as though there were only three valence quarks, with enlarged masses but unchanged quantum numbers. We will return to this point in Chaps. 13–15. These "effective valence quarks" are called *constituent quarks*.

In interpreting deep inelastic scattering, we neglected the rest masses of the bare u- and d-quarks. This is justified, since they are small: $m_u=1.5-5$ MeV/c^2, $m_d=3-9$ MeV/c^2 [PD98]. These masses are commonly called *current quark* masses. However, these are not the masses obtained from hadron spectroscopy; e. g., from calculations of magnetic moments and hadron excitation energies. The constituent quark masses are much larger (300 MeV/c^2). The constituent masses must be mainly due to the cloud of gluons and sea quarks. Their values for all the quark flavours are compiled in Table 9.1.

The d-quark is heavier than the u-quark, which can be easily understood as follows. The proton (uud) and the neutron (ddu) are isospin symmetric; i. e., they transform into each other under interchange of the u- and d-quarks. Since the strong interaction is independent of quark flavour, the neutron-proton mass difference can only be due to the intrinsic quark masses and to the electromagnetic interaction between them. If we assume that the spatial distribution of the u- and d-quarks in the proton corresponds to the distribution of d- and u-quarks in the neutron, then it is easily seen that the Coulomb energy must be higher in the proton. Despite this, the neutron is heavier than the proton which implies that the mass of the d-quark is larger.

8.2 Quarks in Hadrons

A multitude of unstable hadrons are known in addition to the nucleons. Through the study of these hadrons the diverse properties of the strong interaction are revealed. Hadrons can be classified in two groups: the *baryons*, fermions with half-integral spin, and the *mesons*, bosons with integral spin. The hadronic spectrum was uncovered step by step: initially from analyses of photographic plates which had been exposed to cosmic radiation and later in experiments at particle accelerators. Many short-lived particles were thus detected, including excited states of the nucleon. This led to the conclusion that nucleons themselves are composed of smaller structures. This conclusion was then extended to all known hadrons.

Baryons. The lowest mass baryons are the proton and the neutron. They are the "ground states" of a rich excitation spectrum of well-defined energy (or mass) states. This will be discussed further in Chap. 15. In this respect, baryon spectra have many parallels to atomic and molecular spectra. Yet, there is an important difference. The energy (or mass) gaps between individual states are of the same order of magnitude as the nucleon mass. These gaps are then relatively much larger than those of atomic or molecular physics. Consequently these states are also classified as individual particles with corresponding lifetimes.

Like the proton and neutron, other *baryons* are also composed of three quarks. Since quarks have spin $1/2$, baryons have half-integral spin.

When baryons are produced in particle reactions, the same number of antibaryons are simultaneously created. To describe this phenomenon, a new additive quantum number is introduced: *baryon number B*. We assign $B = 1$ to baryons and $B = -1$ to antibaryons. Accordingly, baryon number $+1/3$ is attributed to quark, and baryon number $-1/3$ to antiquarks. All other particles have baryon number $B = 0$. Experiments indicate that baryon number is conserved in all particle reactions and decays. Thus, the quark minus antiquark number is conserved. This would be violated by, e. g., the hypothetical decay of the proton:

$$\mathrm{p} \rightarrow \pi^0 + \mathrm{e}^+ \,.$$

Without baryon number conservation, this decay mode would be energetically favoured. Yet, it has not been observed. The experimental limit of the partial lifetime is given by $\tau(\mathrm{p} \rightarrow \pi^0 + \mathrm{e}^+) > 5.5 \cdot 10^{32}$ years [Be90a].

Mesons. The lightest hadrons are the pions. Their mass, about $140 \, \mathrm{MeV}/c^2$, is much less than that of the nucleon. They are found in three different charge states: π^-, π^0 and π^+. Pions have spin 0. It is therefore natural to assume that they are composed of two quarks, or, more exactly, of a quark and an antiquark: this is the only way to build the three charge states out of quarks. The pions are the lightest systems of quarks. Hence, they can only decay into the even lighter leptons or into photons. Pions have the following quark structure:

$$|\pi^+\rangle = |u\overline{d}\rangle \qquad |\pi^-\rangle = |\overline{u}d\rangle \qquad |\pi^0\rangle = \frac{1}{\sqrt{2}} \left\{ |u\overline{u}\rangle - |d\overline{d}\rangle \right\} .$$

The $|\pi^0\rangle$ is a mixed state of $|u\overline{u}\rangle$ and $|d\overline{d}\rangle$. The above expression includes the correct symmetry and normalisation. The pion mass is considerably smaller than the constituent quark mass described above. This is another indication that the interquark interaction energy has a substantial effect on hadron masses.

Hadrons composed of quark–antiquark pairs are called *mesons*. Mesons have integer spin: their total spin results from a vector sum of the quark and antiquark spins, including a possible integer orbital angular momentum contribution. Mesons eventually decay into electrons, neutrinos and/or photons; there is no "meson number conservation", in contrast to baryon number conservation. This is understood in the quark model: mesons are quark–antiquark combinations $|q\overline{q}\rangle$ and so the number of quarks minus the number of antiquarks is zero. Hence any number of mesons may be produced or annihilated. It is just a matter of convention which mesons are called particles and which antiparticles.

8.3 The Quark–Gluon Interaction

Colour. Quarks have another important property called *colour* which we have previously neglected. This is needed to ensure that quarks in hadrons obey the Pauli principle. Consider the Δ^{++}-resonance which consists of three u-quarks. The Δ^{++} has spin $J = 3/2$ and positive parity; it is the lightest baryon with $J^P = 3/2^+$. We therefore can assume that its orbital angular momentum is $\ell = 0$; so it has a symmetric spatial wave function. In order to yield total angular momentum 3/2, the spins of all three quarks have to be parallel:

$$|\Delta^{++}\rangle = |u^\uparrow u^\uparrow u^\uparrow\rangle .$$

Thus, the spin wave function is also symmetric. The wave function of this system is furthermore symmetric under the interchange of any two quarks, as only quarks of the same flavour are present. Therefore the total wave function appears to be symmetric, in violation of the Pauli principle.

Including the *colour* property, a kind of quark charge, the Pauli principle may be salvaged. The quantum number colour can assume three values, which may be called *red, blue* and *green*. Accordingly, antiquarks carry the anticolours *anti-red, anti-blue*, and *anti-green*. Now the three u-quarks may be distinguished. Thus, a colour wave function antisymmetric under particle interchange can be constructed, and we so have antisymmetry for the total wave function. The quantum number colour was introduced for theoretical reasons; yet, experimental clues indicate that this hypothesis is correct, as will be discussed in Sect. 9.3.

Gluons. The interaction binding quarks into hadrons is called the *strong interaction*. Such a fundamental interaction is, in our current understanding, always connected with a particle exchange. For the strong interaction, *gluons* are the exchange particles that couple to the colour charge. This is analogous to the electromagnetic interaction in which photons are exchanged between electrically charged particles.

The experimental findings of Sect. 8.1 led to the development of a field theory called *quantum chromodynamics* (QCD). As its name implies, QCD is modelled upon quantum electrodynamics (QED). In both, the interaction is mediated by exchange of a massless field particle with $J^P = 1^-$ (a vector boson).

The gluons carry simultaneously colour and anticolour. According to group theory, the 3×3 colour combinations form two multiplets of states: a singlet and an octet. The octet states form a basis from which all other colour states may be constructed. They correspond to an octet of gluons. The way in which these eight states are constructed from colours and anticolours is a matter of convention. One possible choice is:

$$r\bar{g}, \quad r\bar{b}, \quad g\bar{b}, \quad g\bar{r}, \quad b\bar{r}, \quad b\bar{g}, \quad \sqrt{1/2}\,(r\bar{r} - g\bar{g}), \quad \sqrt{1/6}\,(r\bar{r} + g\bar{g} - 2b\bar{b})\,.$$

The colour singlet:

$$\sqrt{1/3}\,(r\bar{r} + g\bar{g} + b\bar{b}),$$

which is symmetrically constructed from the three colours and the three anticolours is invariant with respect to a re-definition of the colour names (rotation in colour space). It therefore has no effect in colour space and cannot be exchanged between colour charges.

By their exchange the eight gluons mediate the interaction between particles carrying colour charge, i.e., not only the quarks but also the gluons themselves. This is an important difference to the electromagnetic interaction, where the photon field quanta have no charge, and therefore cannot couple with each other.

In analogy to the elementary processes of QED (emission and absorption of photons, pair production and annihilation); emission and absorption of gluons (Fig. 8.3a) take place in QCD, as do production and annihilation of quark–antiquark pairs (Fig. 8.3b). In addition, however, three or four gluons can couple to each other in QCD (Fig. 8.3c,d).

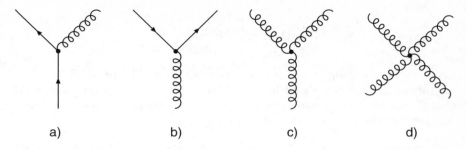

a)   b)   c)   d)

Fig. 8.3. The fundamental interaction diagrams of the strong interaction: emission of a gluon by a quark (**a**), splitting of a gluon into a quark–antiquark pair (**b**) and "self-coupling" of gluons (**c, d**).

Hadrons as colour-neutral objects. With colour, quarks gain an additional degree of freedom. One might therefore expect each hadron to exist in a multitude of versions which, depending upon the colours of the constituent quarks involved, would have different total (net) colours but would be equal in all other respects. In practice only one type of each hadron is observed (one π^-, p, Δ^0 etc.). This implies the existence of an additional condition: only colourless particles, i. e., with no net colour, can exist as free particles.

This condition explains why quarks are not observed as free particles. A single quark can be detached from a hadron only by producing at least two free objects carrying colour: the quark, and the remainder of the hadron. This phenomenon is therefore called *confinement*. Accordingly, the potential acting on a quark limitlessly increases with increasing separation, in sharp contrast to the Coulomb potential. This phenomenon is due to the inter-gluonic interactions.

The combination of a colour with the corresponding anticolour results in a colourless ("white") state. Putting the three different colours together results in a colourless ("white") state as well. This can be graphically depicted by three vectors in a plane symbolizing the three colours, rotated with respect to each other by 120°.

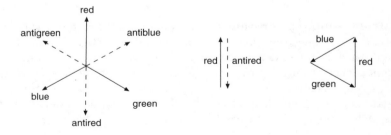

Hence, e.g., the π^+ meson has three possible colour combinations:

$$|\pi^+\rangle = \begin{cases} |u_r \overline{d_{\overline{r}}}\rangle \\ |u_b \overline{d_{\overline{b}}}\rangle \\ |u_g \overline{d_{\overline{g}}}\rangle \,, \end{cases}$$

where the index designates the colour or anticolour. The physical pion is a mixture of these states. By exchange of gluons, which by themselves simultaneously transfer colour and anticolour, the colour combination continuously changes; yet the net-colour "white" is preserved.

In baryons, the colours of the three quarks also combine to yield "white". Hence, to obtain a colour neutral baryon, each quark must have a different colour. The proton is a mixture of such states:

$$|p\rangle = \begin{cases} |u_b u_r d_g\rangle \\ |u_r u_g d_b\rangle \\ \vdots \end{cases} \,.$$

From this argument, it also becomes clear why no hadrons exist which are $|qq\rangle$, or $|qq\overline{q}\rangle$ combinations, or the like. These states would not be colour neutral, no matter what combination of colours were chosen.

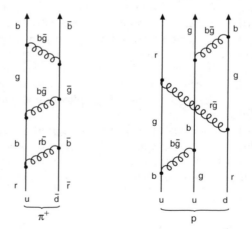

The strong coupling constant α_s. In quantum field theory, the coupling "constant" describing the interaction between two particles is an effective constant which is in fact dependent on Q^2. In the electromagnetic interaction this dependence is only a weak one; in the strong interaction, however, it is very strong. The reason is that gluons as the field quanta of the strong interaction carry colour themselves, and therefore can also couple to other gluons. A first-order perturbative calculation in QCD yields:

$$\alpha_s(Q^2) = \frac{12\pi}{(33 - 2n_f) \cdot \ln(Q^2/\Lambda^2)} \,. \tag{8.6}$$

Here, n_f denotes the number of quark types involved. Since a heavy virtual quark–antiquark pair has a very short lifetime and range, it can be resolved only at very high Q^2. Hence, n_f depends on Q^2, with $n_f \approx 3$–6. The parameter Λ is the only free parameter of QCD. It is determined by comparing predictions with experimental data to be $\Lambda \approx 250$ MeV/c. The application of perturbative expansion procedures in QCD is valid only if $\alpha_s \ll 1$. This is satisfied for $Q^2 \gg \Lambda^2 \approx 0.06$ (GeV/c)^2.

From (7.15), the Q^2-dependence of the coupling strength corresponds to a dependence on separation. For very small distances and corresponding high values of Q^2, the interquark coupling decreases, vanishing asymptotically. In the limit $Q^2 \to \infty$, quarks can be considered to be "free", this is called *asymptotic freedom*. By contrast, at large distances, the interquark coupling increases so strongly, that it is impossible to detach individual quarks from hadrons (confinement).

8.4 Scaling Violations of the Structure Functions

In Sect. 7.2 we showed that the structure function F_2 depends on the quantity x only. We thereby concluded that the nucleon is composed of point-like, charged constituents. Yet, high precision measurements show that to a small degree, F_2 does depend on Q^2. Figure 8.4 shows the experimental results of F_2^{d} as a function of Q^2 at several fixed values of x. The data cover a large kinematic range of x and Q^2. We see that the structure function increases with Q^2 at small values of x; and decreases with increasing Q^2 at large values of x. This behaviour, called scaling violation, is sketched once more in Fig. 8.5. This means that at increasing values of Q^2, there are fewer quarks with large momentum fractions in the nucleon; rather, quarks with small momentum fractions predominate.

This violation of scaling is not caused by any finite size of the quarks. In the framework of QCD, the above violation can be traced back to fundamental processes in which the constituents of the nucleon continuously interact with each other (Fig. 8.3). Quarks can emit or absorb gluons, gluons may split into q$\bar{\text{q}}$ pairs, or emit gluons themselves. Thus, the momentum distribution between the constituents of the nucleon is continually changing.

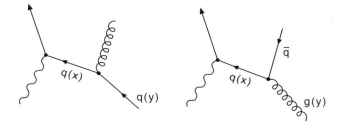

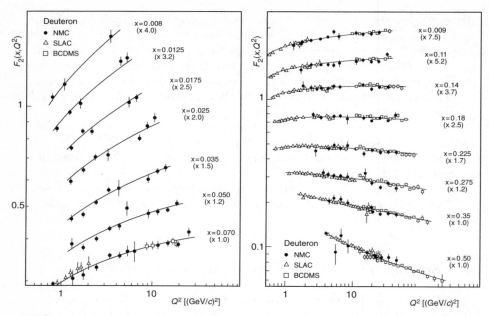

Fig. 8.4. Structure function F_2 of the deuteron as a function of Q^2 at different values of x on a logarithmic scale. The results shown are from muon scattering at CERN (NMC and BCDMS collaboration) [Am92a, Be90b] and from electron scattering at SLAC [Wh92]. For clarity, the data at the various values of x are multiplied by constant factors. The solid line is a QCD fit, taking into account the theoretically predicted scaling violation. The gluon distribution and the strong coupling constant are free parameters here.

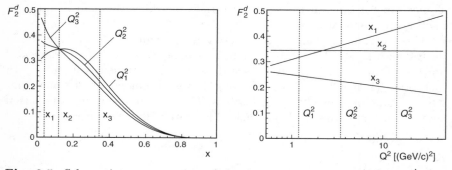

Fig. 8.5. Schematic representation of the deuteron structure function F_2^d as a function of x at various values of Q^2 (*left*); and as a function of Q^2 at constant x (*right*).

Figure 8.6 is an attempt to illustrate how this alters the measurements of structure functions at different values of Q^2. A virtual photon can resolve

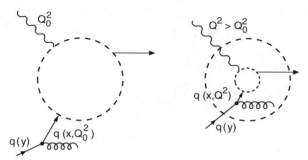

Fig. 8.6. A quark with momentum fraction x can be emitted from a quark or gluon that carries a larger momentum fraction $(x < y < 1)$, *(left)*. The resolving power of the virtual photon increases with Q^2 *(right)*. The diagram shows the interaction of a photon with a quark after it has emitted a gluon. At small $Q^2 = Q_0^2$, the quark and the gluon are seen as a unit. At larger $Q^2 > Q_0^2$, the resolution increases and the momentum fraction of the quark alone is measured, i.e., without that of the gluon; hence, a smaller value is obtained.

dimensions of the order of $\hbar/\sqrt{Q^2}$. At small $Q^2 = Q_0^2$, quarks and any emitted gluons cannot be distinguished and a quark distribution $q(x, Q_0^2)$ is measured. At larger Q^2 and higher resolution, emission and splitting processes must be considered. Thus, the number of partons seen to share the momentum of the nucleon, increases. Therefore, the quark distribution $q(x, Q^2)$ at small momentum fractions x is larger than $q(x, Q_0^2)$; whereas the effect is reversed for large x. This is the origin of the increase of the structure function with Q^2 at small values of x and its decrease at large x. The gluon distribution $g(x, Q^2)$ shows a Q^2-dependence as well, which originates from processes of gluon emission by a quark or by another gluon.

The change in the quark distribution and in the gluon distribution with Q^2, at fixed values of x, is proportional to the strong coupling constant $\alpha_s(Q^2)$ and depends upon the size of the quark and gluon distributions at all larger values of x. The mutual dependence of the quark and gluon distributions can be described by a system of coupled integral-differential equations [Gr72, Li75, Al77]. If $\alpha_s(Q^2)$ and the shape of $q(x, Q_0^2)$ and $g(x, Q_0^2)$ are known at a given value Q_0^2, then $q(x, Q^2)$ and $g(x, Q^2)$ can be predicted from QCD for all other values of Q^2. Alternatively, the coupling $\alpha_s(Q^2)$ and the gluon distribution $g(x, Q^2)$, which cannot be directly measured, can be determined from the observed scaling violation of the structure function $F_2(x, Q^2)$.

The solid lines in Fig. 8.4 show a fit to the scaling violation of the measured structure functions from a QCD calculation [Ar93]. The fit value of $\Lambda \approx 250$ MeV/c corresponds to a coupling constant:

$$\alpha_s(Q^2 = 100\,(\text{GeV}/c)^2) \approx 0.16\,. \tag{8.7}$$

Figure 8.7 is a three dimensional representation of the Q^2-dependence of the structure function F_2^d, here given as $xq(x) \equiv \sum_f x(q_f(x) + \bar{q}_f(x)) \approx \frac{18}{5} F_2^d$,

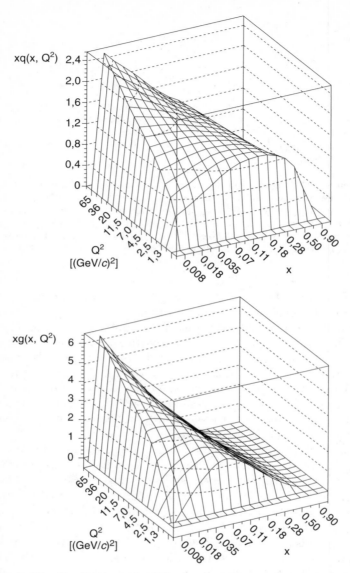

Fig. 8.7. The distributions of quarks (*top*) and gluons (*bottom*) in the nucleon, as functions of x and Q^2 [Bi92]. While $q(x, Q^2)$ and $g(x, Q^2)$ respectively describe the probability of finding a quark or a gluon with momentum fraction x; $xq(x, Q^2)$ and $xg(x, Q^2)$ measure the share of the momentum carried by the quarks or the gluons in the particular x-interval.

and of the gluon structure function $G(x, Q^2) = xg(x, Q^2)$ as they emerge from this QCD analysis. The drastic change in the shape of these distributions with

Q^2 is clearly seen. At small values of x, F_2 increases rapidly with Q^2. At small values of Q^2, the shape approaches that of the valence quark distribution, since the sea quark contribution becomes less and less significant. Similarly, the gluon structure function $xg(x)$ increases rapidly with Q^2 at small values of x. It assumes a shape similar to that of a valence quark distribution at small values of Q^2. For large values of Q^2 the gluon distribution drops rapidly with x. The *change* in the structure function with Q^2 can be calculated. It has however so far proven impossible to predict the x-dependence of $F_2(x, Q_0^2)$ in theory; it is known only from experiment.

Summary. Scaling violation in the structure functions is a highly interesting phenomenon. It is not unusual that particles that appear point-like turn out to be composite when studied more closely (e.g., atomic nuclei in Rutherford scattering with low-energy α particles or high-energy electrons). In deep inelastic scattering, however, a new phenomenon is observed. With increasing resolution, quarks and gluons turn out to be composed of quarks and gluons; which themselves, at even higher resolutions, turn out to be composite as well (Fig. 8.6). The quantum numbers (spin, flavour, colour, ...) of these particles remain the same; only the mass, size, and the effective coupling α_s change. Hence, there appears to be in some sense a self similarity in the internal structure of strongly interacting particles.

Problems

1. **Partons**

   Consider deep inelastic scattering of muons with energy 600 GeV off protons at rest. The data analysis is to be carried out at $Q^2 = 4 \, \text{GeV}^2/c^2$.

   a) What is the smallest value of x which can be attained under these circumstances? You may assume that the minimal scattering energy is $E' = 0$.

   b) How many partons may be resolved with $x > 0.3$, $x > 0.03$ and in the full measurable range of x if we parameterise the parton distribution as follows:

   $$\begin{aligned} q_v(x) &= A(1-x)^3/\sqrt{x} &&\text{for the valence quarks,} \\ q_s(x) &= 0.4(1-x)^8/x &&\text{for the sea quarks and} \\ g(x) &= 4(1-x)^6/x &&\text{for the gluons.} \end{aligned}$$

   The role of the normalisation constant, A, is to take into account that there are 3 valence quarks.

9. Particle Production in e⁺e⁻ Collisions

So far, we have only discussed the light quarks, u and d, and those hadrons composed of these two quarks. The easiest way to produce hadrons with heavier quarks is in e⁺e⁻ collisions. Free electrons and positrons may be produced rather easily. They can be accelerated, stored and made to collide in accelerators. In an electron–positron collision process, all particles which interact electromagnetically and weakly can be produced, as long as the energy of the beam particles is sufficiently high. In an electron–positron electromagnetic annihilation, a virtual photon is produced, which immediately decays into a pair of charged elementary particles. In a weak interaction, the exchanged particle is the heavy vector boson Z^0 (cf. the diagram and see Chap. 11). The symbol f denotes an elementary fermion (quark or lepton) and $\bar{f}$ its antiparticle. The $f\bar{f}$ system must have the quantum numbers of the photon or the Z^0, respectively. In these reactions all fundamental, charged particle–antiparticle pairs can be produced; lepton–antilepton and quark–antiquark pairs. Neutrinos are electrically neutral; hence, neutrino–antineutrino pairs can only be produced by Z^0 exchange.

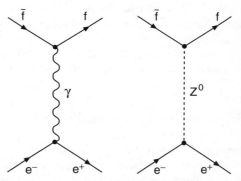

Colliding beams. Which particle–antiparticle pairs can be produced only depends upon the energy of the electrons and positrons. In a storage ring, electrons and positrons orbit with the same energy E, but in opposite directions, and collide head-on. It is conventional to use the Lorentz-invariant energy variable s, the square of the centre of mass energy:

$$
\begin{aligned}
s &= (p_1 c + p_2 c)^2 \\
&= m_1^2 c^4 + m_2^2 c^4 + 2 E_1 E_2 - 2 \boldsymbol{p}_1 \boldsymbol{p}_2 c^2 .
\end{aligned}
\tag{9.1}
$$

In a storage ring with colliding particles of energy E,

$$s = 4E^2 \,. \tag{9.2}$$

Hence, particle–antiparticle pairs with masses of up to $2m = \sqrt{s}/c^2$ can be produced. To discover new particles, the storage ring energy must be raised. One then looks for an increase in the reaction rate, or for resonances in the cross-section.

The great advantage of colliding beam experiments is that the total beam energy is available in the centre of mass system. In a fixed target experiment, with m satisfying $mc^2 \ll E$, s is related to E by:

$$s \approx 2mc^2 \cdot E \,. \tag{9.3}$$

Here, the centre of mass energy only increases proportionally to the square root of the beam energy.

Particle detection. To detect the particles produced in e$^+$e$^-$ annihilation one requires a detector set up around the collision point which covers as much as possible of the the total 4π solid angle. The detector should permit us to trace the tracks back to the interaction point and to identify the particles themselves. The basic form of such a detector is sketched in Fig. 9.1.

9.1 Lepton Pair Production

Before we turn to the creation of heavy quarks, we want to initially consider the leptons. *Leptons* are elementary spin $1/2$ particles which feel the weak and, if they are charged, the electromagnetic interaction — but not, however, the strong interaction.

Muons. The lightest particles which can be produced in electron-positron collisions are muon pairs:

$$\mathrm{e}^+ + \mathrm{e}^- \to \mu^+ + \mu^- \,.$$

The muon μ^- and its antiparticle[1] the μ^+ both have a mass of only 105.7 MeV/c^2 and they are produced in all usual e$^+$e$^-$ storage ring experiments. They penetrate matter very easily[2], whereas electrons because of their small mass and hadrons because of the strong interaction have much smaller ranges. After that of the neutron, theirs is the longest lifetime (2 μs) of any unstable particle. This means that experimentally they may easily be identified. Therefore the process of muon pair production is often used as a reference point for other e$^+$e$^-$ reactions.

---

[1] Antiparticles are generally symbolised by a bar (e.g., $\overline{\nu}_e$). This symbol is generally skipped over for charged leptons since knowledge of the charge alone tells us whether we have a particle or an antiparticle. We thus write e$^+$, μ^+, τ^+.

[2] Muons from cosmic radiation can still be detected in underground mines!

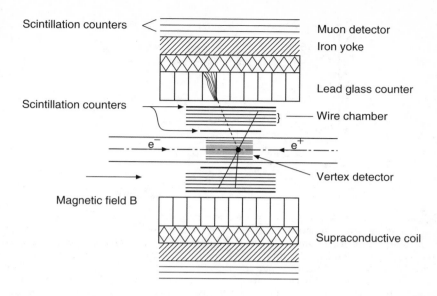

Scintillation counters

Muon detector

Iron yoke

Lead glass counter

Scintillation counters

Wire chamber

e^- e^+

Vertex detector

Magnetic field B

Supraconductive coil

Fig. 9.1. Sketch of a 4π-detector, as used in e^+e^- collision experiments. The detector is inside the coil of a solenoid, which typically produces a magnetic field of around 1T along the beam direction. Charged particles are detected in a vertex detector, mostly composed of silicon microstrip counters, and in wire chambers. The vertex detector is used to locate the interaction point. The curvature of the tracks in the magnetic field tell us the momenta. Photons and electrons are detected as shower formations in electromagnetic calorimeters (of, e.g., lead glass). Muons pass through the iron yoke with little energy loss. They are then seen in the exterior scintillation counters.

Tau leptons. If the centre of mass energy in an e^+e^- reaction suffices, a further lepton pair, the τ^- and τ^+, may be produced. Their lifetime, $3 \cdot 10^{-13}$ s, is much shorter. They may weakly decay into muons or electrons as will be discussed in Sect 10.1f.

The tau was discovered at the SPEAR e^+e^- storage ring at SLAC when oppositely charged electron-muon pairs were observed whose energy was much smaller than the available centre of mass energy [Pe75].

These events were interpreted as the creation and subsequent decay of a heavy lepton–antilepton pair:

$$e^+ + e^- \longrightarrow \tau^+ + \tau^-$$
$$\hookrightarrow \mu^- + \overline{\nu}_\mu + \nu_\tau \quad \text{or} \quad e^- + \overline{\nu}_e + \nu_\tau$$
$$\longrightarrow e^+ + \nu_e + \overline{\nu}_\tau \quad \text{or} \quad \mu^+ + \nu_\mu + \overline{\nu}_\tau \,.$$

The neutrinos which are created are not detected.

The threshold for $\tau^+\tau^-$-pair production, and hence the mass of the τ-lepton, may be read off from the increase of the cross-section of the e^+e^-

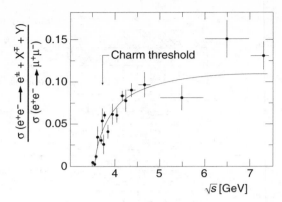

Fig. 9.2. Ratio of the cross-sections for the production of two particles with opposite charges in the reaction $e^+ + e^- \rightarrow e^{\pm} + X^{\mp} + Y$, to the cross-sections for the production of $\mu^+\mu^-$ pairs [Ba78, Ba88]. Here $X^{\mp}$ denotes a charged lepton or meson and Y symbolises the unobserved, neutral particles. The sharp increase at $\sqrt{s} \approx 3.55$ GeV is a result of τ-pair production, which here becomes energetically possible. The threshold for the creation of mesons containing a charmed quark (*arrow*) is only a little above that for τ-lepton production. Both particles have similar decay modes which makes it more difficult to detect τ-leptons.

reaction with the centre of mass energy. One should use as many leptonic and hadronic decay channels as possible to provide a good signature for τ-production (Fig. 9.2). The experimental threshold at $\sqrt{s} = 2m_\tau c^2$ implies that the tau mass is 1.777 GeV/c^2.

Cross-section. The creation of charged lepton pairs may, to a good approximation, be viewed as a purely electromagnetic process (γ exchange). The exchange of Z^0 bosons, and interference between photon and Z^0 exchange, may be neglected if the energy is small compared to the mass of the Z^0. The cross-section may then be found relatively easily. The most complicated case is the elastic process $e^+e^- \rightarrow e^+e^-$, *Bhabha scattering*. Here two processes must be taken into account: the annihilation of the electron and positron into a virtual photon (with subsequent e^+e^--pair creation) and secondly the scattering of the electron and positron off each other.

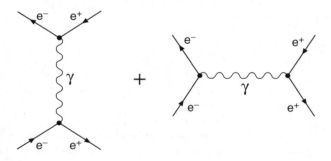

These processes lead to the same final state and so their amplitudes must be added in order to obtain the cross-section.

Muon pair creation is more easily calculated. Other e^+e^- reactions are therefore usually normalised with respect to it. The differential cross-section for this reaction is:

$$\frac{d\sigma}{d\Omega} = \frac{\alpha^2}{4s}\,(\hbar c)^2 \cdot \left(1 + \cos^2\theta\right)\ . \tag{9.4}$$

Integrating over the solid angle Ω yields the total cross-section:

$$\sigma = \frac{4\pi\alpha^2}{3s}(\hbar c)^2\ , \tag{9.5}$$

and one finds

$$\sigma(e^+e^- \to \mu^+\mu^-) = 21.7\,\frac{\text{nbarn}}{(E^2/\text{GeV}^2)}\ . \tag{9.6}$$

The formal derivation of (9.4) may be found in many standard texts [Go86, Na90, Pe87], we will merely try to make it plausible: The photon couples to two elementary charges. Hence the matrix element contains two powers of e and the cross-section, which is proportional to the square of the matrix element, is proportional to e^4 or α^2. The length scale is proportional to $\hbar c$, which enters twice over since cross-sections have the dimension of area. We must further divide by a quantity with dimensions of [energy^2]. Since the masses of the electron and the muon are very small compared to s, this last is the only reasonable choice. The cross-section then falls off with the square of the storage ring's energy. The $(1 + \cos^2\theta)$ angular dependence is typical for the production of two spin 1/2 particles such as muons. Note that (9.4) is, up to this angular dependence, completely analogous to the equation for Mott scattering (5.39) once we recognise that $Q^2c^2 = s = 4E^2 = 4E'^2$ holds here.

Figure 9.3 shows the cross-section for $e^+e^- \to \mu^+\mu^-$ and the prediction of quantum electrodynamics. One sees an excellent agreement between theory and experiment. The cross-section for $e^+e^- \to \tau^+\tau^-$ is also shown in the figure. If the centre of mass energy $\sqrt{s}$ is large enough that the difference in the μ and τ rest masses can be neglected, then the cross-sections for $\mu^+\mu^-$ and $\tau^+\tau^-$ production are identical. One speaks of *lepton universality*, which means that the electron, the muon and the tau behave, apart from their masses and associated effects, identically in all reactions. The muon and the tau may to a certain extent be viewed as being heavier copies of the electron.

Since (9.6) describes the experimental cross-section so well, the form factors of the μ and τ are unity — which according to Table 5.1 means they are point-like particles. No spatial extension of the leptons has yet been seen. The upper limit for the electron is 10^{-18} m. Since the hunt for excited leptons so far has also been unsuccessful, it is currently believed that leptons are indeed elementary, point-like particles.

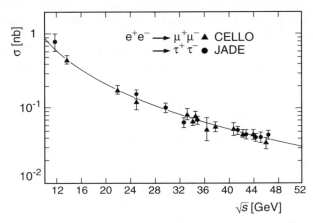

Fig. 9.3. Cross-sections of the reactions e$^+$e$^-$ → $\mu^+\mu^-$ and e$^+$e$^-$ → $\tau^+\tau^-$ as functions of the centre of mass energy $\sqrt{s}$ (from [Ba85] and [Be87]). The solid line shows the cross-section (9.6) predicted by quantum electrodynamics.

9.2 Resonances

If the cross-sections for the production of muon pairs and hadrons in e$^+$e$^-$ scattering are plotted as a function of the centre of mass energy $\sqrt{s}$, one finds in both cases the $1/s$-dependence of (9.5). In the hadronic final state channels this trend is broken by various strong peaks which are sketched in Fig. 9.4. These so-called *resonances* are short lived states which have a fixed mass and well-defined quantum numbers such as angular momentum. It is therefore reasonable to call them particles.

Breit-Wigner formula. The energy dependence of the cross-section of a reaction between two particles a and b close to a resonance energy E_0 is generally described by the *Breit-Wigner formula* (see, e. g., [Pe87]). In the case of elastic scattering, it is approximately given by:

$$\sigma(E) = \frac{\pi \lambda^2 \, (2J + 1)}{(2s_{\mathrm{a}} + 1)(2s_{\mathrm{b}} + 1)} \cdot \frac{\Gamma^2}{(E - E_0)^2 + \Gamma^2/4} \, . \qquad (9.7)$$

Here λ is the reduced wave-length in the centre of mass system, s_{a} and s_{b} are the spins of the reacting particles and Γ is the *width* (half width) of the resonance. The lifetime of such a resonance is $\tau = \hbar/\Gamma$. This formula is similar to that for the resonance of a forced oscillator with large damping. E corresponds to the excitation frequency ω, E_0 to the resonance frequency ω_0 and the width Γ to the damping.

For an inelastic reaction like the case at hand, the cross-section depends upon the *partial widths* Γ_i and Γ_f in the initial and final channels and on the total width Γ_{tot}. The latter is the sum of the partial widths of all possible final channels. The result for an individual decay channel f is

$$\sigma_f(E) = \frac{3\pi\lambdabar^2}{4} \cdot \frac{\Gamma_i \Gamma_f}{(E - E_0)^2 + \Gamma_{\text{tot}}^2/4}, \qquad (9.8)$$

where we have replaced s_a and s_b by the spins of the electrons (1/2) and J by the spin of the photon (1).

The resonances ϱ, ω, and ϕ. First, we discuss resonances at low energies. The width Γ of these states varies between 4 and 150 MeV, corresponding to lifetimes from about 10^{-22} s to 10^{-24} s. These values are typical of the strong interaction. These resonances are therefore interpreted as quark–antiquark bound states whose masses are just equal to the total centre of mass energy of the reaction. The quark–antiquark states must have the same quantum numbers as the virtual photon; in particular, they must have total angular momentum $J = 1$ and negative parity. Such quark–antiquark states are called *vector mesons*; they decay into lighter mesons. The sketch depicts schematically the production and the decay of a resonance.

The analysis of the peak at 770–780 MeV reveals that it is caused by the interference of two resonances, the ϱ^0-meson ($m_{\varrho^0} = 770$ MeV$/c^2$) and the ω-meson ($m_\omega = 782$ MeV$/c^2$). These resonances are produced via the creation of $u\bar{u}$ and $d\bar{d}$ pairs. Since u-quarks and d-quarks have nearly identical masses, the $u\bar{u}$- and $d\bar{d}$-states are approximately degenerate. The ϱ^0 and ω are mixed states of $u\bar{u}$ and $d\bar{d}$.

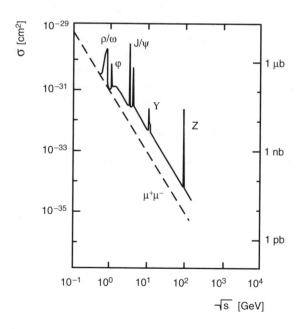

Fig. 9.4. Cross-section of the reaction $e^+e^- \to hadrons$ as a function of the centre of mass energy $\sqrt{s}$ (sketch) [Gr91]. The cross-section for direct muon pair production (9.5) is denoted by a dashed curve.

These two mesons undergo different decays and may be experimentally identified by them (cf. Sect. 14.3):

$$\varrho^0 \rightarrow \pi^+\pi^- ,$$
$$\omega \rightarrow \pi^+\pi^0\pi^- .$$

At an energy of 1019 MeV, the ϕ-resonance is produced. It has a width of only $\Gamma = 4.4$ MeV, and hence a relatively long lifetime compared to other hadrons. The main decay modes ($\approx 85\%$) of the ϕ are into two kaons, which have masses of 494 MeV/c^2 (K$^\pm$) and 498 MeV/c^2 (K^0):

$$\phi \rightarrow K^+ + K^- ,$$
$$\phi \rightarrow K^0 + \overline{K}^0 .$$

Kaons are examples of the so-called *strange particles*. This name reflects the unusual fact that they are produced by the strong interaction, but only decay by the weak interaction; this despite the fact that their decay products include hadrons, i.e., strongly interacting particles.

This behaviour is explained by the fact that kaons are quark–antiquark combinations containing an s or "strange" quark:

$$|K^+\rangle = |u\overline{s}\rangle \qquad |K^0\rangle = |d\overline{s}\rangle$$
$$|K^-\rangle = |\overline{u}s\rangle \qquad |\overline{K}^0\rangle = |\overline{d}s\rangle .$$

The constituent mass attributed to the s-quark is 450 MeV/c^2. In a kaon decay, the s-quark must turn into a light quark which can only happen in weak interaction processes. Kaons and other "strange particles" can be produced in the strong interaction, as long as equal numbers of s-quarks and $\overline{s}$-antiquarks are produced. At least two "strange particles" must therefore be produced simultaneously. We introduce the quantum number S (the *strangeness*), to indicate the number of $\overline{s}$-antiquarks minus the number of s-quarks. This quantum number is conserved in the strong and electromagnetic interactions, but it can be changed in weak interactions.

The ϕ meson decays mainly into two kaons because it is an s$\overline{s}$ system When it decays a u$\overline{u}$ pair or a d$\overline{d}$ pair are produced in the colour field of the strong interaction. The kaons are produced by combining these with the s$\overline{s}$ quarks, as shown in the sketch.

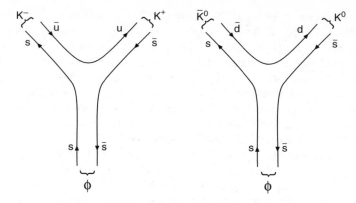

Because of the small mass difference $m_\phi - 2m_K$, the phase space available to this decay is very small. This accounts for the narrow width of the ϕ resonance.

One could ask: why doesn't the ϕ decay mainly into light mesons? The decay into pions is very rare (2.5 %), although the phase space available is much larger. Such a decay is only possible if the s and $\bar{s}$ first annihilate, producing two or three quark–antiquark pairs. According to QCD, this proceeds through a virtual intermediate state with at least three gluons. Hence, this process is suppressed with respect to the decay into two kaons which can proceed through the exchange of one gluon. The enhancement of processes with continuous quark lines is called the *Zweig rule*.

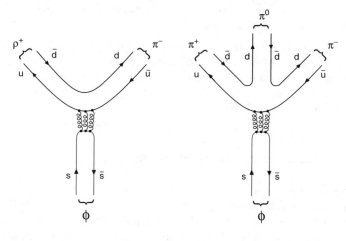

The resonances J/ψ and Υ. Although the s-quarks were known from hadron spectroscopy, it was a surprise when in 1974 an extremely narrow resonance whose width was only 87 keV was discovered at a centre of mass energy of 3097 MeV. It was named J/ψ.[3] The resonance was attributed to the production of a new heavy quark. There were already theoretical suggestions that such a c quark ("charmed" quark) exists. The long lifetime of the J/ψ is explained by its c$\overline{\text{c}}$ structure. The decay into two mesons each containing a c- (or $\overline{\text{c}}$)-quark plus a light quark (in analogy to the decay $\phi \rightarrow$ K$+\overline{\text{K}}$) would be favoured by the Zweig rule, but is impossible for reasons of energy. This is because the mass of any pair of D mesons (c$\overline{\text{u}}$, c$\overline{\text{d}}$ etc.), which were observed in later experiments, is larger than the mass of the J/ψ. More resonances were found at centre of mass energies some 100 MeV higher. They were called ψ', ψ'' etc., and were interpreted as excited states of the c$\overline{\text{c}}$ system. The J/ψ is the lowest c$\overline{\text{c}}$ state with the quantum numbers of the photon $J^P = 1^-$. A c$\overline{\text{c}}$ state, the η_c, exists at a somewhat lower energy, it has quantum numbers 0^- (cf. Sect. 13.2 ff) and cannot be produced directly in e^+e^- annihilation.

A similar behaviour in the cross-section was found at about 10 GeV. Here the series of Upsilon (Υ) resonances was discovered [He77, In77]. These b$\overline{\text{b}}$ states are due to the even heavier b-quark ("bottom" quark). The lowest-lying state at 9.46 GeV also has an extremely narrow width (only 52 keV) and hence a long lifetime.

The t-quark ("top" quark) was found in 1995 in two p$\overline{\text{p}}$ collision experiments at the Tevatron (FNAL) [Ab95a, Ab95b]. These and further experiments imply a t-quark mass of 173.8 ± 5.2 GeV/c^2. Present day e^+e^- accelerators can only attain centre of mass energies of up to around 172 GeV, which is not enough for t$\overline{\text{t}}$ pair production.

The Z^0 resonance. At $\sqrt{s} = 91.2$ GeV, an additional resonance is observed with a width of 2490 MeV. It decays into lepton and quark pairs. The properties of this resonance are such that it is thought to be a real Z^0, the vector boson of the weak interaction. In Sect. 11.2, we will describe what we can learn from this resonance.

9.3 Non-resonant Hadron Production

Up to now we have solely considered resonances in the cross-sections of electron-positron annihilation. Quark–antiquark pairs can, naturally, also be produced among the resonances. Further quark–antiquark pairs are then produced and form hadrons, around the primarily produced quark (or antiquark). This process is called *hadronisation*. Of course only those quarks

---

[3] This particle was discovered nearly simultaneously in two differently conceived experiments (pp collision and e^+e^- annihilation). One collaboration called it J [Au74a], the other ψ [Au74b].

Table 9.1. Charges and masses of the quarks: b, g, r denote the colours blue, green and red. Listed are the masses of "bare" quarks (current quarks) which would be measured in the limit $Q^2 \rightarrow \infty$ [PD98] as well as the masses of constituent quarks, i. e., the effective masses of quarks bound in hadrons. The masses of the quarks, in particular those of the current quarks, are strongly model dependent. For heavy quarks, the relative difference between the two masses is small.

Quark	Colour	Electr. Charge	Mass [MeV/c^2] Bare Quark	Const. Quark
down	b, g, r	−1/3	3 − 9	≈ 300
up	b, g, r	+2/3	1.5 − 5	≈ 300
strange	b, g, r	−1/3	60 − 170	≈ 450
charm	b, g, r	+2/3	1 100 − 1 400	
bottom	b, g, r	−1/3	4 100 − 4 400	
top	b, g, r	+2/3	$168 \cdot 10^3 - 179 \cdot 10^3$	

can be produced whose masses are less than half the centre of mass energy available.

In hadron production, a quark–antiquark pair is initially produced. Hence the cross-section is given by the sum of the individual cross-sections of quark–antiquark pair production. The production of the primary quark–antiquark pair by an electromagnetic interaction can be calculated analogously to muon pair production. Unlike muons, quarks do not carry a full elementary charge of $1 \cdot e$; but rather a charge $z_f \cdot e$ which is $-1/3e$ or $+2/3e$, depending on the quark flavour f. Hence the transition matrix element is proportional to $z_f e^2$, and the cross-section is proportional to $z_f^2 \alpha^2$. Since quarks (antiquarks) carry colour (anticolour), a quark–antiquark pair can be produced in three different colour states. Therefore there is an additional factor of 3 in the cross-section formula. The cross-section is given by:

$$\sigma(\text{e}^+\text{e}^- \rightarrow \text{q}_f\bar{\text{q}}_f) = 3 \cdot z_f^2 \cdot \sigma(\text{e}^+\text{e}^- \rightarrow \mu^+\mu^-)\,, \qquad (9.9)$$

and the ratio of the cross-sections by

$$R := \frac{\sigma(\text{e}^+\text{e}^- \rightarrow \text{hadrons})}{\sigma(\text{e}^+\text{e}^- \rightarrow \mu^+\mu^-)} = \frac{\sum_f \sigma(\text{e}^+\text{e}^- \rightarrow \text{q}_f\bar{\text{q}}_f)}{\sigma(\text{e}^+\text{e}^- \rightarrow \mu^+\mu^-)} = 3 \cdot \sum_f z_f^2\,. \qquad (9.10)$$

Here only those quark types f which can be produced at the centre of mass energy of the reaction contribute to the sum over the quarks.

Figure 9.5 shows schematically the ratio R as a function of the centre of mass energy $\sqrt{s}$. Many experiments had to be carried out at different particle accelerators, each covering a specific region of energy, to obtain such a picture. In the non-resonant regions R increases step by step with increasing energy $\sqrt{s}$. This becomes plausible if we consider the contributions of the individual quark flavours. Below the threshold for J/ψ production, only u$\bar{\text{u}}$, d$\bar{\text{d}}$, and s$\bar{\text{s}}$ pairs can be produced. Above it, c$\bar{\text{c}}$ pairs can also be produced; and at even

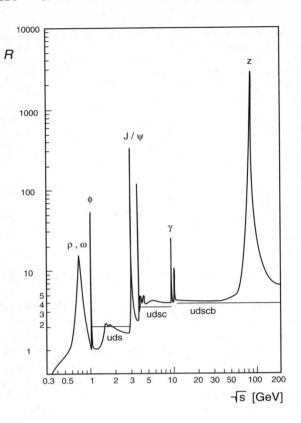

Fig. 9.5. Cross-section of the reaction e$^+$e$^-$ → *hadrons*, normalised to e$^+$e$^-$ → $\mu^+\mu^-$, as a function of the centre of mass energy $\sqrt{s}$ (sketch). The horizontal lines correspond to $R = 6/3$, $R = 10/3$ and $R = 11/3$, the values we expect from (9.10), depending upon the number of quarks involved. The value $R = 15/3$ which is expected if the t-quark participates lies outside the plotted energy range. *(Courtesy of G. Myatt, Oxford)*

higher energies, b$\bar{\text{b}}$ pairs are produced. The sum in (9.10) thus contains at higher energies ever more terms. As a corollary, the increase in R tells us about the charges of the quarks involved. Depending on the energy region, i.e., depending upon the number of quark flavours involved, one expects:

$$R = 3 \cdot \sum_f z_f^2 = 3 \cdot \left\{ \underbrace{(\tfrac{2}{3})^2}_{\text{u}} + \underbrace{(-\tfrac{1}{3})^2}_{\text{d}} + \underbrace{(-\tfrac{1}{3})^2}_{\text{s}} + \underbrace{(\tfrac{2}{3})^2}_{\text{c}} + \underbrace{(-\tfrac{1}{3})^2}_{\text{b}} \right\} \cdot$$

$$\underbrace{\qquad\qquad 3 \cdot 6/9 \qquad\qquad}$$
$$\underbrace{\qquad\qquad\qquad 3 \cdot 10/9 \qquad\qquad\qquad}$$
$$\underbrace{\qquad\qquad\qquad\qquad 3 \cdot 11/9 \qquad\qquad\qquad\qquad}$$

$$(9.11)$$

These predictions are in good agreement with the experimental results. The measurement of R represents an additional way to determine the quark charges and is simultaneously an impressive confirmation of the existence of exactly three colours.

9.4 Gluon Emission

Using e^+e^- scattering it has proven possible to experimentally establish the existence of gluons and to measure the value of α_s, the strong coupling constant.

The first indications for the existence of gluons were provided by deep inelastic scattering of leptons off protons. The integral of the structure function F_2 was only half the expected value. The missing half of the proton momentum was apparently carried by electrically neutral particles which were also

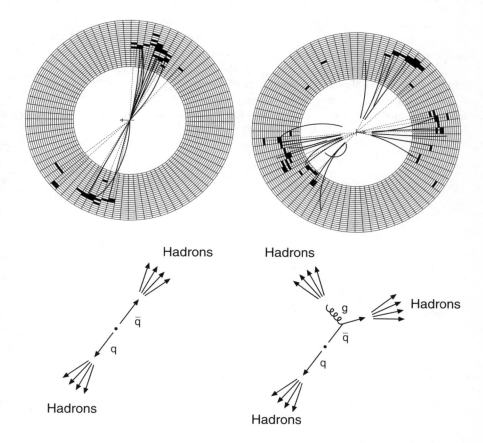

Fig. 9.6. Typical 2-jet and 3-jet events, measured with the JADE detector at the PETRA e^+e^- storage ring. The figures show a projection perpendicular to the beam axis, which is at the centre of the cylindrical detector. The tracks of charged particles (*solid lines*) and of neutral particles (*dotted lines*) are shown. They were reconstructed from the signals in the central wire chamber and in the lead glass calorimeter surrounding the wire chamber. In this projection, the concentration of the produced hadrons in two or three particle jets is clearly visible. (*Courtesy of DESY*)

not involved in weak interactions. They were identified with the gluons. The coupling constant α_s was determined from the scaling violation of the structure function F_2 (Sect. 8.4).

A direct measurement of these quantities is possible by analysing "jets". At high energies, hadrons are typically produced in two jets, emitted in opposite directions. These jets are produced in the hadronisation of the primary quarks and antiquarks (left side of Fig. 9.6).

In addition to simple $q\bar{q}$ production, higher-order processes can occur. For example, a high-energy ("hard") gluon can be emitted, which can then manifest itself as a third jet of hadrons. This corresponds to the emission of a photon in electromagnetic bremsstrahlung. Emission of a hard photon, however, is a relatively rare process, as the electromagnetic coupling constant α is rather small. By contrast, the probability of gluon bremsstrahlung is given by the coupling constant α_s. Such 3-jet events are indeed detected. Figure 9.6 (right) shows a particularly nice example. The coupling constant α_s may be deduced directly from a comparison of the 3- and 2-jet event rates. Measurements at different centre of mass energies also demonstrate that α_s decreases with increasing $Q^2 = s/c^2$ as (8.6) predicts.

_____ **Problems**

1. **Electron-positron collisions**

   a) Electrons and positrons each with a beam energy E of 4 GeV collide head on in a storage ring. What production rate of $\mu^+\mu^-$-pairs would you expect at a luminosity of $10^{32}\,\mathrm{cm}^{-2}\,\mathrm{s}^{-1}$? What production rate for events with hadronic final states would you expect?

   b) It is planned to construct two linear accelerators aimed at each other (a linear collider) from whose ends electrons and positrons will collide head on with a centre of mass energy of 500 GeV. How big must the luminosity be if one wants to measure the hadronic cross-section within two hours with a 10 % statistical error?

2. **Υ resonance**
   Detailed measurements of the $\Upsilon(1S)$ resonance, whose mass is roughly 9460 MeV, are performed at the CESR electron-positron storage ring.

   a) Calculate the uncertainty in the beam energy E and the centre of mass energy W if the radius of curvature of the storage ring is $R = 100\,\mathrm{m}$. We have:
   $$\delta E = \left(\frac{55}{32\sqrt{3}} \frac{\hbar c \, m_e c^2}{2R} \gamma^4 \right)^{1/2}$$
   What does this uncertainty in the energy tell us about the experimental measurement of the Υ (Use the information given in Part b)?

   b) Integrate the Breit-Wigner formula across the region of energy where the $\Upsilon(1S)$ resonance is found. The experimentally observed value of this integral for hadronic final states is $\int \sigma(\mathrm{e}^+\mathrm{e}^- \to \Upsilon \to \mathrm{hadrons})\,\mathrm{d}W \approx 300\,\mathrm{nb\,MeV}$. The decay probabilities for $\Upsilon \to \ell^+\ell^-$ ($\ell = \mathrm{e},\,\mu,\,\tau$) are each around 2.5 %. How large is the total natural decay width of the Υ? What cross-section would one expect at the resonance peak if there was no uncertainty in the beam energy (and the resonance was not broadened by radiative corrections)?

10. Phenomenology of the Weak Interaction

The discovery and the first theories of the weak interaction were based on the phenomenology of β-decay. Bound states formed by the weak interaction are not known, in contrast to those of the electromagnetic, strong and gravitational interactions. The weak interaction is in this sense somewhat foreign. We cannot, for example, base its description on any analogous phenomena in atomic physics. The weak interaction is, however, responsible for the decay of quarks and leptons.

In scattering experiments weak interaction effects are difficult to observe. Reactions of particles which are solely subject to the weak interaction (neutrinos) have extremely tiny cross-sections. In scattering experiments involving charged leptons and hadrons the effects of the weak interaction are overshadowed by those of the strong and electromagnetic interactions. Thus, most of our knowledge of the weak interaction has been obtained from particle decays.

The first theoretical description of β-decay, due to Fermi [Fe34], was constructed analogously to that of the electromagnetic interaction. With some modifications, it is still applicable to low-energy processes. Further milestones in the investigation of the weak interaction were the discovery of parity violation [Wu57], of different neutrino families [Da62] and of CP violation in the K^0 system [Ch64].

Quarks and leptons are equally affected by the weak interaction. In the previous chapter we discussed the quarks at length. We now want to treat the leptons in more detail before we turn to face the phenomena of the weak interaction.

10.1 The Lepton Families

Charged leptons. In our treatment of e^+e^--scattering we encountered the charged leptons: the electron (e), the muon (μ) and the tau (τ) as well as their antiparticles (the e^+, μ^+ and τ^+) which have the same masses as their partners but are oppositely charged.

The electron and the muon are the lightest electrically charged particles. Charge conservation thus ensures that the electron is stable and that an electron is produced when a muon decays. Muon decay proceeds via

$$\mu^- \rightarrow e^- + \overline{\nu}_e + \nu_\mu.$$

In a very few cases an additional photon or e^+e^- pair is produced. The energetically allowed process

$$\mu^- \not\rightarrow e^- + \gamma \,,$$

is, on the other hand, never observed. The muon is therefore not just an excited state of the electron.

The τ-lepton is much heavier than the muon and, indeed, more so than many hadrons. Thus it does not have to decay into lighter leptons

$$\tau^- \rightarrow e^- + \bar{\nu}_e + \nu_\tau \qquad \tau^- \rightarrow \mu^- + \bar{\nu}_\mu + \nu_\tau \,,$$

but can also turn into hadrons, e.g., into a pion and a neutrino

$$\tau^- \rightarrow \pi^- + \nu_\tau \,.$$

In fact more than half of all τ decays follow the hadronic route [Ba88].

Neutrinos. We have already seen several processes in which neutrinos are produced: nuclear β-decay and the decays of charged leptons. Neutrinos are electrically neutral leptons and, as such, do not feel the electromagnetic or strong forces. As a rule they may only be indirectly detected, for example, the sums of the energies and angular momenta of the observed particles in β-decays indicate that another particle as well as the electron must be being emitted.

Experiment has made it completely clear that neutrinos and antineutrinos are distinct particles. The antineutrinos produced in a β-decay

$$n \rightarrow p + e^- + \bar{\nu}_e$$

for example, only induce further reactions in which positrons are produced and do not lead to electrons being created:

$$\bar{\nu}_e + p \rightarrow n + e^+$$
$$\bar{\nu}_e + n \not\rightarrow p + e^- \,.$$

Neutrinos and antineutrinos produced in charged pion decays

$$\pi^- \rightarrow \mu^- + \bar{\nu}_\mu$$
$$\pi^+ \rightarrow \mu^+ + \nu_\mu$$

also behave differently. They induce reactions in which muons are created but never produce electrons [Da62]. This implies that there must be at least two different sorts of neutrinos: an electron-neutrino ν_e, which is associated with the creation and annihilation of electrons, and a muon-neutrino ν_μ, which we similarly associate with the muon. No reactions induced by a tau-neutrino have yet been seen, but, as a result of theoretical work, it is firmly believed

Table 10.1. The leptons and their most important properties [PD98]. Leptons may be divided into three families according to various conservation laws.

Family		Elec. charge	Mass (MeV/c^2)	Spin	Lifetime
1	ν_e	0	$< 15 \cdot 10^{-6}$	1/2	stable
	e	−1	0.511	1/2	stable
2	ν_μ	0	< 0.17	1/2	stable?
	μ	−1	105.7	1/2	$2.197 \cdot 10^{-6}$ s
3	ν_τ	0	< 18.2	1/2	stable?
	τ	−1	1777.1	1/2	$2.900 \cdot 10^{-13}$ s

that the ν_τ neutrino, which is associated with the τ-lepton, differs from the other two neutrinos and that there are therefore three different neutrinos.

It has also been demonstrated for the muon-neutrino that it and its antineutrino are distinct particles: neutrinos from π^+ decays only generate μ^-'s, while antineutrinos from π^- decays only produce μ^+'s.

Only experimental upper bounds exist for the rest masses of the neutrinos. All previous experiments are compatible with a vanishing mass. Whether or not the neutrino masses actually vanish is a question which is currently hotly disputed. We will return to measurements of neutrino masses in Sect. 17.6.

The lepton families. Despite intensive searches at ever higher energies, no further leptons have yet been found. The lower bound for the mass of any further charged lepton is currently 42.8 GeV/c^2. In Sect. 11.2 we will see that there cannot be more than three light neutrinos ($m_\nu \ll 10$ GeV/c^2). To summarize: we now know six different leptons . These are three electrically charged particles (e, μ, τ) and three neutral ones (ν_e, ν_μ, ν_τ). Their most important properties are listed in Table 10.1.

Just like the quarks, the leptons fall into three families, each of which is made up of two particles whose charges differ by one unit. The charged leptons have, like the quarks, very different masses ($m_\mu/m_e \approx 207$, $m_\tau/m_\mu \approx 17$). We still do not have a generally accepted reason for why the fundamental fermions come in three families and we do not understand their masses.

Lepton number conservation. In all the reactions we have mentioned above, the creation or annihilation of a lepton was always associated with the creation or annihilation of an antilepton of the same family. According to our current knowledge this is always the case. As with the baryons, we therefore have a conservation law: in all reactions the number of leptons of a particular family minus the number of the corresponding antileptons is conserved. We write

$$L_\ell = N(\ell) - N(\overline{\ell}) + N(\nu_\ell) - N(\overline{\nu}_\ell) = \text{const.} \quad \text{where} \quad \ell = \text{e}, \, \mu, \, \tau. \quad (10.1)$$

The sum $L = L_e + L_\mu + L_\tau$ is called *lepton number* and the L_ℓ's are individually referred to as *lepton family numbers*. Each one of the three L_e, L_μ and L_τ

numbers are separately conserved. In consequence the following reactions are allowed or forbidden:

Allowed			Forbidden		
$p + \mu^-$	$\to$	$\nu_\mu + n$	$p + \mu^-$	$\not\to$	$\pi^0 + n$
$e^+ + e^-$	$\to$	$\nu_\mu + \bar{\nu}_\mu$	$e^+ + e^-$	$\not\to$	$\nu_e + \nu_\mu$
π^-	$\to$	$\mu^- + \bar{\nu}_\mu$	π^-	$\not\to$	$e^- + \nu_e$
μ^-	$\to$	$e^- + \bar{\nu}_e + \nu_\mu$	μ^-	$\not\to$	$e^- + \bar{\nu}_\mu + \nu_e$
τ^-	$\to$	$\pi^- + \nu_\tau$	τ^-	$\not\to$	$\pi^- + \nu_e$.

Experimentally the upper limits for any violation of these conservation laws in electromagnetic or weak processes are very small. For example we have [PD94]

$$\frac{\Gamma(\mu^\pm \to e^\pm \gamma)}{\Gamma(\mu^\pm \to \text{all channels})} < 5 \cdot 10^{-11}$$

$$\frac{\Gamma(\mu^\pm \to e^\pm e^+ e^-)}{\Gamma(\mu^\pm \to \text{all channels})} < 1 \cdot 10^{-12}. \tag{10.2}$$

All the allowed reactions that we have listed above proceed exclusively through the weak interaction, since in all these cases neutrinos are involved and these particles are only subject to the weak interaction. The opposite conclusion is, however, incorrect. We will see in the following section that there are indeed weak processes which involve neither neutrinos nor any other leptons.

10.2 The Types of Weak Interactions

Recall that the weak interaction can transform a charged lepton into its family's neutrino and that it can produce a charged lepton (antilepton) and its antineutrino (neutrino). In just the same manner quarks of one flavour can be transformed into quarks with another flavour in weak interactions: a typical example of this is the transformation of a d-quark into a u-quark — this takes place in the β-decay of a neutron. In all such reactions the identity of the quarks and leptons involved changes and, simultaneously, the charge changes by $+1e$ or $-1e$. The term *charged current* was coined to describe such reactions. They are mediated by charged particles, the W^+ and W^-.

For a long time only this sort of weak interaction was known. Nowadays we know that weak interactions may also proceed via the exchange of an additional, electrically neutral particle, the Z^0. In this case the quarks and leptons are not changed. One refers to *neutral currents*.

The $W^\pm$ and the Z^0 are vector bosons, i.e., they have spin one. Their masses are large: 80 GeV/c^2 ($W^\pm$) and 91 GeV/c^2 (Z^0). We will return to their experimental detection in Sect. 11.1. In this chapter we will, following the

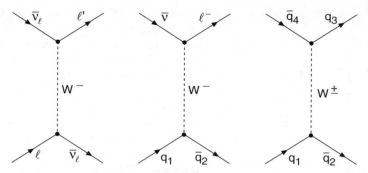

Fig. 10.1. The three sorts of charged current reactions: a leptonic process (*left*), a semileptonic process (*middle*) and a non-leptonic process (*right*).

historical development, initially concern ourselves with the *charged currents*. These may be straightforwardly divided up into three categories (Fig. 10.1): *leptonic processes*, *semileptonic processes* and *non-leptonic processes*.

Leptonic processes. If the W boson only couples to leptons, one speaks of a leptonic process. The underlying reaction is

$$\ell + \bar{\nu}_\ell \longleftrightarrow \ell' + \bar{\nu}_{\ell'}.$$

Examples of this are the leptonic decay of the τ-lepton:

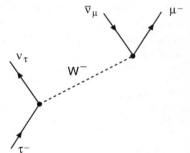

$$\tau^- \;\rightarrow\; \mu^- + \bar{\nu}_\mu + \nu_\tau$$
$$\tau^- \;\rightarrow\; e^- + \bar{\nu}_e + \nu_\tau$$

and the scattering process

$$\nu_\mu + e^- \;\rightarrow\; \mu^- + \nu_e.$$

Semileptonic processes. *Semileptonic processes* are those where the exchanged W boson couples to both leptons and quarks. The fundamental process here is

$$q_1 + \bar{q}_2 \longleftrightarrow \ell + \bar{\nu}_\ell.$$

Examples of this are charged pion decay, the decay of the K^- or the β-decay of the neutron:

Hadron description	Quark description
$\pi^- \rightarrow \mu^- + \bar{\nu}_\mu$	$d + \bar{u} \rightarrow \mu^- + \bar{\nu}_\mu$
$K^- \rightarrow \mu^- + \bar{\nu}_\mu$	$s + \bar{u} \rightarrow \mu^- + \bar{\nu}_\mu$
$n \;\;\rightarrow p + e^- + \bar{\nu}_e$	$d \rightarrow u + e^- + \bar{\nu}_e.$

The β-decay of a neutron may be reduced to the decay of a d-quark in

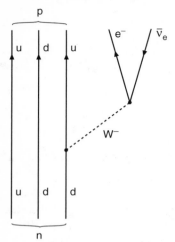

which the two other quarks are not involved. The latter are called *spectator quarks*. Inverse reactions are processes such as K capture $p+e^- \rightarrow n+\nu_e$ and inverse β-decay $\bar{\nu}_e+p \rightarrow n+e^+$ or $\nu_e+n \rightarrow p+e^-$. (Anti-) Neutrinos were directly detected for the first time in the second of these reactions [Co56a] — antineutrinos from the β^--decay of neutron-rich fission products were seen to react with hydrogen. The second reaction may be used to detect solar and stellar neutrinos emanating from β^+-decays of proton-rich nuclei produced in fusion reactions. A further example of a semileptonic process is deep inelastic neutrino scattering, which we will treat in more detail in Sect. 10.6.

Non-leptonic processes. Finally non-leptonic processes do not involve leptons at all. The basic reaction is

$$q_1 + \bar{q}_2 \longleftrightarrow q_3 + \bar{q}_4 \,.$$

Charge conservation requires that the only allowed quarks combinations have a total charge $\pm 1e$. Examples are the hadronic decays of baryons and mesons with strangeness, such as the decay of the Λ^0 hyperon into a nucleon and a pion, or that of $K^+(u\bar{s})$ into two pions:

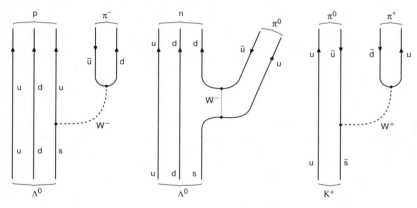

10.3 Coupling Strength of the Charged Current

We now want to deal with charged currents in a more quantitative manner. We will treat leptonic processes in what follows since leptons, in contrast to quarks, exist as free particles which simplifies matters.

As with Mott scattering or e^+e^--annihilation, the transition matrix element for such processes is proportional to the square of the *weak charge g* to which the W Boson couples and to the propagator (4.23) of a massive spin-one particle:

$$\mathcal{M}_{fi} \propto g \cdot \frac{1}{Q^2c^2 + M_{\mathrm{W}}^2c^4} \cdot g \xrightarrow{Q^2 \to 0} \frac{g^2}{M_{\mathrm{W}}^2c^4}. \tag{10.3}$$

The difference to an electromagnetic interaction is seen in the finite mass of the exchange particle. Instead of the photon propagator $(Qc)^{-2}$, we see a propagator which is almost a constant for small enough momenta $Q^2 \ll M_{\mathrm{W}}^2c^2$. We will see in Sect. 11.2 that the weak charge g and the electric charge e are of a similar size. The very large mass of the exchange boson means that at small Q^2 the weak interaction appears to be much weaker than the electromagnetic interaction. It also means that its range $\hbar/M_{\mathrm{W}}c \approx 2.5 \cdot 10^{-3}$ fm is very limited.

In the approximation of the small four-momentum transfer one may then describe this interaction as a point-like interaction of the four particles involved. This was in fact the original description of the weak interaction before the idea of the W and Z bosons was brought in. The coupling strength of this interaction is described by the *Fermi constant* G_{F}, which is proportional to the square of the weak charge g, very much as the electromagnetic coupling constant $\alpha = e^2/(4\pi\varepsilon_0\hbar c)$ is proportional to the electric charge e. It is so defined that $G_{\mathrm{F}}/(\hbar c)^3$ has dimensions of $[1/\mathrm{energy}^2]$ and is related to g by

$$\frac{G_{\mathrm{F}}}{\sqrt{2}} = \frac{\pi\alpha}{2} \cdot \frac{g^2}{e^2} \cdot \frac{(\hbar c)^3}{M_{\mathrm{W}}^2c^4}. \tag{10.4}$$

The decay of the muon. The most exact value for the Fermi constant is obtained from muon decay. The muon decays, as explained in Sect. 10.1, by

$$\mu^- \to e^- + \bar{\nu}_e + \nu_\mu \,, \qquad \mu^+ \to e^+ + \nu_e + \bar{\nu}_\mu \,.$$

Since the muon mass is tiny compared to that of the W boson, it is reasonable to treat this interaction as point-like and describe the coupling via the Fermi constant.

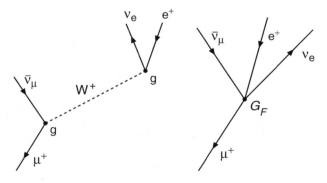

In this approximation the lifetime of the muon may be calculated with the help of the golden rule, if we use the Dirac equation and take into account the amount of phase space available to the three outgoing leptons. One finds that the decay width is:

$$\Gamma_\mu = \frac{\hbar}{\tau_\mu} = \frac{G_F^2}{192\pi^3 (\hbar c)^6} \cdot (m_\mu c^2)^5 \cdot (1+\varepsilon) \ . \tag{10.5}$$

The correction term ε, which reflects higher order (radiative) corrections and phase space effects resulting from the finite electron mass, is small (see Eq. 5 in [Ma91]). It should be noted that the transition rate is proportional to the fifth power of the energy and hence the mass of the decaying muon. In Sect. 15.5 we will show in detail how the phase space may be calculated and how the E^5-dependence can be derived (in the example of the β-decay of the neutron).

The muon mass and lifetime have been measured to a high precision:

$$\begin{aligned} m_\mu &= (105.658\,389 \pm 0.000\,034)\ \mathrm{MeV}/c^2\,, \\ \tau_\mu &= (2.197\,035 \pm 0.000\,040)\cdot 10^{-6}\ \mathrm{s}\,. \end{aligned} \tag{10.6}$$

This yields a value for the Fermi constant

$$\frac{G_F}{(\hbar c)^3} = (1.166\,39 \pm 0.000\,01)\cdot 10^{-5}\,\mathrm{GeV}^{-2}. \tag{10.7}$$

Neutrino-electron scattering. Neutrino-electron scattering is a reaction between free, elementary particles. It proceeds exclusively through the weak interaction. We can discuss the effects of the effective coupling strength G_F on the cross-section of this reaction and show why the weak interaction is called "weak".

In the diagram below the scattering of muon-neutrinos off electrons in which the ν_μ is changed into a μ^- is shown.

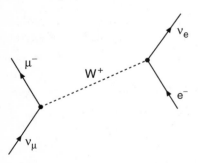

We have chosen this process as our example since it can only take place via W-exchange. Calculating ν_e-e^- scattering is more complicated since both Z- and W-exchange lead to the same final state and thus interfere with each other.

For small four-momenta the total cross-section for neutrino-electron scattering is proportional to the square of the effective coupling constant G_F. Similarly to our discussion of the total cross-section in e^+e^- annihilation in Sect. 9.1, the characteristic length and energy scales of the reaction (the constants $\hbar c$ and the centre of mass energy $\sqrt{s}$) must enter the cross-section in such a way as to yield the correct dimensions ([area]):

$$\sigma = \frac{G_{\mathrm{F}}^2}{\pi(\hbar c)^4} \cdot s \,, \qquad (10.8)$$

where s may be found in the laboratory frame from (9.3) to be $s = 2m_e c^2 E_\nu$. From (10.7) one finds that the cross-section in the laboratory frame is:

$$\sigma_{\mathrm{lab}} = 1.7 \cdot 10^{-41} \, \mathrm{cm}^2 \cdot E_\nu / \mathrm{GeV}. \qquad (10.9)$$

This is an extremely tiny cross-section. To illustrate this point we now estimate the distance L which a neutrino must traverse until it weakly interacts with an electron. The electron density in iron is

$$n_e = \frac{Z}{A} \varrho N_A \approx 22 \cdot 10^{23} \, \mathrm{cm}^{-3} \,. \qquad (10.10)$$

Neutrinos produced in the sun via hydrogen fusion have typically an energy of around 1 MeV. Their mean free path is therefore $L = (n_e \cdot \sigma)^{-1} = 2.6 \cdot 10^{17}$ m, which is about 30 light years![1]

At very high energies the simple formula (10.9) is no longer valid, since the cross-section would limitlessly grow with the neutrino energy. This of course will not happen in practice: at large four momentum transfers $Q^2 \gg M_{\mathrm{W}}^2 c^2$ the propagator term primarily determines the energy dependence of the cross-section. A point-like interaction approximation no longer holds. At a fixed centre of mass energy $\sqrt{s}$ the cross-section falls off, as in electromagnetic scattering, as $1/Q^4$. The total cross-section is on the other hand [Co73]:

$$\sigma = \frac{G_{\mathrm{F}}^2}{\pi(\hbar c)^4} \cdot \frac{M_{\mathrm{W}}^2 c^4}{s + M_{\mathrm{W}}^2 c^4} \cdot s \,. \qquad (10.11)$$

The cross-section does not then increase linearly with s, as the point-like approximation implies, rather it asymptotically approaches a constant value.

Neutral currents. Up to now we have only considered neutrino-electron scattering via W^+ exchange, i.e., through charged currents. Neutrinos and electrons can, however, interact via Z^0-exchange, i.e., neutral current interactions are possible.

Elastic muon-neutrino scattering off electrons

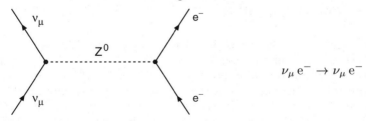

---

[1] The absorption of neutrinos by the atomic nuclei is neglected here. This is a reasonable approximation for neutrino energies less than 1 MeV, but would need to be modified for higher energies.

is particularly suitable for investigating the weak interaction via Z^0-exchange. This is because conservation of lepton family number precludes W-exchange. Reactions of this kind were first seen in 1973 at CERN [Ha73]. This was the first experimental signal for weak neutral currents.

Universality of the weak interaction. If we assume that the weak charge g is the same for all quarks and leptons, then (10.5) must hold for all possible charged decays of the fundamental fermions into lighter leptons or quarks. All the decay channels then contribute equally, up to a phase space correction coming from the different masses, to the total decay width.

We choose to consider the example of the decay of the τ-lepton. This particle has essentially three routes open to it[2]

$$
\begin{aligned}
\tau^- &\rightarrow \nu_\tau + \bar{\nu}_e + e^- \\
\tau^- &\rightarrow \nu_\tau + \bar{\nu}_\mu + \mu^- \\
\tau^- &\rightarrow \nu_\tau + \bar{u} + d
\end{aligned}
\tag{10.12}
$$

whose widths are $\Gamma_{\tau e} \approx \Gamma_{\tau\mu}$ and $\Gamma_{\tau d\bar{u}} \approx 3\Gamma_{\tau\mu}$. The factor of three follows from the $\bar{u}d$-pair appearing in three different colour combinations (r$\bar{r}$, b$\bar{b}$, g$\bar{g}$).

From the mass term in (10.5) we have:

$$
\Gamma_{\tau e} = (m_\tau/m_\mu)^5 \cdot \Gamma_{\mu e} ,
\tag{10.13}
$$

and the lifetime is thus predicted to be:

$$
\tau_\tau = \frac{\hbar}{\Gamma_{\tau e} + \Gamma_{\tau\mu} + \Gamma_{\tau d\bar{u}}} \approx \frac{\tau_\mu}{5 \cdot (m_\tau/m_\mu)^5} \approx 3.1 \cdot 10^{-13}\,\text{s} .
\tag{10.14}
$$

Experimentally we find [PD98]

$$
\tau_\tau^{\text{exp}} = (2.900 \pm 0.012) \cdot 10^{-13}\,\text{s} .
\tag{10.15}
$$

This good agreement confirms that quarks occur in three different colours and is strongly supportive of the quark and lepton weak charges being identical.

10.4 The Quark Families

We have claimed that the weak charge is universal, and that all the weak reactions which proceed through W exchange can therefore be calculated using the one coupling constant g or G_F. The lifetime of the τ-lepton seemed to illustrate this point: our expectations, based on the assumption that the W boson couples with the same strength to both quarks and leptons were fulfilled. However, the lifetime does not contain the decay widths for leptonic

---

[2] The appearance of further hadronic decay channels will be treated in the next section.

and hadronic processes separately, but only their sum. Furthermore it is very sensitive to the mass of the τ-lepton. Hence, this is not a particularly precise test of weak charge universality.

The coupling to quarks can be better determined from semi-leptonic hadron decays. This yields a smaller value for the coupling than that obtained from muon data. If a d-quark is transformed into a u-quark, as in the β-decay of the neutron, the coupling constant appears to be about 4 % smaller. In processes in which an s-quark is transformed into a u-quark, as in Λ^0 decay, it even appears to be 20 times smaller.

The Cabibbo angle. An explanation of these findings was proposed by Cabibbo as early as 1963 [Ca63], at a time at which quarks had not been introduced. We will re-express Cabibbo's hypothesis in modern terms. In weak interactions with charged currents, leptons can only be transformed into their "partners" in the same doublet; e.g., $e^- \leftrightarrow \nu_e$ and $\mu^- \leftrightarrow \nu_\mu$. Similarly we may group the quarks into families, according to their charges and masses:

$$\begin{pmatrix} u \\ d \end{pmatrix} \begin{pmatrix} c \\ s \end{pmatrix} \begin{pmatrix} t \\ b \end{pmatrix} .$$

Quark transitions are observed not only within a family but, to a lesser degree, from one family to another. For charged currents, the "partner" of the flavour eigenstate $|u\rangle$ is therefore not the flavour eigenstate $|d\rangle$, but a linear combination of $|d\rangle$ and $|s\rangle$. We call this linear combination $|d'\rangle$. Similarly the partner of the c-quark is a linear combination of $|s\rangle$ und $|d\rangle$, orthogonal to $|d'\rangle$, which we call $|s'\rangle$.

The coefficients of these linear combinations can be written as the cosine and sine of an angle called the *Cabibbo angle* θ_C. The quark eigenstates $|d'\rangle$ and $|s'\rangle$ of W exchange are related to the eigenstates $|d\rangle$ and $|s\rangle$ of the strong interaction, by a rotation through θ_C:

$$\begin{aligned} |d'\rangle &= \cos\theta_C\,|d\rangle + \sin\theta_C\,|s\rangle \\ |s'\rangle &= \cos\theta_C\,|s\rangle - \sin\theta_C\,|d\rangle , \end{aligned} \qquad (10.16)$$

which may be written as a matrix:

$$\begin{pmatrix} |d'\rangle \\ |s'\rangle \end{pmatrix} = \begin{pmatrix} \cos\theta_C & \sin\theta_C \\ -\sin\theta_C & \cos\theta_C \end{pmatrix} \cdot \begin{pmatrix} |d\rangle \\ |s\rangle \end{pmatrix} . \qquad (10.17)$$

Whether the state vectors $|d\rangle$ and $|s\rangle$ or the state vectors $|u\rangle$ and $|c\rangle$ are rotated, or indeed both pairs simultaneously, is a matter of convention alone. Only the difference in the rotation angles is of physical importance. Usually the vectors of the charge $-1/3$ quarks are rotated while those of the charge $+2/3$ quarks are left untouched.

Experimentally, θ_C is determined by comparing the lifetimes and branching ratios of the semi-leptonic and hadronic decays of various particles as shown in the sketch. This yields:

$$\sin \theta_C \approx 0.22 , \quad \text{and} \quad \cos \theta_C \approx 0.98 . \quad (10.18)$$

The transitions $c \leftrightarrow d$ and $s \leftrightarrow u$, as compared to $c \leftrightarrow s$ and $d \leftrightarrow u$, are therefore suppressed by a factor of

$$\sin^2 \theta_C : \cos^2 \theta_C \approx 1 : 20 . \quad (10.19)$$

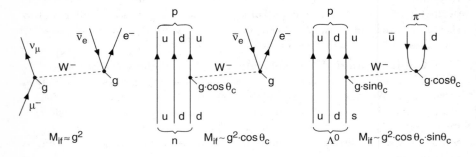

We can now make our treatment of τ decay more precise. In (10.12), we stated that $\tau \to \nu_\tau + \bar{u} + d$ is "essentially" the only hadronic decay of the τ. But $\tau \to \nu_\tau + \bar{u} + s$ is also energetically possible. Whereas the former decay is only slightly suppressed by a factor of $\cos^2 \theta_C$, the latter is faced with a factor of $\sin^2 \theta_C$. However, since $\cos^2 \theta_C$ and $\sin^2 \theta_C$ add to one our conclusion concerning the lifetime of the τ-lepton is not affected, as long as we ignore the difference in the quark masses.

The Cabibbo-Kobayashi-Maskawa matrix. Adding the third generation of quarks, the 2×2 matrix of (10.17) is replaced by a 3×3 matrix [Ko73]. This is called the *Cabibbo-Kobayashi-Maskawa matrix* (CKM matrix):

$$\begin{pmatrix} |d'\rangle \\ |s'\rangle \\ |b'\rangle \end{pmatrix} = \begin{pmatrix} V_{ud} & V_{us} & V_{ub} \\ V_{cd} & V_{cs} & V_{cb} \\ V_{td} & V_{ts} & V_{tb} \end{pmatrix} \cdot \begin{pmatrix} |d\rangle \\ |s\rangle \\ |b\rangle \end{pmatrix} . \quad (10.20)$$

The probability for a transition from a quark q to a quark q' is proportional to $|V_{qq'}|^2$, the square of the magnitude of the matrix element.

The matrix elements are by now rather well known [Ma91]. They are correlated since the matrix is unitary. The total number of independent parameters is four: three real angles and an imaginary phase. The phase affects weak processes of higher order via the interference terms. *CP violation* (cf. Sect. 14.4) is attributed to the existence of this imaginary phase [Pa89].

The following numbers represent the 90%–confidence limits on the magnitudes of the matrix elements [PD98]:

$$\left(|V_{ij}| \right) = \begin{pmatrix} 0.9745 \cdots 0.9760 & 0.217 \cdots 0.224 & 0.0018 \cdots 0.0045 \\ 0.217 \cdots 0.224 & 0.9737 \cdots 0.9753 & 0.036 \cdots 0.042 \\ 0.004 \cdots 0.013 & 0.035 \cdots 0.042 & 0.9991 \cdots 0.9994 \end{pmatrix} .$$

$$(10.21)$$

The diagonal elements of this matrix describe transitions within a family; they deviate from unity by only a few percent. The values of the matrix elements V_{cb} and V_{ts} are nearly one order of magnitude smaller than those of V_{us} and V_{cd}. Accordingly, transitions from the third to the second generation (t → s, b → c) are suppressed by nearly two orders of magnitude compared to transitions from the second to the first generation. This applies to an even higher degree for transitions from the third to the first generation. The direct transition b → u was detected in the semi-leptonic decay of B mesons into non-charmed mesons [Fu90, Al90, Al91].

Weak quark decays only proceed through W exchange. Neutral currents which change the quark flavour (e. g., c → u) have thus far not been observed.

■ Whether transitions from one family to another are also possible for leptons has not yet been fully clarified. So far, no experiment has detected any particle decay or reaction in which the lepton family numbers were not conserved. A necessary (but not sufficient) condition for a CKM mixing of the leptons would be for the neutrinos to have a finite rest mass. If neutrinos really are exactly massless: then any mixture of neutrinos will also be an eigenstate of the mass operator and the mass eigenstates $|\nu_e'\rangle$, $|\nu_\mu'\rangle$ and $|\nu_\tau'\rangle$ can be *defined* via $|\nu_e\rangle$, $|\nu_\mu\rangle$ and $|\nu_\tau\rangle$ as the exact "partners" of $|e\rangle, |\mu\rangle$ and $|\tau\rangle$. The conservation of lepton family number then necessarily follows.

Direct measurements of the neutrino masses (e. g., from the kinematical analysis of β-decay, see Sect. 17.6) have up to now been compatible with a vanishing mass. Experiments which seek to indirectly detect any neutrino mass, look for transitions (oscillations) between the eigenstates of the flavour families $|\nu_e\rangle, |\nu_\mu\rangle$ and $|\nu_\tau\rangle$ either in neutrino beams from reactors and accelerators or in solar/atmospheric neutrinos [Ku89].

Recent experiments with solar neutrinos (see Sec. 19.5 for how neutrinos are created in the sun) indicate that such oscillations may be taking place: the measured flux of solar neutrinos [Ha96] is only about half the size which is predicted by solar models which describe in detail the elementary particle reactions and energy production in the sun. An anomaly has also been observed in the measurement of atmospheric neutrinos: the measured rate in terrestrial detectors for the production of ν_μ's in the atmosphere depends strongly upon whether the neutrinos only pass through the atmosphere or also through the entire body of the earth. Experimentally a difference of a factor of 2 between the rates is measured [Fu98] although the earth is so transparent to neutrinos that the flux of neutrinos should not be noticeably weakened by it.

Both results can be interpreted as implying that in-flight neutrinos transform into another variety. Should these results be confirmed, this would be a proof of CKM lepton mixing and thus of a finite neutrino mass.

10.5 Parity Violation

A property unique to the weak interaction is parity violation. This means that weak interaction reactions are not invariant under space inversion.

An example of a quantity which changes under a spatial inversion is *helicity*

$$h = \frac{s \cdot p}{|s| \cdot |p|}, \qquad (10.22)$$

which we introduced in Sect. 5.3. The numerator is a scalar product of an axial vector (spin) and a vector (momentum). Whereas spin preserves its orientation under mirror reflection, the direction of the momentum is reversed. Thus helicity is a pseudoscalar, changing sign when the parity operator is applied to it. An interaction which depends upon helicity is therefore not invariant under spatial reflections.

In general, the operator of an interaction described by the exchange of a spin-1 particle can have a vector or an axial vector nature. In order for an interaction to conserve parity, and therefore to couple identically to both right and left-handed particles, it must be either purely vectorial or purely axial. In electromagnetic interactions, for example, it is experimentally observed that only a vector part is present. But in parity violating interactions, the matrix element has a vector part as well as an axial vector part. Their strengths are described by two coefficients, c_V und c_A. The closer the size of the two parts the stronger is the parity violation. *Maximum parity violation* occurs if both contributions are equal in magnitude. A $(V + A)$–interaction, i. e., a sum of vector and axial interactions of equal strength $(c_V = c_A)$, couples exclusively to right-handed fermions and left-handed antifermions. A $(V-A)$– interaction $(c_V = -c_A)$ only couples to left-handed fermions and right-handed antifermions.

As we will show, the angular distribution of electrons produced in the decay of polarised muons exhibits parity violation. This decay can be used to measure the ratio c_V/c_A. Such experiments yield $c_V = -c_A = 1$ for the coupling strength of W bosons to leptons. One therefore speaks of a *V-minus-A theory* of charged currents. Parity violation is maximal. If a neutrino or an antineutrino is produced by W exchange, the neutrino helicity is negative, while the antineutrino helicity is positive. In fact all experiments are consistent with *neutrinos being always left-handed and antineutrinos right-handed*. We will describe such an experiment in Sect. 17.6.

For massive particles $\beta = v/c < 1$ and the above considerations must be modified. On the one hand, massive fermions can be superpositions of right-handed and of left-handed particles. On the other hand, right-handed and left-handed states receive contributions with the opposite helicity, which increase the more β decreases. This is because helicity is only Lorentz-invariant for massless particles. For particles with a non-vanishing rest mass it is always possible to find a reference frame in which the particle is "overtaken", i.e., in which its direction of motion and thus its helicity are reversed.

CP conservation. It may be easily seen that if the helicity of the neutrinos is fixed, then *C-parity* ("charge conjugation") is simultaneously violated. Application of the C-parity operator replaces all particles by their antiparticles. Thus, left-handed neutrinos would be transformed into left-handed antineutrinos, which are not found in nature. Therefore physical processes which involve neutrinos, and in general all weak processes, a priori violate C-parity. The combined application of space inversion (P) and of charge conjugation (C), however, yields a process which is physically possible. Here, left-handed fermions are transformed into right-handed antifermions, which interact with equal strength. This is called the *CP conservation* property of the weak interaction. The only known case in which CP symmetry is not conserved (CP violation) will be discussed in Sect. 14.4.

Parity violation in muon decay. An instructive example of parity violation is the muon decay $\mu^- \rightarrow e^- + \nu_\mu + \overline{\nu}_e$. In the rest frame of the muon, the momentum of the electron is maximised if the momenta of the neutrinos are parallel to each other, and antiparallel to the momentum of the electron. From the sketch it is apparent that the spin of the emitted electron must be in the same direction as that of the muon since the spins of the $(\nu_e, \overline{\nu}_\mu)$ pair cancel.

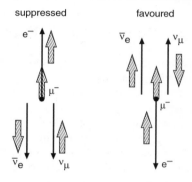

Experimentally it is observed that electrons from polarised muon decays are preferentially emitted with their spins opposite to their momentum; i. e., they are left-handed. This left-right asymmetry is a manifestation of parity violation. The ratio of the vector to axial vector fractions can be determined from the angular distribution [Bu85].

Helicity suppressed pion decay. Our second example is the decay of the charged pion. The lightest hadron with electric charge, the π^-, can only decay in a semi-leptonic weak process, i.e., through a charged current, according to:

$$\pi^- \rightarrow \mu^- + \overline{\nu}_\mu \,,$$
$$\pi^- \rightarrow e^- + \overline{\nu}_e \,.$$

The second process is suppressed, compared to the first one, by a factor of $1 : 8000$ [Br92] (cf. Table 14.3). From the amount of phase space available, however, one would expect the pion to decay about 3.5 times more often into an electron than into a muon. This behaviour may be explained from helicity considerations.

The particles emitted in such two-particle pion decays depart, in the centre of mass system, in opposite directions. Since the pion has spin zero, the spins of the two leptons must be opposite to each other. Thus, the projections on the direction of motion are either $+1/2$ for both, or $-1/2$ for both. The latter case is impossible as the helicity of antineutrinos is fixed. Therefore, the spin projection of the muon (electron) is $+1/2$.

If electrons and muons were massless, two-body pion decays would be forbidden. A massless electron, or muon, would have have to be 100 % rigth-handed, but W bosons only couple to left-handed leptons. Because of their finite mass, electrons and muons with their spins pointing in their directions of motion actually also have a left-handed component, which is proportional to $1-\beta$. The W boson couples to this component. Since the electron mass is so small, $1-\beta_e = 2.6 \cdot 10^{-5}$ is very small in pion decay, compared to $1-\beta_\mu = 0.72$. Hence, the left-handed component of the electron is far smaller than that of the muon, and the electron decay is accordingly strongly suppressed.

10.6 Deep Inelastic Neutrino Scattering

Deep inelastic scattering of neutrinos off nucleons gives us information about the quark distributions in the nucleon which cannot be obtained from electron or muon scattering alone. In contrast to photon exchange, the exchange of W bosons (charged currents) in neutrino scattering distinguishes between the helicity and charged states of the fermions involved. This is then exploited to separately determine the quark and antiquark distributions in the nuclei.

In deep inelastic neutrino scattering experiments, muon (anti)neutrinos are generally used, which, as discussed in Sect. 10.5, stem from weak pion and kaon decays. These latter particles can be produced in large numbers by bombarding a solid block with highly energetic protons. Since (anti)neutrinos have very small cross-sections the targets that are used (e.g., iron) are generally many metres long. The deep inelastic scattering takes place off both the protons and the neutrons in the target.

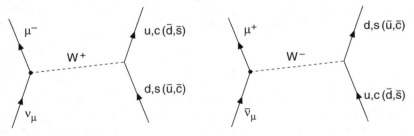

When left handed neutrinos scatter off nucleons, the exchanged W^+ can only interact with the negatively charged, left handed quarks (d_L, s_L) and negatively charged, right handed antiquarks ($\bar{u}_R$, $\bar{c}_R$) which are thereby transformed into the corresponding (anti)quarks of the same family. In analogy to our description of τ decay, we can neglect complications due to Cabibbo mixing if the energies are large enough that we can ignore the differences in the quark masses. Equivalently for the scattering of right handed antineutrinos, the W^- which is exchanged can only interact with the positively charged, left handed quarks (u_L, c_L) and positively charged, right handed antiquarks ($\bar{d}_R$, $\bar{s}_R$).

The scattering off the quarks and antiquarks is characterised by different angle and energy distributions for the outgoing leptons. This becomes plausible if one (analogously to our considerations in the case of Mott scattering in Sect. 5.3) considers the extreme case of scattering through $\theta_{\text{c.m.}} = 180°$ in the centre of mass frame for the neutrino and the quark. We choose the quantisation axis $\hat{z}$ to be the direction of the incoming neutrino's momentum. Since the W boson only couples to left handed fermions, both the neutrino and the quark have in the high energy limit negative helicities and the projection of the total spin on the $\hat{z}$ axis is, both before and after scattering through 180° $S_3 = 0$.

This also holds for all other scattering angles, i.e., the scattering is isotropic. On the other hand if a left handed neutrino interacts with a right handed antiquark, the spin projection before the scattering is $S_3 = -1$ but after being scattered through 180° it is $S_3 = +1$. Hence scattering through 180° is forbidden by conservation of angular momentum. An angular dependence, proportional to $(1 + \cos\theta_{\text{c.m.}})^2$, is found in the cross-section. In the laboratory frame this corresponds to an energy dependence proportional to $(1 - y)^2$ where

$$y = \frac{\nu}{E_\nu} = \frac{E_\nu - E'_\mu}{E_\nu} \qquad (10.23)$$

is that fraction of the neutrino's energy which is transferred to the quark. Completely analogous considerations hold for antineutrino scattering.

The cross-section for neutrino-nucleon scattering may be written analogously to the cross-section for neutrino-electron scattering (10.9) if we take into account the fact that the interacting quark only carries a fraction x of the momentum of the nucleon and that the centre of mass energy in the neutrino-quark centre of mass system is x times smaller than in the neutrino-nucleon system. One finds:

$$\frac{\mathrm{d}^2\sigma}{\mathrm{d}x\mathrm{d}y} = \frac{G_F^2}{\pi(\hbar c)^4} \cdot \left(\frac{M_W^2 c^4}{Q^2 c^2 + M_W^2 c^4}\right)^2 \cdot 2M_p c^2 E_\nu \cdot x \cdot K \qquad (10.24)$$

where

$$K = \begin{cases} d(x) + s(x) + (\overline{u}(x) + \overline{c}(x))(1-y)^2 & \text{for } \nu\text{-p scattering,} \\ \overline{d}(x) + \overline{s}(x) + (u(x) + c(x))(1-y)^2 & \text{for } \overline{\nu}\text{-p scattering.} \end{cases} \qquad (10.25)$$

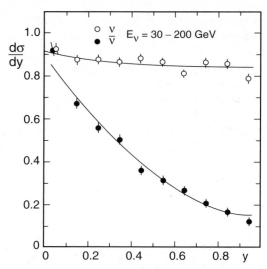

Fig. 10.2. Differential cross-sections $d\sigma/dy$ for neutrino and antineutrino scattering off nucleons as a function of y (in arbitrary units).

Figure 10.2 shows the dependence of the integrated (over x) cross-section as a function of y. For neutrino scattering we have two contributions: a large constant contribution from scattering off the quarks, and a small contribution from scattering off the antiquarks which falls off as $(1 - y)^2$. In antineutrino scattering one observes a strong $(1 - y)^2$ dependence from the interaction with the quarks and a small energy independent part from the antiquarks.

Suitable combinations of the data from neutrino and antineutrino scattering off protons and neutrons can be used to separate the distributions of valence and sea quarks shown in Fig. 7.7.

_____ **Problems**

1. **Particle reactions**
   Show whether the following particle reactions and decays are possible or not.
   State which interaction is concerned and sketch the quark composition of the
   hadrons involved.

   $$
   \begin{aligned}
   p + \overline{p} &\rightarrow \pi^+ + \pi^- + \pi^0 + \pi^+ + \pi^- \\
   p + K^- &\rightarrow \Sigma^+ + \pi^- + \pi^+ + \pi^- + \pi^0 \\
   p + \pi^- &\rightarrow \Lambda^0 + \overline{\Sigma}^0 \\
   \overline{\nu}_\mu + p &\rightarrow \mu^+ + n \\
   \nu_e + p &\rightarrow e^+ + \Lambda^0 + K^0 \\
   \Sigma^0 &\rightarrow \Lambda^0 + \gamma
   \end{aligned}
   $$

2. **Parity and C-parity**
   a) Which of the following particle states are eigenstates of the charge conjugation operator $\mathcal{C}$ and what are their respective eigenvalues?

   $|\gamma\rangle; \; |\pi^0\rangle; \; |\pi^+\rangle; \; |\pi^-\rangle; \; |\pi^+\rangle - |\pi^-\rangle; \; |\nu_e\rangle; \; |\Sigma^0\rangle.$

   b) How do the following quantities behave under the parity operation? (Supply
   a brief explanation.)

position vector r	momentum p
angular momentum L	spin σ
electric field E	magnetic field B
electric dipole moment $\sigma \cdot E$	magnetic dipole moment $\sigma \cdot B$
helicity $\sigma \cdot p$	transversal polarisation $\sigma \cdot (p_1 \times p_2)$

3. **Parity and C-parity of the f_2-mesons**
   The $f_2(1270)$-meson has spin 2 and decays, amongst other routes, into $\pi^+\pi^-$.
   a) Use this decay to find the parity and C-parity of the f_2.
   b) Investigate whether the decays $f_2 \rightarrow \pi^0\pi^0$ and $f_2 \rightarrow \gamma\gamma$ are allowed.

4. **Pion decay and the Golden Rule**
   Calculate the ratio of the partial decay widths

   $$
   \frac{\Gamma(\pi^+ \rightarrow e^+\nu)}{\Gamma(\pi^+ \rightarrow \mu^+\nu)}
   $$

   and so verify the relevant claims in the text. From the Golden Rule it holds that
   $\Gamma(\pi \rightarrow \ell\nu) \propto |\mathcal{M}_{\pi\ell}|^2 \varrho(E_0)$, where $|\mathcal{M}_{\pi\ell}|$ is the transition matrix element and
   $\varrho(E_0) = dn/dE_0$ is the density of states (ℓ denotes the charged lepton). The
   calculation may be approached as follows:
   a) Derive formulae for the momenta and energies of the charged leptons ℓ^+ as
   functions of m_ℓ and m_π and so find numerical values for $1 - v/c$.
   b) We have $|\mathcal{M}_{\pi\ell}|^2 \propto 1 - v/c$. Use this to express the ratio of the squares of
   the matrix elements as a function of the particle masses involved and find its
   numerical value.
   c) Calculate the ratio of the densities of states $\varrho_e(E_0)/\varrho_\mu(E_0)$ as a function
   of the masses of the particles involved. Exploit the fact that the density of
   states in momentum space is $dn/d|p| \propto |p|^2$ ($|p| = |p_{\ell+}| = |p_\nu|$) and that
   $E_0 = E_{\ell+} + E_\nu$. For which of the two decays is the "phase space" bigger?
   d) Combine the results from b) and c) to obtain the ratio of the partial decay
   widths as a function of the masses of the particles involved. Find its numerical
   value and compare it with its experimental value of $(1.230 \pm 0.004) \cdot 10^{-4}$.

11. Exchange Bosons of the Weak Interaction

The idea that the weak interaction is mediated by very heavy exchange bosons was generally accepted long before they were discovered. The structure of the Fermi theory of β-decay implies that the interaction is pointlike, which in turn implies that the exchange bosons have to be very heavy particles. Quantitatively, however, this was confirmed only when the W and the Z bosons were detected experimentally [Ar83, Ba83a] and their properties could be measured. The Z^0 boson's properties imply a mixing of the electromagnetic and weak interactions. The electroweak unification theory due to Glashow, Salam and Weinberg from the early seventies was thus confirmed. Today it is the basis of the standard model of elementary particle physics.

11.1 Real W and Z Bosons

The production of a real W or Z boson requires that a lepton and antilepton or a quark and antiquark interact. The centre of mass energy necessary for this is $\sqrt{s} = M_{W,Z}\,c^2$. This energy is most easily reached using colliding particle beams.

In e^+e^- colliders, a centre of mass energy of $\sqrt{s} = 2E_e = M_Z c^2$ is necessary for the production of Z^0 particles via:

$$e^+ + e^- \rightarrow Z^0\,.$$

This became technically possible in 1989, when the SLC (Stanford Linear Collider) and the LEP became operational; now large numbers of Z^0 bosons can be produced. W bosons can also be produced in e^+e^- reactions, but only in pairs:

$$e^+ + e^- \rightarrow W^+ + W^-\,.$$

Hence, significantly higher energies are necessary for their production: $\sqrt{s} > 2M_W c^2$.

In 1996 the beam energy at LEP was upgraded to 86 GeV. This made a precise measurement of the W-mass of the decay products of the W^+W^- pairs possible.

For many years the production of $W^\pm$ or Z^0 bosons was only possible with the help of quarks in proton beams via the reactions:

$$u + \overline{u} \to Z^0, \qquad d + \overline{u} \to W^-,$$
$$d + \overline{d} \to Z^0, \qquad u + \overline{d} \to W^+.$$

For these reactions, however, it is insufficient to collide two proton beams each with half the rest energy of the vector bosons. Rather, the quarks which participate have to carry enough centre of mass energy $\sqrt{\hat{s}}$ to produce the bosons. In a fast moving system, quarks carry only a fraction xP_p of the proton momentum P_p (cf. Sect. 7.3). About half the total momentum is carried by gluons; the rest is distributed among several quarks, with the mean x for valence quarks and sea quarks given by:

$$\langle x_v \rangle \approx 0.12 \qquad \langle x_s \rangle \approx 0.04. \tag{11.1}$$

One can produce a Z^0 boson in a head-on collision of two protons according to:

$$u + \overline{u} \to Z^0.$$

But the proton beam energy E_p must be close to $E_p \approx 600$ GeV in order to satisfy:

$$M_Z c^2 = \sqrt{\hat{s}} \approx \sqrt{\langle x_u \rangle \langle x_{\overline{u}} \rangle \cdot s} = 2 \cdot \sqrt{0.12 \cdot 0.04} \cdot E_p. \tag{11.2}$$

Proton-antiproton collisions are more favourable, since the momentum distributions of the $\overline{u}$- and $\overline{d}$-valence quarks in antiprotons are equal to those of the u- and d-valence quarks in protons. Consequently, only about half the energy is necessary. Since a p and a $\overline{p}$ have opposite charges, it is also not necessary to build two separate accelerator rings; both beams can in fact be injected in opposite directions into the same ring. At the SPS (Super Proton Synchrotron) at CERN, which was renamed Sp$\overline{p}$S (Super Proton Antiproton Storage ring) for this, protons and antiprotons of up to 318 GeV were stored; at the Tevatron (FNAL), 900 GeV beam energies are attained.

The bosons were detected for the first time in 1983 at CERN at the UA1 [Ar83] and UA2 [Ba83a, An87] experiments in the decays:

$$Z^0 \to e^+ + e^-, \qquad\qquad W^+ \to e^+ + \nu_e,$$
$$Z^0 \to \mu^+ + \mu^-, \qquad\qquad W^+ \to \mu^+ + \nu_\mu.$$

The Z^0 boson has a very simple experimental signature. One observes a high-energy e^+e^- or $\mu^+\mu^-$ pair flying off in opposite directions. Figure 11.1 shows a so-called "lego diagram" of one of the first events. The figure shows the transverse energy measured in the calorimeter cells plotted against the polar and azimuthal angles of the leptons relative to the incoming proton beam. The height of the "lego bars" measures the energy of the leptons. The total energy of both leptons corresponds to the mass of the Z^0.

The detection of the charged vector bosons is somewhat more complicated, since only the charged lepton leaves a trail in the detector and the neutrino is not seen. The presence of the neutrino may be inferred from the

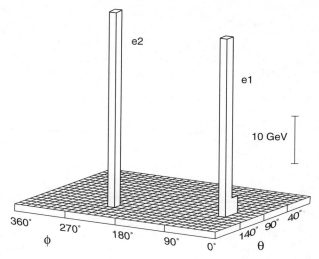

Fig. 11.1. "Lego diagram" of one of the first events of the reaction $q\overline{q} \to Z^0 \to e^+e^-$, in which the Z^0 boson was detected at CERN. The transverse energies of the electron and positron detected in the calorimeter elements are plotted as a function of the polar and azimuthal angles [Ba83b].

momentum balance. When the transverse momenta (the momentum components perpendicular to the beam direction) of all the detected particles are added together the sum is found to be different from zero. This *missing* (transverse) *momentum* is ascribed to the neutrino.

Mass and width of the W boson. The distribution of the transverse momenta of the charged leptons may also be used to find the mass of the $W^{\pm}$. Consider a W^+ produced at rest and then decaying into an e^+ and a ν_e, as shown in Fig. 11.2a. The transverse momentum of the positron is roughly given by:

$$p_t^{e^+} \approx \frac{M_W \cdot c}{2} \sin \theta \,, \tag{11.3}$$

where θ is the angle at which the positron is emitted with respect to the beam axis. We now consider the dependence of the cross-section on p_t or on $\cos \theta$. We have:

$$\frac{\mathrm{d}\sigma}{\mathrm{d}p_t} = \frac{\mathrm{d}\sigma}{\mathrm{d}\cos\theta} \cdot \frac{\mathrm{d}\cos\theta}{\mathrm{d}p_t} \,, \tag{11.4}$$

from which follows:

$$\frac{\mathrm{d}\sigma}{\mathrm{d}p_t} = \frac{\mathrm{d}\sigma}{\mathrm{d}\cos\theta} \cdot \frac{2p_t}{M_W c} \cdot \frac{1}{\sqrt{(M_W c/2)^2 - p_t^2}} \,. \tag{11.5}$$

The cross-section should have a maximum at $p_t = M_W c/2$ (because of the transformation of variables, also called a *Jacobian peak*) and should then

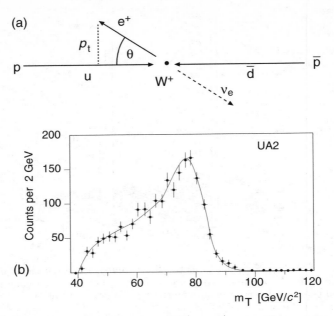

Fig. 11.2. (a) Kinematics of the decay $W^+ \to e^+ + \nu_e$. The maximum possible transverse momentum p_t of the e^+ is $M_W c/2$. (b) Distribution of the "transverse mass" $m_t = 2p_t/c$ of e^+ and e^- in the reaction $q_1 + \bar{q}_2 \to e^{\pm} +$ "nothing", from the UA2 experiment at CERN [Al92b].

drop off rapidly, as shown by the solid line in Fig. 11.2b. Since the W is not produced at rest and has a finite decay width the distribution is smeared out. The most precise figures to date for the width and mass of the W are:

$$
\begin{aligned}
M_W &= 80.41 \pm 0.10 \text{ GeV}/c^2, \\
\Gamma_W &= 2.06 \pm 0.07 \text{ GeV}.
\end{aligned}
\tag{11.6}
$$

Mass and width of the Z boson. Since the cross-section for creating Z-bosons in e^+e^- collisions is much larger than the cross-section for creating W bosons, in either e^+e^- or $p\bar{p}$ collisions, the mass and width of the Z^0 boson have been much more precisely determined than their W boson counterparts. Furthermore, the energies of the e^+ and e^- beams are known to an accuracy of a few MeV, which means that the measurements are extremely good. The experimental values of the Z^0 parameters and width are [PD98]:

$$
\begin{aligned}
M_Z &= 91.187 \pm 0.007 \text{ GeV}/c^2 \\
\Gamma_Z &= 2.490 \pm 0.007 \text{ GeV}.
\end{aligned}
\tag{11.7}
$$

Decays of the W boson. When we dealt with the charged current decays of hadrons and leptons we saw that the W boson only couples to left-handed fermions (maximum parity violation) and that the coupling is always the same (universality). Only the Cabibbo rotation causes a small correction in the coupling to the quarks.

If this universality of the weak interaction holds, then all types of fermion–antifermion pairs should be equally likely to be produced in the decay of real W bosons. The colour charges mean that an extra factor of 3 is expected for quark-antiquark production. The production of a t-quark is impossible because of its larger mass. Thus, if we neglect the differences between the fermion masses, a ratio of $1:1:1:3:3$ is expected for the production of the pairs $e^+\nu_e$, $\mu^+\nu_\mu$, $\tau^+\nu_\tau$, $u\bar{d}'$, and $c\bar{s}'$, in the decay of the W^+ boson. Here, the states $\bar{d}'$ and $\bar{s}'$ are the Cabibbo-rotated eigenstates of the weak interaction.

Because of the process of hadronisation, it is not always possible in an experiment to unequivocally determine the type of quark–antiquark pair into which a W boson decays. Leptonic decay channels can be identified much more easily. According to the above estimate, a decay fraction of $1/9$ is expected for each lepton pair. The experimental results are [PD98]:

$$
\begin{aligned}
W^\pm \;\to\; & e^\pm + \overset{(-)}{\nu_e} && 10.9 \pm 0.4\,\% \\
& \mu^\pm + \overset{(-)}{\nu_\mu} && 10.2 \pm 0.5\,\% \\
& \tau^\pm + \overset{(-)}{\nu_\tau} && 11.3 \pm 0.8\,\%,
\end{aligned}
\tag{11.8}
$$

in very good agreement with our prediction.

Decays of the Z boson. If the Z boson mediates the weak interaction in the same way as the W boson does, it should also couple with the same strength to all lepton-antilepton pairs and to all quark–antiquark pairs. One therefore should expect a ratio of $1:1:1:1:1:1:3:3:3:3:3$ for the six leptonic channels and the five hadronic channels which are energetically accessible; i.e., $1/21$ for each lepton–antilepton pair, and $1/7$ for each quark–antiquark pair.

To determine the branching ratios, the various pairs of charged leptons and hadronic decays must be distinguished with appropriate detectors. The different quark–antiquark channels cannot always be separated. Decays into neutrino–antineutrino pairs cannot be directly detected. In order to measure their contribution, the cross-sections for all other decays are measured, and compared to the total width of the Z^0 boson. Treating the spin dependences correctly [Na90], we rewrite the Breit-Wigner formula (9.8) in the form:

$$
\sigma_{i \to f}(s) = 12\pi(\hbar c)^2 \cdot \frac{\Gamma_i \cdot \Gamma_f}{(s - M_Z^2 c^4)^2 + M_Z^2 c^4 \Gamma_{\text{tot}}^2}.
\tag{11.9}
$$

Here, Γ_i is the partial width of the initial channel (the partial width for the decay $Z^0 \to e^+ e^-$) and Γ_f is the partial width of the final channel. The total width of the Z^0 is the sum of the partial widths of all the possible decays into fermion–antifermion pairs:

$$
\Gamma_{\text{tot}}(Z^0) = \sum_{\text{all fermions f}} \Gamma(Z^0 \to f\bar{f}).
\tag{11.10}
$$

Each final channel thus yields a resonance curve with a maximum at $\sqrt{s} = M_Z c^2$, and a total width of Γ_{tot}. Its height is proportional to the partial width Γ_f. The partial width Γ_f can experimentally be determined from the ratio of the events of the corresponding channel to the total number of all Z^0 events.

Analyses of the experiments at LEP and SLC yield the following branching ratios [PD98]:

$$
\begin{aligned}
Z^0 \longrightarrow\ & e^+ + e^- & 3.366 &\pm 0.008\,\% \\
& \mu^+ + \mu^- & 3.367 &\pm 0.013\,\% \\
& \tau^+ + \tau^- & 3.360 &\pm 0.015\,\% \\
& \nu_{e,\mu,\tau} + \bar{\nu}_{e,\mu,\tau} & 20.01 &\pm 0.16\ \% \\
& \text{hadrons} & 69.90 &\pm 0.15\ \% .
\end{aligned}
\tag{11.11}
$$

Thus, the probability for a decay into charged leptons is significantly different from the decay probability into neutrinos. The coupling of the Z^0 boson apparently depends on the electric charge. Hence the Z^0 cannot simply be a "neutral W boson" coupling with the same strength to all fermions; rather it mediates a more complicated interaction.

11.2 Electroweak Unification

The properties of the Z^0 boson are attractively described in the theory of the *electroweak interaction*. In this framework, developed by Salam and Weinberg, the electromagnetic and weak interactions are understood as two aspects of the same interaction.

Weak isospin. The electroweak interaction theory can be elegantly described by introducing a new quantum number, the *weak isospin* T, in analogy to the isospin of the strong interaction. Each family of left-handed quarks and leptons forms a doublet of fermions which can transform into each other by emitting (or absorbing) a W boson. The electric charges $z_f \cdot e$ of the two fermions in a doublet always differ by one unit. The weak isospin ascribed to them is $T = 1/2$, and the third component is $T_3 = \pm 1/2$. For right-handed antifermions, the signs of T_3 and z_f are inverted. By contrast, right-handed fermions (and left-handed antifermions) do not couple to W bosons. They are described as singlets ($T = T_3 = 0$). Hence, the left-handed leptons and the (Cabibbo-rotated) left-handed quarks of each family form two doublets and there are additionally three right-handed fermion singlets.

The Weinberg angle. We now continue our description of the weak isospin formalism. One requires conservation of T_3 in reactions with charged currents. The W^- boson must then be assigned the quantum number $T_3(W^-) = -1$ and the W^+ boson $T_3(W^+) = +1$. A third state should therefore exist with $T = 1$, $T_3 = 0$, coupling with the same strength g as the $W^\pm$ to the fermion

Table 11.1. Multiplets of the electroweak interaction. The quarks d′, s′ and b′ emerge from the mass eigenstates through a generalised Cabibbo rotation (Kobayashi-Maskawa matrix). Weak isospin T doublets are joined in parentheses. The electric charges of the two states of each doublet always differ by one unit. The sign of the third component T_3 is defined so that that the difference $z_f - T_3$ is constant within each doublet.

		Fermion Multipletts		T	T_3	z_f
Leptons	$\begin{pmatrix} \nu_e \\ e \end{pmatrix}_L$	$\begin{pmatrix} \nu_\mu \\ \mu \end{pmatrix}_L$	$\begin{pmatrix} \nu_\tau \\ \tau \end{pmatrix}_L$	$1/2$	$+1/2$ $-1/2$	0 -1
	e_R	μ_R	τ_R	0	0	-1
Quarks	$\begin{pmatrix} u \\ d' \end{pmatrix}_L$	$\begin{pmatrix} c \\ s' \end{pmatrix}_L$	$\begin{pmatrix} t \\ b' \end{pmatrix}_L$	$1/2$	$+1/2$ $-1/2$	$+2/3$ $-1/3$
	u_R	c_R	t_R	0	0	$+2/3$
	d_R	s_R	b_R	0	0	$-1/3$

doublets. This state is denoted by W^0; and together with the W^+ and the W^- it forms a weak isospin triplet.

The W^0 cannot be identical to the Z^0, since we saw that the coupling of the latter also depends on the electric charge. One now postulates the existence of an additional state B^0, a singlet of the weak isospin ($T = 0$, $T_3 = 0$). Its coupling strength does not have to be equal to that of the triplet ($W^\pm, W^0$). The corresponding weak charge is denoted by g'. The B^0 and W^0 couple to fermions without changing their weak isospin and hence without changing their type.

Experimentally two neutral vector bosons, the photon and the Z^0, are indeed known. The basic idea of the electroweak unification is to describe the photon and the Z^0 as mutually orthogonal, linear combinations of the B^0 and the W^0. This mixing is, analogously to the description of quark mixing in terms of the Cabibbo angle (10.16), expressed as a rotation through the so-called *electroweak mixing angle* θ_W (also called the *Weinberg angle*):

$$|\gamma\rangle = \cos\theta_W|B^0\rangle + \sin\theta_W|W^0\rangle$$

$$|Z^0\rangle = -\sin\theta_W|B^0\rangle + \cos\theta_W|W^0\rangle . \tag{11.12}$$

The connection between the Weinberg angle θ_W, the weak charges g und g' and the electric charge e is given by demanding that the photon couples to the charges of the left and right handed fermions but not to the neutrinos. One so obtains [Na90]

$$\tan\theta_W = \frac{g'}{g} , \quad \sin\theta_W = \frac{g'}{\sqrt{g^2 + g'^2}} , \quad \cos\theta_W = \frac{g}{\sqrt{g^2 + g'^2}} . \tag{11.13}$$

The electromagnetic charge is given by:

$$e = g \cdot \sin \theta_W. \tag{11.14}$$

The Weinberg angle can be determined, for example, from ν–e scattering, from electroweak interference in e^+e^- scattering, out of the width of the Z^0, or from the ratio of the masses of the $W^\pm$ and the Z^0 [Am87, Co88]. A combined analysis of such experiments gives the following result [PD98]:

$$\sin^2 \theta_W = 0.231\,24 \pm 0.000\,24\,. \tag{11.15}$$

Hence, the weak coupling constant ($\alpha_w \propto g \cdot g$) is about four times stronger than the electromagnetic one ($\alpha \propto e \cdot e$). It is the propagator term in the matrix element (10.3), which is responsible for the tiny effective strength of the weak interaction at low energies.

This Weinberg mixing somewhat complicates the interaction. The W boson couples with equal strength to all the quarks and leptons (universality) but always to only left-handed particles and right-handed antiparticles (maximal parity violation). In the coupling of the Z boson, however, the electric charges of the fundamental fermions play a part as well. The coupling strength of the Z^0 to a fermion f is:

$$g_Z(f) = \frac{g}{\cos \theta_W} \cdot \hat{g}(f) \qquad \text{where} \quad \hat{g}(f) = T_3 - z_f \sin^2 \theta_W\,, \tag{11.16}$$

and z_f is the electric charge of the fermion in units of the elementary charge e.
The ratio of the masses of the W and Z bosons. The electroweak unification theory could be used to predict the absolute masses of the W and the Z fairly well before their actual discovery. According to (10.4) and (11.14), the electromagnetic coupling constant α, the Fermi constant G_F and the mass of the W boson are related by

$$M_W^2 c^4 = \frac{4\pi\alpha}{8 \sin^2 \theta_W} \cdot \frac{\sqrt{2}\,(\hbar c)^3}{G_F}\,. \tag{11.17}$$

It is important to realise that in in quantum field theory the "constants" α and $\sin^2 \theta_W$ are in fact weakly dependent upon the energy range (renormalisation) [El82, Fa90]. For the mass region of 11.17, we have $\alpha \approx 1/128$ and $\sin^2 \theta_W \approx 0.231$. The mass of the Z boson is fixed by the relation:

$$\frac{M_W}{M_Z} = \cos \theta_W \approx 0.88\,. \tag{11.18}$$

This is in good agreement with the ratio calculated from the experimentally measured masses (11.6) and (11.7):

$$\frac{M_W}{M_Z} = 0.8818 \pm 0.0011\,. \tag{11.19}$$

The resulting value of $\sin^2 \theta_W$ is in very good agreement with the results of other experiments. The value given in (11.15) is from the combined analysis of all experiments.

Interpretation of the width of the Z^0. A detailed study of the producti-on of Z^0 bosons in electron–positron annihilation delivers a very precise check of the predictions of the standard model of electroweak unification.

The coupling of a Z^0 to a fermion f is proportional to the quantity $\hat{g}(f)$ defined in (11.16). The partial width Γ for a decay $Z^0 \to f\bar{f}$ is a superposition of two parts, one for each helicity state:

$$\Gamma_f = \Gamma_0 \cdot \left[\hat{g}_L^2(f) + \hat{g}_R^2(f) \right], \tag{11.20}$$

where

$$\Gamma_0 = \frac{G_F}{3\pi\sqrt{2}\,(\hbar c)^3} \cdot M_Z^3 c^6 \approx 663 \text{ MeV}. \tag{11.21}$$

For left-handed neutrinos, $T_3 = 1/2$, $z_f = 0$; hence,

$$\hat{g}_L(\nu) = \frac{1}{2}. \tag{11.22}$$

We believe that right-handed neutrinos are not found in nature. They would have $T_3 = z_f = \hat{g}_R = 0$ and would not be subject to the interactions of the standard model. The contribution of each $\nu\bar{\nu}$ pair to the total width is therefore:

$$\Gamma_\nu \approx 165.8 \text{ MeV}. \tag{11.23}$$

The d, s and b quarks have $T_3 = -1/2$ (left-handed) or $T_3 = 0$ (right-handed) and $z_f = -1/3$. This yields:

$$\hat{g}_L(d) = -\frac{1}{2} + \frac{1}{3}\sin^2\theta_W, \qquad \hat{g}_R(d) = \frac{1}{3}\sin^2\theta_W. \tag{11.24}$$

Recalling that quark–antiquark pairs can be produced in three colour com-binations (r$\bar{\text{r}}$, g$\bar{\text{g}}$, b$\bar{\text{b}}$), the total contribution of these quarks is:

$$\Gamma_d = \Gamma_s = \Gamma_b = 3 \cdot 122.4 \text{ MeV}. \tag{11.25}$$

Similarly the contribution of the u and c quarks is:

$$\Gamma_u = \Gamma_c = 3 \cdot 94.9 \text{ MeV}, \tag{11.26}$$

and the contribution of the charged leptons is:

$$\Gamma_e = \Gamma_\mu = \Gamma_\tau = 83.3 \text{ MeV}. \tag{11.27}$$

Decays into $\nu\bar{\nu}$ pairs cannot be directly detected in an experiment, but they manifest themselves in their contributions to the total width. Taking account of the finite masses of the quarks and charged leptons only produces small corrections, as these masses are small compared to the mass of the Z boson.

Including all known quarks and leptons in the calculations, one finds that the total width is 2418 MeV. After incorporating quantum field theoretical

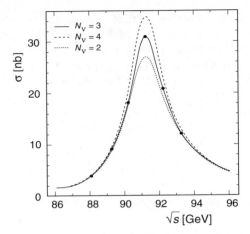

Fig. 11.3. Cross-section of the reaction $e^+e^- \to hadrons$ close to the Z^0 resonance. The data shown are the results of the OPAL experiment at CERN [Bu91]. According to (11.9) the measured width of the resonance yields the total cross-section. The more types of light leptons exist, the smaller the fraction of the total cross-section that remains for the production of hadrons. The lines show the theoretical predictions, based on the measured width of the resonance, assuming that 2, 3, or 4 massless neutrinos exist.

corrections due to higher-order processes (radiation corrections) the width predicted is [La95]:

$$\Gamma_{\text{tot}}^{\text{theor.}} = (2497 \pm 6) \text{ MeV} . \tag{11.28}$$

This is in very good agreement with the experimental value (11.7) of:

$$\Gamma_{\text{tot}}^{\text{exp.}} = (2490 \pm 7) \text{ MeV} . \tag{11.29}$$

The proportion of the total number of decays into pairs of charged leptons is equal to the ratio of the widths (11.27) and (11.28):

$$\frac{\Gamma_{e,\mu,\tau}}{\Gamma_{\text{tot}}} = 3.37 \% . \tag{11.30}$$

The experimental branching ratios (11.11) are in excellent agreement with this theoretical value.

If a fourth type of light neutrino were to couple to the Z^0 in the same way, then the total width would be larger by 166 MeV. We thus can deduce from the experimental result that exactly three types of *light* neutrinos exist (Fig. 11.3). This may be interpreted as implying that the total number of generations of quarks and leptons is three (and three only).

Symmetry breaking. Notwithstanding the successes of electroweak unification, the theory is aesthetically flawed: the mixture of states described by the Weinberg rotation (11.12) should only occur for states with similar energies (masses). Yet, the photon is massless and the W and Z bosons have very large masses. How this can happen is a central and, as yet, not really answered question in particle physics.

A possible answer is associated with *spontaneous symmetry breaking,* a concept known from the physics of phase transitions. This assumes an asymmetric vacuum ground state. The best-known examples of this idea are the magnetic properties of iron, and the Meissner effect (or Meissner-Ochsenfeld effect) in superconductivity.

■ To illustrate how symmmetry breaking can generate a mass, we now consider the analogy of ferromagnetism. Above the Curie temperature, iron is paramagnetic and the spins of the valence electrons are isotropically distributed. No force is required to alter spin orientations. The fields that carry the magnetic interaction may, as far as spatial rotations are concerned, be considered massless. When the temperature drops below the Curie point, a phase transition takes place and iron becomes ferromagnetic. The spins, or the magnetic moments of the valence electrons turn spontaneously to point in a common direction which is not fixed a priori. The space within the ferromagnet is no longer isotropic, rather it has a definite preferred direction. Force must be used to turn the spins away from the preferred direction. Thus the carriers of the magnetic interaction now have a mass as far as rotations are concerned. This process is called *spontaneous symmetry breaking.*

The Meissner effect, the absence of external magnetic fields in superconductors, provides an even better analogy to particle production by symmetry breaking. Above the transition temperature of the superconductor, magnetic fields propagate freely within the conductor. With the transition to the superconducting phase, however, they are expelled from the superconductor. They can only penetrate the superconductor at its surface and drop off exponentially inside. An observer within the superconductor could explain this effect by a finite range of the magnetic field in the superconductor. In analogy to the discussion of the Yukawa force (Sect. 16.3) he therefore would ascribe a finite mass to the photon.

Where is the spontaneous symmetry breaking in this process? This is what actually happens in superconductivity: below the critical temperature, Cooper pairs are formed out of the conduction electrons which organise themselves into a correlated state of definite energy; the energy of the superconducting ground state. For an observer within the superconductor, the ground state of the superconductor is the ground state of the vacuum. As the temperature sinks a current is induced in the superconductor which compensates the external magnetic field and expels it from the superconductor. The correlated Cooper pairs are responsible for this current. Just as in the case of the ferromagnet where the spins are no longer free to choose their orientation, the phase of a Cooper pair is here fixed by the phase of the other Cooper pairs. This effect corresponds to a symmetry breaking of the ground state.

In a theoretical model, proposed independently by Englert and Brout [En64] and by Higgs [Hi64], the masses of the Z^0 and of the $W^\pm$ bosons are explained in analogy to the Meissner effect. In this model, so-called Higgs fields are postulated, which — compared to our example — correspond to the ground state of correlated Cooper pairs in superconductivity. At sufficiently high temperatures (or energies) the Z^0 and $W^\pm$ bosons are massless like the photon. Below the energy of the phase transition, the boson masses are produced by the Higgs fields, just as the "photon mass" is in the Meissner effect.

The masses of the Z^0 and the $W^\pm$ bosons must be independent of their location and orientation in the universe. Hence, the Higgs fields must be scalars. In the theory of electroweak unification, there are thus four Higgs fields, one for each boson. During the cooling of the system, three *Higgs bosons,* the quanta of the Higgs field, are absorbed by the Z^0 and by the $W^\pm$. This generates their masses. Since the photon remains massless there must still be a free Higgs boson.

The existence of these Higgs fields is fundamental to the modern interpretation of elementary particle physics. The search for non-absorbed Higgs bosons is the main motivation for the construction of a new accelerator and storage ring at CERN, the *Large Hadron Collider* (LHC). The experimental proof of their existence would be a complete confirmation of the Glashow-Salam-Weinberg theory of electroweak unification. The non-existence of the Higgs bosons, however, would require completely new theoretical concepts. One could compare this situation with that at the end of the nineteenth century, when the existence of the aether had a similar importance for the interpretation of physics.

Problems

1. **Number of neutrino generations**

   At the LEP storage ring at CERN Z^0-bosons are produced in electron-positron annihilations at a centre of mass energy of about 91 GeV before decaying into fermions: $e^+ e^- \rightarrow Z^0 \rightarrow f\bar{f}$. Use the following measurements from the OPAL experiment to verify the statement that there are exactly three sorts of light neutrinos (with $m_\nu < m_{Z^0}/2$). The measurement of the resonance curve (11.9) yielded: $\sigma_{\text{had}}^{\text{max}} = 41.45 \pm 0.31$ nb, $\Gamma_{\text{had}} = 1738 \pm 12$ MeV, $\Gamma_\ell = 83.27 \pm 0.50$ MeV, $M_Z = 91.182 \pm 0.009$ GeV/c^2. All quark final states are here combined into a single width Γ_{had} and Γ_ℓ is the decay width of the Z^0 into (single) charged leptons. Derive a formula for the number of neutrino species N_ν and use the ratio Γ_ℓ/Γ_ν from the text to calculate N_ν. Estimate the error in N_ν from the experimental errors.

12. The Standard Model

Se non è vero, è ben trovato.

Giordano Bruno
Gli eroici furori

Die Wissenschaft hat ewig Grenzen,
aber keine ewigen Grenzen.

P. du Bois-Reymond
Über die Grenzen
des Naturerkennens

The *standard model* of elementary particle physics comprises the unified theory of the electroweak interaction and quantum chromodynamics. In the following, we will once more summarize what we have learned in previous chapters about the different particles and interactions.

– As well as gravitation, we know of three elementary interactions which have very similar structures. Each of them is mediated by the exchange of vector bosons.

Interaction	couples to	Exchange particle(s)	Mass (GeV/c^2)	J^P
strong	colour charge	8 gluons (g)	0	1^-
electromagn.	electric charge	photon (γ)	0	1^-
weak	weak charge	$\mathrm{W}^\pm$, Z^0	$\approx 10^2$	1

Gluons carry colour and therefore interact with each other. The bosons of the weak interaction themselves carry weak charge and couple with each other as well.

– As well as the exchange bosons, the known fundamental particles are the quarks and the leptons. They are fermions with spin-1/2. They are grouped, according to their masses, into three "families", or "generations".

Fermions	Family 1 2 3	Electr. charge	Colour	Weak Isospin left-hd. right-hd.	Spin
Leptons	ν_e ν_μ ν_τ e μ τ	0 −1	—	1/2 — 0	1/2
Quarks	u c t d s b	+2/3 −1/3	r, b, g	1/2 0 0	1/2

Each fermion has an associated antifermion. It has the same mass as the fermion, but opposite electric charge, colour and third component of weak isospin.

From the measured width of the Z^0 resonance, one can deduce that no further (fourth) massless neutrino exists. Thus, the existence of a fourth generation of fermions (at least one with a light neutrino) can be excluded.

– The range of the electromagnetic interaction is infinite since photons are massless. Because of the large mass of the exchange bosons of the weak interaction, its range is limited to 10^{-3} fm. Gluons have zero rest mass. Yet, the effective range of the strong interaction is limited by the mutual interaction of the gluons. The energy of the colour field increases with increasing distance. At distances $\gtrsim 1$ fm, it is sufficiently large to produce real quark–antiquark pairs. "Free" particles always have to be colour neutral.

– The electromagnetic interaction and the weak interaction can be interpreted as two aspects of a single interaction: the electroweak interaction. The corresponding charges are related by the Weinberg angle, cf. (11.14).

– Different conservation laws apply to the different interactions:

  • The following physical quantities are conserved in all three interactions: energy (E), momentum ($\boldsymbol{p}$), angular momentum ($\boldsymbol{L}$), charge (Q), colour, baryon number (B) and the three lepton numbers (L_e, L_μ, L_τ).

  • The P and C parities are conserved in the strong and in the electromagnetic interaction; but not in the weak interaction. For the charged current of the weak interaction, parity violation is maximal. The charged current only couples to left-handed fermions and right-handed antifermions. The neutral weak current is partly parity violating. It couples to left-handed and right-handed fermions and antifermions, but with different strengths. One case is known in which the combined CP parity is not conserved.

  • Only the charged current of the weak interaction transforms one type of quark into another type (quarks of a different flavour) and one type of lepton into another. Thus, the quantum numbers determining the quark flavour (third component of isospin (I_3), strangeness (S), charm (C) etc.) are conserved in all other interactions.

  • The magnitude of the isospin (I) is conserved in strong interactions.

The possible transitions within a lepton doublet, or a quark doublet, as well as transitions between families are shown in Fig. 12.1. This picture is possibly the forerunner of a new type of spectroscopy, more elementary than the atomic, nuclear or hadronic spectroscopies.

In summary, experiments are in astoundingly good quantitative agreement with the assumptions of the standard model. These include the grouping of the fermions into left-handed doublets and right-handed singlets of weak isospin, the strength of the coupling of the Z^0 to left-handed and right-handed fermions, the three-fold nature of the quark families because of colour and

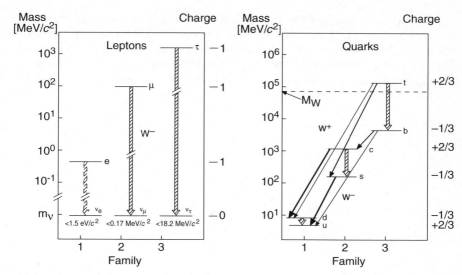

Fig. 12.1. Transitions between quark and lepton states via charged currents. The strength of the coupling is reflected in the width of the arrows. As a result of energy and charge conservation, the transition $e^- \rightarrow \nu_e$ is observed in reactions such as lepton scattering or K-electron capture, but not as the decay of a free particle. The mass of the t quark is so large, that it decays by emission of a *real* W^+ boson.

the ratio of the masses of the $W^\pm$ and Z^0. We thus possess a self-contained picture of the fundamental building blocks of matter and of their interactions.

And yet today's standard model is unsatisfactory in many respects. A large number of free parameters remain, as many as 18 or more, depending on the counting scheme [Na90]. These are the masses of the fermions and bosons, the coupling constants of the interactions and the coefficients of the Kobayashi-Maskawa matrix. These parameters are not given by the standard model; they must be determined experimentally and have then to be incorporated ad hoc into the model.

Many questions are still completely open. Why do exactly three families of fermions exist? Do neutrinos have finite masses? What is the origin of the masses of all the other fermions and of the W and Z boson? Does the Higgs boson exist? Is it a coincidence that that within every family the fermions which carry more charge (strong, electromagnetic, weak) have larger masses? Are baryon number and lepton number strictly conserved? What is the origin of CP violation? What is the orgin of the mixture of quark families, described by the Kobayashi-Maskawa matrix? Will such a mixture eventually be found for leptons as well? Why are there just four interactions? What determines the magnitudes of the coupling constants of the different interactions? Is it possible to unify the strong and electroweak interactions, as one has unified the electromagnetic and weak interactions? Will it be possible to include gravitation in a complete unification?

Such questions reflect the experience physicists have gained in analysing the building blocks of matter. On their journey from solid bodies to quarks via molecules, atoms, nuclei, and hadrons, they have constantly found new, fundamental particles. The question "Why?" implicitly assumes that more fundamental reasons exist for observed phenomena — new experiments are the only way to check this assumption.

Nature has always looked like a horrible mess, but as we go along we see patterns and put theories together; a certain clarity comes and things get simpler.

Richard P. Feynman [Fe85]

Part II

Synthesis:
Composite Systems

Naturam expelles furca, tamen usque recurret.

Horace, epist. I,XX

13. Quarkonia

Analogy is perhaps the physicist's most powerful conceptual tool for understanding new phenomena or opening new areas of investigation. Early in this century, for example, Ernest Rutherford and Niels Bohr conceived the atom as a miniature solar system in which electrons circle the nucleus as planets circle the sun.

V. L. Telegdi [Te62]

In the following we are going to consider hadronic bound-states. The simplest example are heavy quark-antiquark ($c\bar{c}$ and $b\bar{b}$) pairs, which are known as *quarkonia*. Due to the large quark masses they may be approximately treated in a nonrelativistic manner. The *hydrogen atom* and *positronium* will serve as electromagnetic analogues.

13.1 The Hydrogen Atom and Positronium Analogues

The simplest atomic bound-state is the hydrogen atom, which is composed of a proton and an electron. To a first approximation the bound-states and energy levels may be calculated from the nonrelativistic Schrödinger equation. The static Coulomb potential $V_C \propto 1/r$ is then incorporated into the Hamiltonian

$$\left(-\frac{\hbar^2}{2m}\triangle - \frac{\alpha\hbar c}{r} \right) \psi(\boldsymbol{r}) = E\,\psi(\boldsymbol{r}). \tag{13.1}$$

The eigenstates are characterised by the number of nodes N in the radial wave functions and the orbital angular momentum ℓ. For the particular case of the Coulomb potential, states with identical $n = N + \ell + 1$ are degenerate and n is therefore called the *principal quantum number*. The allowed energy levels E_n are found to be

$$E_n = -\frac{\alpha^2 mc^2}{2n^2}\,, \tag{13.2}$$

where α is the electromagnetic coupling constant and m is the reduced mass of the system:

$$m = \frac{M_{\mathrm{p}}m_{\mathrm{e}}}{M_{\mathrm{p}} + m_{\mathrm{e}}} \approx m_{\mathrm{e}} = 0.511\,\mathrm{MeV}/c^2\,. \tag{13.3}$$

The binding energy of the hydrogen ground state ($n = 1$) is $E_1 = -13.6$ eV. The Bohr radius r_{b} is given by

$$r_{\mathrm{b}} = \frac{\hbar \cdot c}{\alpha \cdot mc^2} \approx \frac{197\,\mathrm{MeV} \cdot \mathrm{fm}}{137^{-1} \cdot 0.511\,\mathrm{MeV}} = 0.53 \cdot 10^5\,\mathrm{fm} \,. \qquad (13.4)$$

The spin-orbit interaction ("fine structure") and the spin-spin-interaction ("hyperfine structure") split the degeneracy of the principal energy levels as is shown in Fig. 13.1. These corrections to the general $1/n^2$ behaviour of the energy levels are, however, very small. The fine structure correction is of order α^2 while that of the hyperfine structure is of order $\alpha^2 \cdot \mu_{\mathrm{p}}/\mu_{\mathrm{e}}$. The ratio of the hyperfine splitting of the $1\mathrm{s}_{1/2}$ level to the gap between the $n = 1$ and $n = 2$ principal energy levels is therefore merely $E_{\mathrm{HFS}}/E_n \approx 5 \cdot 10^{-7}$. Here we employ the notation $n\ell_j$ for states when fine structure effects are taken into account. The orbital angular momenta quantum numbers $\ell = 0, 1, 2, 3$ are then denoted by the letters $\mathrm{s, p, d, f}$. The quantum number j is the total angular momentum of the electron, $\boldsymbol{j} = \boldsymbol{\ell} + \boldsymbol{s}$. A fourth quantum number f is used to describe the hyperfine effects (see Fig. 13.1 left). This describes the total angular momentum of the atom, $\boldsymbol{f} = \boldsymbol{j} + \boldsymbol{i}$, with the proton's spin $\boldsymbol{i}$ included.

The energy states of positronium, the bound $\mathrm{e}^+\mathrm{e}^-$ system, can be found in an analogous way to the above. The main differences are that the reduced mass $(m = m_{\mathrm{e}}/2)$ is only half the value of the hydrogen case and the spin-spin coupling is much larger than before, since the electron magnetic moment is roughly 650 times larger than that of the proton. The smaller reduced mass means that the binding energies of the bound states are only half the size of those of the hydrogen atom while the Bohr radius is twice its previous value (Fig. 13.2). The stronger spin-spin coupling now means that the positronium spectrum does not display the clear hierarchy of fine and hyperfine structure effects that we know from the hydrogen atom. The spin-orbit and spin-spin forces are of a similar size (Fig. 13.1).

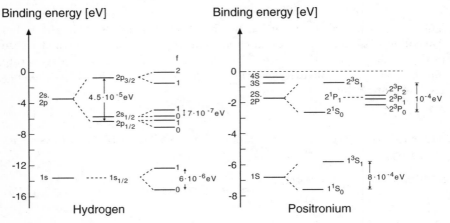

Fig. 13.1. The energy levels of the hydrogen atom and of positronium. The ground states ($n=1$) and the first excited states ($n=2$) are shown together with their fine and hyperfine splitting. The splitting is not shown to scale.

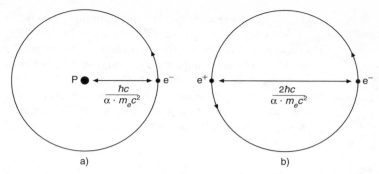

Fig. 13.2. The first Bohr orbits of the hydrogen atom (**a**) and positronium (**b**) (from [Na90]). The Bohr radius describes the average separation of the two bound particles.

Thus for positronium the total spin S and the total angular momentum J as well as the principal quantum number n and the orbital angular momentum L are the useful quantum numbers. S can take on the values 0 (singlet) and 1 (triplet), and J obeys the triangle inequality, $|L - S| \leq J \leq L + S$. The notation $n^{2S+1}L_J$ is commonly employed, where the orbital angular momentum L is represented by the capital letters (S, P, D, F). Thus $2\,^3P_1$ signifies a positronium state with $n = 2$ and $S = L = J = 1$.

Since electrons and positrons annihilate, positronium has a finite lifetime. It primarily decays into 2 or 3 photons, depending upon whether the total spin is 0 or 1. The decay width for the two-photon decay of the 1^1S_0 state is found to be [Na90]

$$\Gamma(1^1S_0 \to 2\gamma) = \frac{4\pi\alpha^2\hbar^3}{m_e^2 c}|\psi(0)|^2 \,. \tag{13.5}$$

Note that $|\psi(0)|^2$ is the square of the wave function at the origin, i. e. the probability that e^+ und e^- meet at a point. Equation (13.5) yields a lifetime of $\approx 10^{-10}\,\mathrm{s}$.

The potential and the coupling constant of the electromagnetic interaction are very well known, and electromagnetic transitions in positronium as well as its lifetime can be calculated to high precision and excellent agreement with experiment is found. Quarkonia, i.e., systems built up of strongly interacting heavy quark-antiquark pairs, can be investigated in an analogous manner. The effective potential and the coupling strength of the strong interaction can thus be determined from the experimental spectrum and transition strengths between the various states.

13.2 Charmonium

Bound states of c and $\bar{c}$ quarks are, in analogy to positronium, called *charmonium*. For historical reasons a somewhat different nomenclature is employed for charmonium states than is used for positronium. The first quantum number is $n_{q\bar{q}} = N+1$, where N is the number of nodes in the radial wave

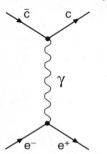

function, while for positronium the atomic convention, according to which the principal quantum number is defined as $n_{atom} = N + \ell + 1$, is used.

$c\bar{c}$ pairs are most easily produced in the decay of virtual photons generated in e^+e^- collisions with a centre of mass energy of around 3–4.5 GeV

$$e^+ + e^- \rightarrow \gamma \rightarrow c\bar{c}.$$

Various resonances may be detected by varying the beam energy and looking for peaks in the cross section. These are then ascribed to the various charmonium states (Fig. 13.3). Because of the intermediate virtual photon, only $c\bar{c}$ states with the quantum numbers of a photon, ($J^P = 1^-$), can be created in this way. The lowest state with such quantum numbers is the 1^3S_1, which is called the J/ψ (see p. 122) and has a mass of 3.097 GeV/c^2. Higher resonances with masses up to 4.4 GeV/c^2 have been detected.

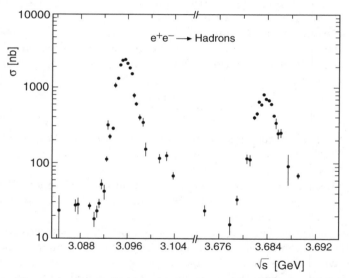

Fig. 13.3. The cross section of the reaction $e^+e^- \rightarrow$ *hadrons*, plotted against the centre of mass energy in two different intervals each of 25 MeV. The two peaks which are both 100 times larger than the continuum represent the lowest charmonium states with $J^P = 1^-$ (the J/ψ (1^3S_1) and the ψ (2^3S_1)). That the experimental width of these resonances is a few MeV is a consequence of the detector's resolution: widths of 87 keV and 277 keV respectively may be extracted from the lifetimes of the resonances. The results shown are early data from the e^+e^- ring SPEAR at Stanford [Fe75].

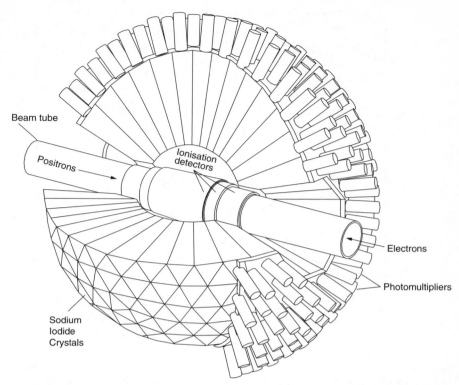

Fig. 13.4. A (crystal ball) detector built out of spherically arranged NaI crystals. High energy photons from electromagnetic $c\bar{c}$ transitions are absorbed by the crystals. This creates a shower of electron-positron pairs which generate many low energy, visible photons. These are then detected by photomultipliers attached to the rear of the crystals. The current measured from the photomultipliers is proportional to the energy of the initial photon (from [Kö86]).

Charmonium states only have a finite lifetime. They predominantly decay via the strong interaction into hadrons. Excited states can, however, by the emission of a photon, decay into lower energy states, just as in atomic physics or for positronium. The emitted photons may be measured with a detector that covers the entire solid angle around the e^+e^- interaction zone (4π detectors). *Crystal balls,* which are composed of spherically arranged scintillators (NaI crystals) are particularly well suited to this task (Fig. 13.4).

If one generates, say, the excited charmonium ψ (2^3S_1) state one then may measure the photon spectrum shown in Fig. 13.5, in which various sharp lines are clearly visible. The photon energy is between 100 and 700 MeV. The stronger lines are electric dipole transitions which obey the selection rules, $\Delta L = 1$ and $\Delta S = 0$. Intermediate states with total angular momentum 0, 1 or 2 and positive parity must therefore be created in such decays. The parity

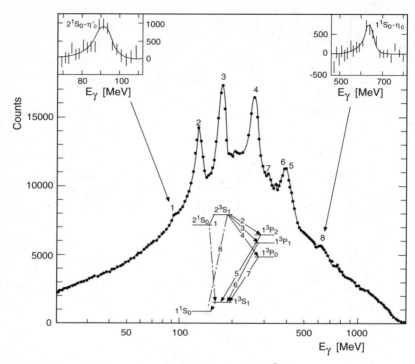

Fig. 13.5. The photon spectrum in the decay of $\psi\,(2^3S_1)$, as measured in a crystal ball, and a sketch of the so extracted charmonium energy levels. The strong peaks in the photon spectrum represent the so numbered transitions in the sketch. The continuous lines in the sketch represent parity changing electric dipole transitions and the dashed lines denote magnetic dipole transitions which do not change parity [Kö86].

of the spatial wave function is just $(-1)^L$, where L is the orbital angular momentum. Furthermore from the Dirac theory fermions and antifermions have opposite intrinsic parity. Thus the parity of $q\bar{q}$ states is generally $(-1)^{L+1}$. Armed with this information we can reconstruct the diagram in Fig. 13.5. We see that after the $\psi\,(2^3S_1)$ state is generated it primarily decays into the 1^3P_J charmonium triplet system which is known as χ_c. These χ_c states then decay into J/ψ's. The spin 0 charmonium states (n^1S_0), which are called η_c, and cannot be produced in e^+e^- collisions, are only produced in magnetic dipole transitions from J/ψ or $\psi\,(2^3S_1)$. These obey the selection rules $\Delta L = 0$ and $\Delta S = 1$ and thus connect states with the same parity. They correspond to a spin flip of one of the c-quarks. Magnetic dipole transitions are weaker than electric dipole transitions. They are, however, observed in charmonium, since the spin-spin interaction for $c\bar{c}$ states is significantly stronger than in atomic systems. This is due to the much smaller separation between the partners compared to atomic systems.

13.3 Quark–Antiquark Potential

If we compare the spectra of charmonium and positronium, we find that the states with $n=1$ and $n=2$ are very similarly arranged once an overall increase in the positronium scale of about 10^8 is taken into account (Fig. 13.6). The higher charmonium states do not, on the other hand, display the $1/n^2$ behaviour we see in positronium.

What can we learn from this about the potential and the coupling constant of the strong interaction? Since the potential determines the relative positions of the energy levels, it is clear that the potential of the strong interaction must, similarly to the electromagnetic one, be of a Coulomb type (at least at very short distances, i.e., for $n = 1, 2$). This observation is supported by quantum chromodynamics which describes the force between the quarks via gluon exchange and predicts a r^{-1} potential at short distances. The absence, in comparison to positronium, of any degeneracy between the 2^3S and 1^3P states suggests that the potential is not of a pure Coulomb form even at fairly small quark-antiquark separations. Since quarks have not been experimentally observed, it is plausible to postulate a potential which is of a

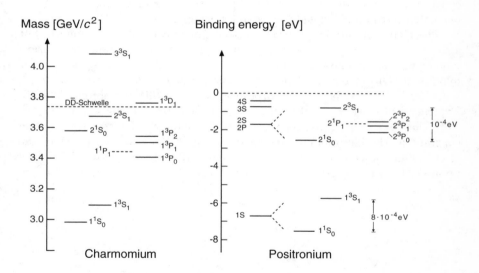

Fig. 13.6. Comparison of the energy levels of positronium and charmonium. The energy scales were chosen such that the 1S and 2S states of the two systems coincide horizontally. As a result of the differences in nomenclature for the first quantum number, the 2P states in positronium actually correspond to the 1P levels in charmonium. The splitting of the positronium states has been magnified. Dashed states have been calculated but not yet experimentally detected. Note that the $n=1$ and $n=2$ level patterns are very similar, while the 2S–3S separations are distinctly different. The dashed, horizontal line marks the threshold where positronium breaks up and charmonium decays into two D mesons (see Sect. 13.6).

Coulomb type at short distances and grows linearly at greater separations, thus leading to the confinement of quarks in hadrons.

An ansatz for the potential is therefore

$$V = -\frac{4}{3}\frac{\alpha_s(r)\hbar c}{r} + k \cdot r, \tag{13.6}$$

which displays the asymptotic behaviour $V(r \to 0) \propto 1/r$ and $V(r \to \infty) \to \infty$. The factor of 4/3 is a theoretical consequence of quarks coming in three different colours. The strong coupling constant α_s is actually not a constant at all, but depends upon the separation r of the quarks (8.6), becoming smaller as the separation increases. This is a direct consequence of QCD and results in the so-called *asymptotic freedom* property of the strong force. This behaviour allows us to view quarks as quasi-free particles at short distances as we have already discussed for deep-inelastic scattering.

While a Coulomb potential corresponds to a dipole field, where the field lines are spread out in space (Fig. 13.7a), the kr term leads to a so-called flux tube. The lines of force between the quarks are "stretched" (Fig. 13.7b) and the field energy increases linearly with the separation of the quarks. The constant k in the second term of the potential determines the field energy per unit length and is called the "string tension".

The charmonium energy levels depend not only upon the potential but also upon the kinetic terms in the Hamiltonian, which contain the a priori unknown c-quark mass m_c. The three unknown quantities α_s, k and m_c may be roughly determined by fitting the principal energy levels of the $c\bar{c}$ states from the nonrelativistic Schrödinger equation with the potential (13.6). Typical results are: $\alpha_s \approx 0.15$–0.25, $k \approx 1$ GeV/fm and $m_c \approx 1.5$ GeV/c^2. Note that m_c is the constituent mass of the c-quark. The strong coupling constant in the charmonium system is about 20–30 times larger than the electromagnetic coupling, $\alpha = 1/137$. Figure 13.8 shows a potential, based upon (13.6), where the calculated radii of the charmonium states are given.

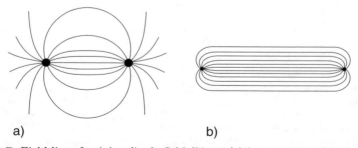

a)   b)

Fig. 13.7. Field lines for (**a**) a dipole field ($V \propto 1/r$) between two electric charges, (**b**) a potential $V \propto r$ between two widely separated quarks.

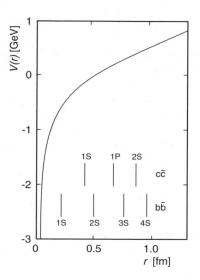

Fig. 13.8. Strong interaction potential versus the separation r of two quarks. This potential is roughly described by (13.6). The vertical lines mark the radii of the $c\bar{c}$ and $b\bar{b}$ states as calculated from such a potential (from [Go84]).

The J/ψ (1^3S_1) has, for example, a radius[1] of approximately $r \approx 0.4\,\mathrm{fm}$, which is five orders of magnitude smaller than that of positronium.

To fully describe the energy levels of Fig. 13.6 one must incorporate further terms into the potential. Similarly to the case of atomic physics, one can describe the splitting of the P states very well through a spin-orbit interaction. The splitting of the S states of charmonium and the related spin-spin interaction will be treated in the next section.

The Coulomb potential describes forces that decrease with distance. The integral of this force is the ionisation energy. The strong interaction potential, (13.6), on the other hand, describes a force between quarks which remains constant at large separations. To remove a coloured particle such as a quark from a hadron would require an infinitely high energy. Thus, since the isolation of coloured objects is impossible, we find only colourless objects in nature. This does not, however, mean that quarks cannot be detached from one another.

Quarks are not liberated in such circumstances, rather fresh hadrons are produced if the energy in the flux tube crosses a specific threshold. The now detached quarks become constituents of these new hadrons. If, for example, a quark is knocked out of a hadron in deep inelastic scattering, the flux tube between this quark and the remainder of the original hadron breaks when the tube reaches a length of about 1–2 fm. The field energy is converted into a quark and an antiquark. These then separately attach themselves to the two ends of the flux tube and thus produce two colour neutral hadrons. This is the previously mentioned *hadronisation* process.

---

[1] By this we mean the average separation between the quark and the antiquark (see Fig. 13.2).

13.4 The Chromomagnetic Interaction

The similarity between the potential of the strong force and that of the electromagnetic interaction is due to the short distance r^{-1} Coulombic term. This part corresponds to 1-gluon (1-photon) exchange. Charmonium displays a strong splitting of the S states, as does positronium, and this is due to a spin-spin interaction. This force is only large at small distances and thus 1-gluon exchange should essentially account for it in quarkonium. The spin-spin interaction splitting, and hence the force itself, is, however, roughly 1000 times larger for charmonium than in positronium.

The spin-spin interaction for positronium takes the form

$$V_{ss}(e^+e^-) = \frac{-2\mu_0}{3}\,\boldsymbol{\mu}_1\cdot\boldsymbol{\mu}_2\,\delta(\boldsymbol{x})\,, \tag{13.7}$$

where μ_0 is the vacuum permeability. This equation describes the point interaction of the magnetic moments $\boldsymbol{\mu}_{1,2}$ of e^+ and e^-. The magnetic moment of the electron (positron) is just

$$\boldsymbol{\mu}_i = \frac{z_i e\hbar}{2m_i}\boldsymbol{\sigma}_i \quad \text{where} \quad z_i = Q_i/e = \pm 1\,, \tag{13.8}$$

and the components of the vector $\boldsymbol{\sigma}$ are the Pauli matrices; $\sigma_x^2 = \sigma_y^2 = \sigma_z^2 = \mathbb{1}$. The potential $V_{ss}(e^+e^-)$ may then be expressed as

$$V_{ss}(e^+e^-) = \frac{-\hbar^2\,\mu_0}{6}\frac{z_1 z_2 e^2}{m_1 m_2}\,\boldsymbol{\sigma}_1\cdot\boldsymbol{\sigma}_2\,\delta(\boldsymbol{x}) = \frac{2\pi\hbar^3}{3\,c}\,\alpha\,\frac{\boldsymbol{\sigma}_1\cdot\boldsymbol{\sigma}_2}{m_e^2}\,\delta(\boldsymbol{x})\,. \tag{13.9}$$

The quark colour charges lead to a spin-spin interaction called the *chromomagnetic* or *colour magnetic interaction*. To generalise the electromagnetic spin-spin force to describe the chromomagnetic spin-spin interaction we have to replace the electromagnetic coupling constant α by α_s and alter the factor to take the three colour charges into account. We thus obtain for the quark-antiquark spin-spin interaction

$$V_{ss}(q\bar{q}) = \frac{8\pi\hbar^3}{9\,c}\alpha_s\frac{\boldsymbol{\sigma}_q\cdot\boldsymbol{\sigma}_{\bar{q}}}{m_q m_{\bar{q}}}\,\delta(\boldsymbol{x})\,. \tag{13.10}$$

The chromomagnetic energy thus depends upon the relative spin orientations of the quark and the antiquark. The expectation value of $\boldsymbol{\sigma}_q\cdot\boldsymbol{\sigma}_{\bar{q}}$ is found to be

$$\boldsymbol{\sigma}_q\cdot\boldsymbol{\sigma}_{\bar{q}} = 4\boldsymbol{s}_q\cdot\boldsymbol{s}_{\bar{q}}/\hbar^2 \;=\; 2\cdot[S(S+1) - s_q(s_q+1) - s_{\bar{q}}(s_{\bar{q}}+1)]$$

$$= \begin{cases} -3 & \text{for} \quad S=0\,, \\ +1 & \text{for} \quad S=1\,, \end{cases} \tag{13.11}$$

where S is the total spin of the charmonium state and we have used the identity, $\boldsymbol{S}^2 = (\boldsymbol{s}_q + \boldsymbol{s}_{\bar{q}})^2$. One thus obtains an energy splitting from this chromomagnetic interaction of the form

$$\Delta E_{\mathrm{ss}} = \langle \psi | V_{\mathrm{ss}} | \psi \rangle = 4 \cdot \frac{8\pi\hbar^3}{9\,c} \frac{\alpha_{\mathrm{s}}}{m_{\mathrm{q}} m_{\overline{\mathrm{q}}}} | \psi(0) |^2 \,. \tag{13.12}$$

This splitting is only important for S states, since only then is the wave function at the origin $\psi(0)$ non-vanishing.

The observed charmonium transition from the state 1^3S_1 to 1^1S_0 (i.e., $J/\psi \to \eta_c$) is a magnetic transition, which corresponds to one of the quarks flipping its spin. The measured photon energy, and hence the gap between the states, is approximately 120 MeV. The colour magnetic force (13.12) should account for this splitting. Although an exact calculation of the wave function is not possible, we can use the values of α_{s} and m_{c} from the last section to see that our ansatz for the chromomagnetic interaction is consistent with the observed splitting of the states. We will see in Chap. 14 that the spin-spin force also plays a role for light mesons and indeed describes their mass spectrum very well.

The c-quark's mass. The c-quark mass which we obtained from our study of the charmonium spectrum is its constituent quark mass, i.e., the effective quark mass in the bound state. This constituent mass has two parts: the intrinsic (or "bare") quark mass and a "dynamical" part which comes from the cloud of sea quarks and gluons that surrounds the quark. The fact that charmed hadrons are 4–10 times heavier than light hadrons implies that the constituent mass of the c-quark is predominantly intrinsic since the dynamical masses themselves should be more or less similar for all hadrons. We should not forget that even if the dynamical masses are small compared to the heavy quark constituent mass, the potential we have used is a phenomenological one which merely describes the interaction between *constituent* quarks.

13.5 Bottonium and Toponium

A further group of narrow resonances are found in e^+e^- scattering at centre of mass energies of around 10 GeV. These are understood as $b\overline{b}$ bound states and are called *bottonium*. The lowest $b\overline{b}$ state which can be obtained from e^+e^- annihilation is called the Υ and has a mass of 9.46 GeV/c^2. Higher $b\overline{b}$ excitations have been found with masses up to 11 GeV/c^2.

Various electromagnetic transitions between the various bottonium states are also observed. As well as a 1^3P_J state, a 2^3P_J state has been observed. The spectrum of these states closely parallels that of charmonium (Fig. 13.9). This indicates that the quark-antiquark potential is independent of quark flavour. The b-quark mass is about 3 times as large as that of the c-quark. The radius of the quarkonium ground state is from (13.4) inversely proportional both to the quark mass and to the strong coupling constant α_{s}. The 1S $b\overline{b}$ state thus has a radius of roughly 0.2 fm (cf. Fig. 13.8), i.e., about half that of the equivalent $c\overline{c}$ state. Furthermore the nonrelativistic treatment of bottonium is better justified than was the case for charmonium. The approximately

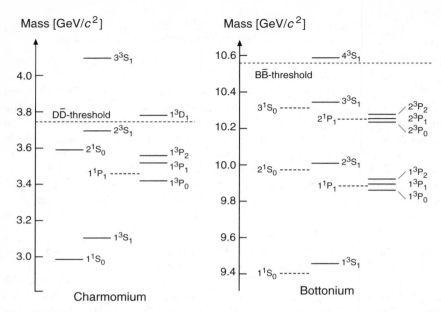

Fig. 13.9. Energy levels of charmonium and bottonium. Dashed levels are theoretically predicted, but not yet experimentally observed. The spectra display a very similar structure. The dashed line shows the threshold beyond which charmonium (bottonium) decays into hadrons containing the initial quarks, i.e., D (B) mesons. Below the threshold electromagnetic transitions from ^{3}S states into ^{3}P and ^{1}S states are observed. For bottonium the first and second excitations ($n = 2, 3$) lie below this threshold, for charmonium only the first does.

equal mass difference between the 1S and 2S states in both systems is, however, astounding. A purely Coulombic potential would cause the levels to be proportional to the reduced mass of the system, (13.2). It is thus clear that the long distance part of the potential kr cancels the mass dependence of the energy levels at the c- and b-quark mass scales.

The t-quark has, due to its large mass, only a fleeting lifetime. Thus no pronounced $t\bar{t}$ states *(toponium)* are expected.

13.6 The Decay Channels of Heavy Quarkonia

Up to now we have essentially dealt with the electromagnetic transitions between various levels of quarkonia. But actually it is astonishing that electromagnetic decays occur at all at an observable rate. One would naively expect a strongly interacting object to decay "strongly". The decays of heavy quarkonia have been in fact investigated very thoroughly [Kö89] so as to obtain the most accurate possible picture of the quark-antiquark interaction.

There are in principle four different ways in which quarkonia can change its state or decay. They are:

a) A change of excitation level via photon emission (electromagnetic), e.g.,

$$\chi_{c1} \, (1^3P_1) \rightarrow J/\psi \, (1^3S_1) + \gamma$$

b) Quark-antiquark annihilation into real or virtual photons or gluons (electromagnetic or strong), e.g.,

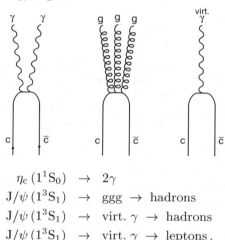

$$
\begin{aligned}
\eta_c \, (1^1S_0) &\rightarrow 2\gamma \\
J/\psi \, (1^3S_1) &\rightarrow ggg \rightarrow \text{hadrons} \\
J/\psi \, (1^3S_1) &\rightarrow \text{virt. } \gamma \rightarrow \text{hadrons} \\
J/\psi \, (1^3S_1) &\rightarrow \text{virt. } \gamma \rightarrow \text{leptons} .
\end{aligned}
$$

The J/ψ decays about 30% of the time electromagnetically into hadrons or charged leptons and about 70% of the time strongly. The electromagnetic route can, despite the smallness of α, compete with the strong one, since in the strong case three gluons must be exchanged to conserve colour and parity. A factor of α_s^3 thus lowers this decay probability (compared to α^2 in the electromagnetic case). States such as η_c, which have $J = 0$, can decay into two gluons or two real photons. The decay of the J/ψ ($J{=}1$) is mediated by three gluons or a single virtual photon.

c) Creation of one or more light $q\bar{q}$ pairs from the vacuum to form light mesons (strong interaction)

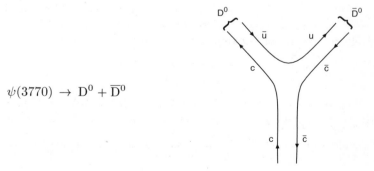

$$\psi(3770) \rightarrow D^0 + \overline{D}^0$$

d) Weak decay of one or both heavy quarks, e.g.,

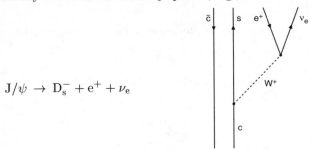

$$J/\psi \;\to\; D_s^- + e^+ + \nu_e$$

In practice the weak decay (d) is unimportant since the strong and electromagnetic decays proceed much more quickly. The strong decay (c) is, in principle, the most likely, but this can only take place above a certain threshold since the light $q\bar{q}$ pairs need to be created from the quarkonia binding energy. Hence only options (a) and (b) are available to quarkonia below this threshold.

Electromagnetic processes like deexcitation via photon emission are relatively slow. Furthermore, although hadronisation via the annihilation (b) into gluons is a strong process such decays are, according to the *Zweig rule* (cf. Sect. 9.2) suppressed relative to those decays (c) where the initial quarks still exist in the final state. For these reasons the width of those quarkonium levels below the mesonic threshold is very small (e.g., $\Gamma = 88\,\text{keV}$ for the J/ψ).

The first charmonium state beyond this threshold is the $\psi\,(1^3D_1)$ which has a mass of $3770\,\text{MeV}/c^2$. It has, compared to the J/ψ, rather a large width, $\Gamma \approx 24\,\text{MeV}$. For the more strongly bound $b\bar{b}$ system the decay channel into mesons with b-quarks is first to the third excitation, the $\Upsilon\,(4^3S_1)$ $(10\,580\,\text{MeV}/c^2)$ (cf. Fig. 13.9).

The lightest quarks are the u- and d-quarks and their pair production opens the mesonic decay channels. Charmonium, say, decays into

$$c\bar{c} \;\to\; c\bar{u} + \bar{c}u\,,$$
$$c\bar{c} \;\to\; c\bar{d} + \bar{c}d\,,$$

where $c\bar{u}$ is called the D^0 meson, $\bar{c}u$ the $\overline{D}^0$, $c\bar{d}$ the D^+ and $\bar{c}d$ the D^-. The masses of these mesons are $1864.6\,\text{MeV}/c^2$ (D^0) and $1869.3\,\text{MeV}/c^2$ $(D^\pm)$. The preferred decays of bottonium are analogously

$$b\bar{b} \;\to\; b\bar{u} + \bar{b}u\,,$$
$$b\bar{b} \;\to\; b\bar{d} + \bar{b}d\,.$$

These mesons are called[2] B^- and B^+ $(m = 5278.9\,\text{MeV}/c^2)$, as well as $\overline{B}^0$ and B^0 $(m = 5279.2\,\text{MeV}/c^2)$. For higher exitations decays into mesons with

---

[2] The standard nomenclature for mesons containing heavy quarks is such that the neutral meson with a b-quark is called a $\overline{B}^0$ and the meson with a $\overline{b}$ is known

s-quarks are also possible:

$$
\begin{aligned}
c\overline{c} &\rightarrow c\overline{s} + \overline{c}s & &(D_s^+ \text{ and } D_s^-)\,, \\
b\overline{b} &\rightarrow b\overline{s} + \overline{b}s & &(\overline{B}_s^0 \text{ and } B_s^0)\,.
\end{aligned}
$$

Such mesons are accordingly heavier. The mass of $D_s^\pm$ meson is, for example, 1968.5 MeV/c^2. All of these mesons eventually decay weakly into lighter mesons such as pions.

13.7 Decay Widths as a Test of QCD

The decays and decay rates of quarkonia can provide us with information about the strong coupling constant α_s. Let us consider the 1^1S_0 charmonium state (η_c) which can decay into either two photons or two gluons. (In the latter case we will only experimentally observe the end products of hadronisation.) Measurements of the ratio of these two decay widths can determine α_s, in principle, in a very elegant way.

The formula for the decay width into 2 *real* photons is essentially just the same as for positronium (13.5), one needs only to recall that the c-quarks have fractional electric charge $z_c = 2/3$ and come in three flavours.

$$
\Gamma(1^1S_0 \rightarrow 2\gamma) = \frac{3 \cdot 4\pi z_c^4 \alpha^2 \hbar^3}{m_c^2 c} |\psi(0)|^2 \, (1 + \varepsilon')\,. \tag{13.13}
$$

The ε' term signifies higher order QCD corrections which can be approximately calculated.

To consider the 2 gluon decay, one must replace α by α_s. In contrast to photons, gluons do not exist as real particles but rather have to hadronise. For this process we set the strong coupling constant to one. The different colour-anticolour combinations also mean we must use a different overall colour factor which takes the various gluon combinations into account:

$$
\Gamma(1^1S_0 \rightarrow 2g \rightarrow \text{hadrons}) = \frac{8\pi}{3} \frac{\alpha_s^2 \hbar^3}{m_c^2 c} |\psi(0)|^2 \, (1 + \varepsilon'')\,. \tag{13.14}
$$

ε'' signifies QCD corrections once again.

The ratio of these decay widths is

$$
\frac{\Gamma(2\gamma)}{\Gamma(2g)} = \frac{8}{9} \frac{\alpha^2}{\alpha_s^2} (1 + \varepsilon)\,. \tag{13.15}
$$

The correction factor ε itself depends upon α_s and is about $\varepsilon \approx -0.5$. From the experimentally determined ratio $\Gamma(2\gamma)/\Gamma(2g) \approx (3.0 \pm 1.2) \cdot 10^{-4}$ [PD98]

---

as a B^0. An electrically neutral $q\overline{q}$ state is marked with a bar, if the heavier quark/antiquark is negatively charged [PD98].

one finds the value $\alpha_\mathrm{s}(m_{\mathrm{J}/\psi}^2 c^2) \approx 0.25 \pm 0.05$. This is consistent with the value from the charmonium spectrum. From (8.6) we see that α_s always depends upon a distance or, equivalently, energy or mass scale. In this case the scale is fixed by the constituent mass of the c-quark or by the J/ψ mass.

The above result, despite the simplicity of the original idea, suffers from both experimental and theoretical uncertainties. As well as QCD corrections, there are further corrections from the relativistic motion of the quarks. For a better determination of α_s from charmonium physics one can investigate other decay channels. The comparison, for instance, of the decay rates

$$\frac{\Gamma(\mathrm{J}/\psi \to 3\mathrm{g} \to \text{hadrons})}{\Gamma(\mathrm{J}/\psi \to \gamma \to 2\,\text{leptons})} \propto \frac{\alpha_\mathrm{s}^3}{\alpha^2}, \tag{13.16}$$

is simpler from an experimental viewpoint. Both here and in studies of other channels one finds $\alpha_\mathrm{s}(m_{\mathrm{J}/\psi}^2 c^2) \approx 0.2 \cdots 0.3$ [Kw87].

The comparison of various bottonium decays yields the coupling strength α in a more accurate way since both QCD corrections and relativistic effects are smaller. From QCD one expects α_s to be smaller, the coupling is supposed to decrease with the separation. This is indeed the case. One finds from the ratio

$$\frac{i\Gamma(\Upsilon \to \gamma\mathrm{gg} \to \gamma + \text{hadrons})}{i\Gamma(\Upsilon \to \mathrm{ggg} \to \text{hadrons})} \propto \frac{\alpha}{\alpha_\mathrm{s}}, \tag{13.17}$$

which is (2.75 ± 0.04), that $\alpha_\mathrm{s}(m_\Upsilon^2 c^2) = 0.163 \pm 0.016$ [Ne97]. The error is dominated by uncertainties in the theoretical corrections.

These examples demonstrate that the annihilation of a $\mathrm{q}\bar{\mathrm{q}}$ pair in both the electromagnetic and strong interactions may formally be described in the same manner. The only essential difference is the coupling constant. This comparison can be understood as a test of the applicability of QCD at short distances, which, after all, is where the $\mathrm{q}\bar{\mathrm{q}}$ annihilation takes place. In this region QCD and QED possess the same structure since both interactions are well described by the exchange of a single vector boson (a gluon or a photon).

Problems

1. **Weak charge**
   Bound states are known to exist for the strong interaction (hadrons, nuclei), electromagnetism (atoms, solids) and gravity (the solar system, stars) but we do not have such states for the weak force. Estimate, in analogy to positronium, how heavy two particles would have to be if the Bohr radius of their bound state would be rougly equal to the range of the weak interaction.

2. **Muonic and hadronic atoms**
   Negatively charged particles that live long enough (μ^-, π^-, K^-, $\overline{p}$, Σ^-, Ξ^-, Ω^-), can be captured by the field of an atomic nucleus. Calculate the energy of atomic ($2p \rightarrow 1s$) transitions in hydrogen-type "atoms" where the electron is replaced by the above particles. Use the formulae of Chap. 13. The lifetime of the 2p state in the H atom is $\tau_{\mathrm{H}} = 1.76 \cdot 10^9$ s. What is the lifetime, as determined from electromagnetic transitions, of the 2p state in a $p\overline{p}$ system (protonium)? Remember to take the scaling of the matrix element and of phase space into account.

3. **Hyperfine structure**
   In a two-fermion system the hyperfine structure splitting between the levels 1^3S_1 and 1^1S_0 is proportional to the product of the magnetic moments of the fermions, $\Delta E \propto |\psi(0)|^2 \mu_1 \mu_2$, where $\mu_i = g_i \frac{e_i}{2m_i}$. The g-factor of the proton is $g_{\mathrm{p}} = 5.5858$ and those of the electron and the muon are $g_{\mathrm{e}} \approx g_{\mu} \approx 2.0023$. In positronium an additional factor of 7/4 arises in the formula for ΔE, which takes the level shifts of the triplet state by pair annihilation graphs into account. In the hydrogen atom, the level splitting corresponds to a transition frequency $f_{\mathrm{H}} = 1420$ MHz. Estimate the values for positronium and muonium ($\mu^+ e^-$). (Hint: $\psi(0) \propto r_{\mathrm{b}}^{-3/2}$; use the reduced mass in the expression for $|\psi(0)|^2$.) Compare your result with the measured values of the transition frequencies, 203.4 GHz for positronium and 4.463 GHz for muonium. How can the (tiny) difference be explained?

4. **B-meson factory**
   Υ-mesons with masses 10.58 GeV/c^2 are produced in the reaction $e^+ e^- \rightarrow \Upsilon(4S)$ at the DORIS and CESR storage rings. The $\Upsilon(4S)$-mesons are at rest in the laboratory frame and decay immediately into a pair of B-mesons: $\Upsilon \rightarrow B^+ B^-$. The mass m_{B} of the B-mesons is 5.28 GeV/c^2 and the lifetime τ is 1.5 psec.
   a) How large is the average decay length of the B-mesons in the laboratory frame?
   b) To increase the decay length, the $\Upsilon(4S)$-mesons need to be given momentum in the laboratory frame. This idea is being employed at SLAC where a "B-factory" is being built where electrons and positrons with different energies collide. What momentum do the B-mesons need to have, if their average decay length is to be 0.2 mm?
   c) What energy do the $\Upsilon(4S)$-mesons, in whose decay the B-mesons are produced, need to have for this?
   d) What energy do the electron and positron beams need to have to produce these $\Upsilon(4S)$-mesons? To simplify the last three questions, without altering the result, assume that the B-mesons have a mass of 5.29 GeV/c^2 (instead of the correct 5.28 GeV/c^2).

14. Mesons Made from Light Quarks

We have seen that the mesons containing the heavy c- and b-quarks may be relatively simply described. In particular since charmonium and bottonium have very different masses they cannot be confused with each other. Furthermore the D and B mesons may be straightforwardly identified with specific quark-antiquark flavour and charge combinations.

Turning now to those mesons that are solely built out of the light flavours (i.e., u, d and s) we encounter a more complicated situation. The constituent masses of these quarks, especially those of the u- and d-quarks, are so similar that we cannot expect to straightforwardly distinguish the mesons according to their quark content but must expect to encounter mixed states of all three light flavours. We shall therefore now consider all of the mesons that are made up of u-, d- and s-quarks.

Another consequence of the light quark masses is that we cannot expect to treat these mesons in a nonrelativistic manner. However, our investigation of the light meson spectrum will lead us to the surprising conclusion that these particles can be at least semi-quantitatively described in a nonrelativistic model. The constituent quark concept is founded upon this success.

14.1 Mesonic Multiplets

Mesonic quantum numbers. We assume that the quarks and antiquarks of the lowest lying mesons do not have any relative orbital angular momentum ($L = 0$). We will only treat such states in what follows. Recall first that quarks and antiquarks have opposite intrinsic parities and so these mesons all have parity, $(-1)^{L+1} = -1$. The quark spins now determine the mesonic total angular momentum. They can add up to either $S = 1$ or $S = 0$. The $J^P = 0^-$ states are called *pseudoscalar mesons* while the $J^P = 1^-$ are the *vector mesons*. One naturally expects 9 different meson combinations from the 3 quarks and 3 antiquarks.

Isospin and strangeness. Let us initially consider just the two lightest quarks. Since the u- and d-quark constituent masses are both around 300 MeV/c^2 (see Table 9.1) there is a natural mixing of degenerate states with the same quantum numbers. To describe $u\bar{u}$- and $d\bar{d}$-quarkonia it is helpful to introduce the idea of *isospin*. The u- and d-quarks form an isospin doublet ($I = 1/2$) with $I_3 = +1/2$ for the u-quark and $I_3 = -1/2$ for the d quark.

This strong isospin is conserved by the strong interaction which does not distinguish between directions in strong isospin space. Quantum mechanically isospin is treated like angular momentum, which reflects itself in isospin addition and the use of ladder operators. The spins of two electrons may combine to form a (spin-)triplet or a singlet, and we can similarly form an (isospin-)triplet or singlet from the 2×2 combinations of a u- or a d-quark with a $\bar{\text{u}}$- or a $\bar{\text{d}}$-quark.

These ideas must be extended to include the s-quark. Its flavour is associated with a further additive quantum number, *strangeness*. The s-Quark has $S = -1$ and the antiquark $S = +1$. Mesons containing one s (anti)quark are eigenstates of the strong interaction, since strangeness can only be changed in weak processes. Zero strangeness $\text{s}\bar{\text{s}}$ states on the other hand can mix with $\text{u}\bar{\text{u}}$ and $\text{d}\bar{\text{d}}$ states since these possess the same quantum numbers. Note that the somewhat larger s-quark constituent mass of about 450 MeV/c^2 implies that this mixing is smaller than that of $\text{u}\bar{\text{u}}$ and $\text{d}\bar{\text{d}}$ states.

Group theory now tells us that the 3×3 combinations of three quarks and three antiquarks form an octet and a singlet. Recall that the 3×3 combinations of colours and anticolours also form an octet and a singlet for the case of the gluons (Sect. 8.3). The underlying symmetry is known as SU(3) in group theory.

We will see below that the larger s-mass leads to this symmetry being less evident in the spectrum. Thus while the mesons inside an isospin triplet have almost identical masses, those of an octet vary noticeably. Were we now to include the c-quark in these considerations we would find that the resulting symmetry was much less evident in the mesonic spectrum.

Vector mesons. Light vector mesons are produced in e^+e^- collisions, just as heavy quarkonia can be. As we saw in Sect. 9.2 (Fig. 9.4) there are three resonances at a centre of mass energy of around 1 GeV. The highest one is at 1019 MeV and is called the ϕ meson. Since the ϕ mostly decays into strange mesons, it is interpreted as the following $\text{s}\bar{\text{s}}$ state:

$$|\phi\rangle = |\text{s}^{\uparrow}\bar{\text{s}}^{\uparrow}\rangle \,,$$

where the arrows signify the 3-component of the quark spins. The pair of light resonances with nearly equal masses, the ϱ and ω mesons, are interpreted as mixed states of u- and d-quarks.

The broad first resonance at 770 MeV is called the ϱ^0 meson. It has two charged partners with almost the same mass. These arise in other reactions. Together they form the isospin triplet: ϱ^+, ϱ^0, ϱ^-. These ϱ mesons are states with isospin 1 built out of the u-, $\bar{\text{u}}$-, d- and $\bar{\text{d}}$-quarks. They may be easily constructed if we recall the quark quantum numbers given in Table 14.1. The charged ϱ mesons are then the states

$$|\varrho^+\rangle = |\text{u}^{\uparrow}\bar{\text{d}}^{\uparrow}\rangle \qquad |\varrho^-\rangle = |\bar{\text{u}}^{\uparrow}\text{d}^{\uparrow}\rangle \,,$$

Table 14.1. The quantum numbers of the light quarks and antiquarks: $B =$ baryon number, $J =$ spin, $I =$ isospin, $I_3 =$ 3-component of the isospin, $S =$ strangeness, $Q/e =$ charge.

	B	J	I	I_3	S	Q/e
u	+1/3	1/2	1/2	+1/2	0	+2/3
d	+1/3	1/2	1/2	−1/2	0	−1/3
s	+1/3	1/2	0	0	−1	−1/3
$\overline{u}$	−1/3	1/2	1/2	−1/2	0	−2/3
$\overline{d}$	−1/3	1/2	1/2	+1/2	0	+1/3
$\overline{s}$	−1/3	1/2	0	0	+1	+1/3

with $I = 1$ and $I_3 = \pm 1$. We may now construct their uncharged partner (for example by applying the ladder operators $I^{\pm}$). We find

$$|\varrho^0\rangle = \frac{1}{\sqrt{2}} \left\{ |u^{\uparrow}\overline{u}^{\uparrow}\rangle - |d^{\uparrow}\overline{d}^{\uparrow}\rangle \right\} .$$

The orthogonal wave function with zero isospin is then just the ω-meson:

$$|\omega\rangle = \frac{1}{\sqrt{2}} \left\{ |u^{\uparrow}\overline{u}^{\uparrow}\rangle + |d^{\uparrow}\overline{d}^{\uparrow}\rangle \right\} .$$

In contradistinction to coupling the angular momentum of two spin half particles there is here a minus sign in the triplet state and a plus in the singlet. The real reason for this is that we have here particle-antiparticle combinations (see, e.g., [Go84]).

Vector mesons with strangeness $S \neq 0$ are called K^* mesons and may be produced by colliding high energy protons against a target:

$$p + p \rightarrow p + \Sigma^+ + K^{*0} .$$

The final state in such experiments must contain an equal number of s-quarks and -antiquarks bound inside hadrons. In this example the K^{*0} contains the $\overline{s}$-antiquark and the Σ^+-baryon contains the s-quark. Strangeness is a conserved quantum number in the strong interaction.

There are four combinations of light quarks which each have just one s- or $\overline{s}$-quark:

$$|K^{*-}\rangle = |s^{\uparrow}\overline{u}^{\uparrow}\rangle \qquad |\overline{K}^{*0}\rangle = |s^{\uparrow}\overline{d}^{\uparrow}\rangle$$
$$|K^{*+}\rangle = |u^{\uparrow}\overline{s}^{\uparrow}\rangle \qquad |K^{*0}\rangle = |d^{\uparrow}\overline{s}^{\uparrow}\rangle .$$

The two pairs K^{*-}, $\overline{K}^{*0}$ and K^{*0}, K^{*+} are both strong isospin doublets.

The ϱ, ω, ϕ and K^* are all of the possible $3 \times 3 = 9$ combinations. They have all been seen in experiments — which is clear evidence of the correctness of the quark model.[1] This classification is made clear in Fig. 14.1. The vector

---

[1] Historically it was the other way around. The quark model was developed so as to order the various mesons into multiplets and hence explain the mesons.

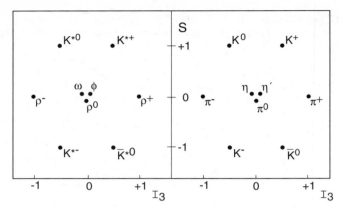

Fig. 14.1. The lightest vector ($J^P = 1^-$) (*left*) and pseudoscalar mesons ($J^P = 0^-$) (*right*), classified according to their isospin I_3 and strangeness S.

mesons are ordered according to their strangeness S and the third component of the isospin I_3. The threefold symmetry of this scheme is due to the three fundamental quark flavours from which the mesons are made. Mesons and antimesons are diagonally opposite to each other and the three mesons at the centre are each their own antiparticles.

Pseudoscalar mesons. The quark and antiquark pair in pseudoscalar mesons have opposite spins and their angular momentum and parity are $J^P = 0^-$. The name "pseudoscalar" arises as follows: spin-0 particles are usually called scalars, while spin-1 particles are known as vectors, but scalar quantities should be invariant under parity transformations. The prefix „pseudo" reflects that these particles possess an unnatural, odd (negative) parity.

The quark structure of the pseudoscalar mesons mirrors that of the vector mesons (Fig. 14.1). The π meson isospin triplet corresponds to the ϱ meson. The pseudoscalars with the quark content of the K^* vector mesons are known as K mesons. Finally, the η' and η correspond to the ϕ and the ω. There are, however, differences in the quark mixings in the isospin singlets. As shown in Fig. 14.1 there are three mesonic states with the quantum numbers $S = I_3 = 0$. These are a symmetric flavour singlet and two octet states. One of these last two has isospin 1 and is therefore a mixture of $u\bar{u}$ and $d\bar{d}$. The π^0 and ϱ^0 occupy this slot in their respective multiplets. The remaining octet state and the singlet can mix with each other since the SU(3) flavour symmetry is broken ($m_s \neq m_{u,d}$). This mixing is rather small for the pseudoscalar case and η and η' are fairly pure octet and singlet states:

$$|\eta\rangle \approx |\eta_8\rangle = \frac{1}{\sqrt{6}}\left\{|u^\uparrow\bar{u}^\downarrow\rangle + |d^\uparrow\bar{d}^\downarrow\rangle - 2|s^\uparrow\bar{s}^\downarrow\rangle\right\},$$

$$|\eta'\rangle \approx |\eta_1\rangle = \frac{1}{\sqrt{3}}\left\{|u^\uparrow\bar{u}^\downarrow\rangle + |d^\uparrow\bar{d}^\downarrow\rangle + |s^\uparrow\bar{s}^\downarrow\rangle\right\}.$$

The vector meson octet and singlet states are, on the other hand, more strongly mixed. It so happens that the mixing angle is roughly $\arctan 1/\sqrt{2}$, which means that the ϕ meson is an almost pure s̄s state and that the ω is a mix of uū and dd̄ whose strange content can safely be neglected [PD98].

14.2 Meson Masses

The masses of the light mesons can be read off from Fig. 14.2. It is striking that the $J = 1$ states have much larger masses than their $J = 0$ partners. The gap between the π and ϱ masses is, for example, about 600 MeV/c^2. This should be contrasted with the splitting of the 1^1S_0 and 1^3S_1 states of charmonium and bottonium, which is only around 100 MeV/c^2.

Just as for the states of heavy quarkonia with total spins $S = 0$ and $S = 1$, the mass difference between the light pseudoscalars and vectors can be traced back to a spin-spin interaction. From (13.10) and (13.11) we find a mass difference of

$$\Delta M_{\mathrm{ss}} = \begin{cases} -3 \cdot \dfrac{8\hbar^3}{9c^3} \dfrac{\pi\alpha_{\mathrm{s}}}{m_{\mathrm{q}} m_{\overline{\mathrm{q}}}} |\psi(0)|^2 & \text{for pseudoscalar mesons}, \\[2ex] +1 \cdot \dfrac{8\hbar^3}{9c^3} \dfrac{\pi\alpha_{\mathrm{s}}}{m_{\mathrm{q}} m_{\overline{\mathrm{q}}}} |\psi(0)|^2 & \text{for vector mesons}. \end{cases} \tag{14.1}$$

Note the dependence of the mass gap on the constituent quark masses. The increase of the gap as the constituent mass decreases is the dominant effect, despite an opposing tendency from the $|\psi(0)|^2$ term (this is proportional to

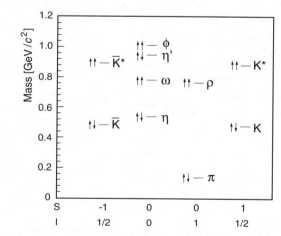

Fig. 14.2. The spectrum of the light pseudoscalar and vector mesons. The multiplets are ordered according to their strangeness S und isospin I. The angular momenta of the various mesons are indicated by arrows. Note that the vector mesons are significantly heavier than their pseudoscalar equivalents.

$1/r_b^3$ and thus grows with the quark mass). Hence this mass gap is larger for the light systems.

The absolute masses of all the light mesons can be described by a phenomenological formula

$$M_{q\bar{q}} = m_q + m_{\bar{q}} + \Delta M_{ss}\,, \tag{14.2}$$

where $m_{q,\bar{q}}$ once again refers to the constituent quark mass. The unknowns in this equation are the constituent masses of the three light quarks. We assume that the u and d masses are the same, and that the product $\alpha_s \cdot |\psi(0)|^2$ is to a rough approximation the same for all of the mesons under consideration here. We may now, with the help of (14.2), extract the quark masses from the experimental results for the meson masses. We thus obtain the following constituent quark masses: $m_{u,d} \approx 310$ MeV/c^2, $m_s \approx 483$ MeV/c^2 [Ga81]. The use of these values yields mesonic masses which only deviate from their true values at the level of a few percent (Table 14.2). These light quark constituent masses are predominantly generated by the cloud of gluons and virtual quark-antiquark pairs that surround the quark. The bare masses are only around 5–10 MeV/c^2 for the u- and d-quarks and about 150 MeV/c^2 for the s. This simple calculation of the mesonic masses demonstrates that the constituent quark concept is valid, even for those quarks with only a tiny bare mass.

It is actually highly surprising that (14.2) describes the mesonic spectrum so very well. After all the equation takes no account of possible mass terms which could depend upon the quark kinetic energy or upon the strong potential (13.6). It appears to be a peculiarity of the potential of the strong interaction that its make up from a Coulombic and a linearly increasing term effectively cancels these mass terms to a very good approximation.

Table 14.2. Light meson masses both from experiment and from (14.2) [Ga81]. The calculations are fitted to the average mass of an isospin multiplet and do not cover those, albeit minor, mass differences arising from electromagnetic effects.

Meson	J^P	I	Mass [MeV/c^2] Calculated	Mass [MeV/c^2] Experiment	
π	0^-	1	140	135.0	π^0
				139.6	$\pi^\pm$
K	0^-	1/2	485	497.7	K^0
				493.7	K^-
η	0^-	0	559	547.3	
η'	0^-	0	—	957.8	
ϱ	1^-	1	780	770.0	
K^*	1^-	1/2	896	896.1	K^{*0}
				891.7	K^{*-}
ω	1^-	0	780	781.9	
ϕ	1^-	0	1032	1019.4	

14.3 Decay Channels

The masses and quantum numbers of the various mesons may also be used to make sense of how these particles decay. The most important decay channels of the pseudoscalar and vector mesons treated here are listed in Table 14.3.

We start with the lightest mesons, the pions. The π^0 is the lightest of all the hadrons and so, although it can decay electromagnetically, it cannot decay strongly. The $\pi^\pm$ can, on the other hand, only decay semileptonically, i.e., through the weak interaction. This is because conservation of charge and of lepton number require that the final state must comprise of a charged lepton and a neutrino. This means that these mesons have long lifetimes. The decay $\pi^- \rightarrow \mathrm{e}^- + \overline{\nu}_\mathrm{e}$ is strongly suppressed compared to $\pi^- \rightarrow \mu^- + \overline{\nu}_\mu$ because of helicity conservation (see p. 143).

The next heaviest mesons are the K mesons (kaons). Since these are the lightest mesons containing an s-quark, their decay into a lighter particle requires the s-quark to change its flavour, which is only possible in weak processes. Kaons are thus also relatively long lived. They decay both non-leptonically (into pions) and semileptonically. The decay of the K^0 is a case for itself and will be treated in Sect. 14.4 in some depth.

As pions and kaons are both long lived and easy to produce it is possible to produce beams of them with a definite momentum. These beams may then be used in scattering experiments. High energy pions and kaons can furthermore be used to produce secondary particle beams of muons or neutrinos if they are allowed to decay in flight.

The strong decays of vector mesons are normally into their lighter pseudoscalar counterparts with some extra pions as a common byproduct. The decays of the ϱ and the K^* are typical here. Their lifetimes are roughly 10^{-23} s.

The ω meson, in contrast to the ϱ, is not allowed to strongly decay into two pions for reasons of isospin and angular momentum conservation. More precisely, this is a consequence of *G-parity* conservation in the strong interaction. G-parity is a combination of C-parity and isospin symmetry [Ga66] and will not be treated here.

How the ϕ decays has already been mentioned in Sect. 9.2 (p. 120). According to the Zweig rule it prefers to decay into a meson with an s-quark and one with an $\overline{\mathrm{s}}$, or, in other words, into a pair of kaons. Since their combined mass is almost as large as that of the original ϕ, the phase space available is small and the ϕ meson consequently has a relatively long lifetime.

The η und η' decay in a somewhat unusual manner. It is easily seen that the η is not allowed to strongly decay into two pions. Note first that the two pion state must have relative angular momentum $\ell = 0$. This follows from angular momentum conservation: both the η and π have spin 0, the pion has odd intrinsic parity and the final two pion state must have total parity $P_{\pi\pi} = (-1)^2 \cdot (-1)^{\ell=0} = +1$. The η has, however, negative parity and so this final state can only be reached by a weak process. A decay into three pions

Table 14.3. The most important decay channels of the lightest pseudoscalar and vector mesons. The resonance's width is often given, instead of the lifetime, for those mesons which can decay in strong processes. The two quantities are related by $\Gamma = \hbar/\tau$ (where $\hbar = 6.6 \cdot 10^{-22}$ MeV s).

	Meson	Lifetime [s]	Most common decay channels		Comments
Pseudoscalar mesons	$\pi^{\pm}$	$2.6 \cdot 10^{-8}$	$\mu^{\pm} \overset{(-)}{\nu_{\mu}}$ $e^{\pm} \overset{(-)}{\nu_{e}}$	$\approx 100\,\%$ $1.2 \cdot 10^{-4}$	(see Sect. 10.5)
	π^{0}	$8.4 \cdot 10^{-17}$	2γ	$99\,\%$	Electromagnetic
	$K^{\pm}$	$1.2 \cdot 10^{-8}$	$\mu^{\pm} \overset{(-)}{\nu_{\mu}}$ $\pi^{\pm}\pi^{0}$ 3π	$64\,\%$ $21\,\%$ $7\,\%$	
	K^{0}_{S}	$8.9 \cdot 10^{-11}$	2π	$\approx 100\,\%$	(K^{0} decay: see Sect. 14.4)
	K^{0}_{L}	$5.2 \cdot 10^{-8}$	3π $\pi\mu\nu$ $\pi e\nu$ 2π	$34\,\%$ $27\,\%$ $39\,\%$ $3 \cdot 10^{-3}$	 CP violating
	η	$5.5 \cdot 10^{-19}$	3π 2γ	$55\,\%$ $39\,\%$	Electromagnetic Electromagnetic
	η'	$3.3 \cdot 10^{-21}$	$\pi\pi\eta$ $\varrho^{0}\gamma$	$65\,\%$ $30\,\%$	 Electromagnetic
Vector mesons	ϱ	$4.3 \cdot 10^{-24}$	2π	$\approx 100\,\%$	
	K^{*}	$1.3 \cdot 10^{-23}$	$K\pi$	$\approx 100\,\%$	
	ω	$7.8 \cdot 10^{-23}$	3π	$89\,\%$	
	ϕ	$1.5 \cdot 10^{-22}$	$2K$ $\varrho\pi$	$83\,\%$ $13\,\%$	 Zweig-suppressed

can conserve parity but not isospin since pions, for reasons of symmetry, cannot couple to zero isospin. The upshot is that the η predominantly decays electromagnetically, as isospin need not then be conserved, and its lifetime is orders of magnitudes greater than those of strongly decaying particles.

The η' prefers to decay into $\pi\pi\eta$ but this rate is still broadly comparable to that of its electromagnetic decay into $\varrho\gamma$. This shows that the strong process must also be suppressed and the η' must have a fairly long lifetime. The story underlying this is a complicated one [Ne91] and will not be recounted here.

14.4 Neutral Kaon Decay

The decays of the K^0 and the $\overline{K}^0$ are of great importance for our understanding of the P- and C-parities (spatial reflection and particle-antiparticle conjugation).

Neutral kaons can decay into either two or three pions. The two pion final state must have positive parity, recall our discussion of the decay of the η, while the three pion system has negative parity. The fact that both decays are possible is a classic example of parity violation.

K^0 and $\overline{K}^0$ mixing. Since the K^0 and $\overline{K}^0$ can decay into the same final states, they can also transform into each other via an intermediate state of virtual pions [Ge55]:

$$K^0 \longleftrightarrow \left\{ \begin{array}{c} 2\pi \\ 3\pi \end{array} \right\} \longleftrightarrow \overline{K}^0 .$$

In terms of quarks this oscillation corresponds to box diagrams:

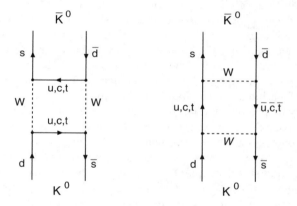

CP conservation. This possible mixing of particles and antiparticles leads to highly interesting effects. In Sect. 10.5 we said that the weak interaction violates parity maximally. This was particularly clear for the neutrino, which only occurs as a left handed particle $|\nu_L\rangle$ and a right handed antiparticle $|\overline{\nu}_R\rangle$. In K^0 decay parity violation shows itself via decays into 2 and 3 pions. For the neutrinos we further saw that the combined application of spatial reflection and charge conjugation ($\mathcal{P}$ and $\mathcal{C}$) lead to a physically allowed state: $\mathcal{CP}|\nu_L\rangle \to |\overline{\nu}_R\rangle$. The V-minus-A theory of the weak interaction may be so formulated that the combined CP quantum number is conserved.

Let us now apply this knowledge to the K^0-$\overline{K}^0$ system. The 2 and 3 pion final states are both eigenstates of the combined $\mathcal{CP}$ operator and have distinct eigenvalues

$$\begin{aligned}
\mathcal{CP}\,|\pi^0\pi^0\rangle &= +1 \cdot |\pi^0\pi^0\rangle & \mathcal{CP}\,|\pi^0\pi^0\pi^0\rangle &= -1 \cdot |\pi^0\pi^0\pi^0\rangle \\
\mathcal{CP}\,|\pi^+\pi^-\rangle &= +1 \cdot |\pi^-\pi^+\rangle & \mathcal{CP}\,|\pi^+\pi^-\pi^0\rangle &= -1 \cdot |\pi^-\pi^+\pi^0\rangle ,
\end{aligned}$$

but neither K^0 nor $\overline{K}^0$ have well-defined CP parity:

$$CP\,|K^0\rangle = -1 \cdot |\overline{K}^0\rangle \qquad CP\,|\overline{K}^0\rangle = -1 \cdot |K^0\rangle \,.$$

The relative phase between the K^0 and the $\overline{K}^0$ can be chosen arbitrarily. We have picked the convention $C|K^0\rangle = +|\overline{K}^0\rangle$ and this together with the kaon's odd parity leads to the minus sign under the CP transformation.

If we suppose that the weak force violates both the P- and C-parities but is invariant under CP then the initial kaon state has to have well-defined CP parity before its decay. Such CP eigenstates can be constructed from linear combinations in the following way:

$$|K_1^0\rangle \;=\; \frac{1}{\sqrt{2}}\left\{|K^0\rangle - |\overline{K}^0\rangle\right\} \qquad \text{where} \quad CP|K_1^0\rangle = +1 \cdot |K_1^0\rangle$$

$$|K_2^0\rangle \;=\; \frac{1}{\sqrt{2}}\left\{|K^0\rangle + |\overline{K}^0\rangle\right\} \qquad \text{where} \quad CP|K_2^0\rangle = -1 \cdot |K_2^0\rangle \,.$$

This assumption, of CP conservation, means that we have to understand the hadronic decay of a neutral kaon as the decay of either a K_1^0 into two pions or of a K_2^0 into three pions. The two decay probabilities must differ sharply from one another. The phase space available to the three pion decay is significantly smaller than for the two pion case, this follows from the rest mass of three pions being nearly that of the neutral kaon, and so the K_2^0 state ought to be much longer lived than its K_1^0 sibling.

Kaons may be produced in large numbers by colliding high energy protons onto a target. An example is the reaction $p+n \to p+\Lambda^0+K^0$. The strong force conserves strangeness S and so the neutral kaons are in an eigenstate of the strong interaction. In the case at hand it is $|K^0\rangle$ which has strangeness $S=+1$. This may be understood in quantum mechanics as a linear combination of the two CP eigenstates $|K_1^0\rangle$ and $|K_2^0\rangle$. In practice both in reactions where K^0 and in those where $\overline{K}^0$ mesons are produced an equal mixture of short and long lived particles are observed. These are called K_S^0 and K_L^0 (for *short* and *long*) respectively (Table 14.3). The short lived kaons decay into two pions and the long lived ones into three.

CP violation. After a time of flight much longer than the lifetime of the K_S^0 these shorter lived particles must have all decayed. Thus at a sufficient distance from the production target we have a pure beam of K_L^0 particles. Precision measurements have shown that these long lived kaons decay with a tiny, but non-vanishing, probability into two, instead of three, pions [Ch64, Kl92b, Gi97]. This must mean either that the K_L^0 mass eigenstate is not identical to the K_2^0 CP eigenstate or that the matrix element for the decay of the K_2^0 contains a term which permits a decay into two pions. In both cases CP symmetry is broken.

Studies of the semileptonic decay of the K_L^0

$$K_L^0 \to \pi^\pm + \mu^\mp + \overset{(-)}{\nu_\mu} \qquad K_L^0 \to \pi^\pm + e^\mp + \overset{(-)}{\nu_e}$$

reveal an asymmetry between the creation of particles and antiparticles: there is a slight preponderance of decays with positively charged leptons in the final state (the ratio is 1.0033 : 1). This is a further, albeit very tiny, case of CP violation.

CP violation has only been experimentally observed in this $K^0 \leftrightarrow \overline{K}^0$ system. It is nevertheless expected that other electrically neutral meson-antimeson systems will display similar behaviour: ($D^0 \leftrightarrow \overline{D}^0$, $B^0 \leftrightarrow \overline{B}^0$, $B_s^0 \leftrightarrow \overline{B}_s^0$). In 1987 B^0-$\overline{B}^0$ mixing was indeed discovered at DESY [Al87a, Al87b, Al92a]. CP violation in this system has, however, not yet been seen.

_____ **Problems**

1. **ϱ^0-decay**
   The ϱ^0 ($J^P = 1^-, I = 0$) almost 100% decays into $\pi^+ + \pi^-$. Why does it not also decay into $2\,\pi^0$?

2. **D^+-decay**
   $D^+(c\overline{d})$ decays into many channels. What value would you expect for the ratio:

   $$R = \frac{\Gamma(D^+ \to K^- + \pi^+ + \pi^+)}{\Gamma(D^+ \to \pi^- + \pi^+ + \pi^+)} \ . \tag{14.3}$$

3. **Pion and kaon decay**
   High energy neutrino beams can be generated using the decay of high energy, charged pions and kaons:

   $$\pi^\pm \quad \to \quad \mu^\pm + \overset{(-)}{\nu}_\mu$$
   $$K^\pm \quad \to \quad \mu^\pm + \overset{(-)}{\nu}_\mu \ .$$

   a) What fraction F of the pions and kaons in a 200 GeV beam decays inside a distance $d = 100$ m? (Use the particle masses and lifetimes given in Tables 14.2 and 14.3)
   b) How large are the minimal and maximal neutrino energies in both cases?

15. The Baryons

The best known baryons are the proton and the neutron. These are collectively referred to as the nucleons. Our study of deep inelastic scattering has taught us that they are composed of three valence quarks, gluons and a "sea" of quark-antiquark pairs. The following treatment of the baryonic spectrum will, analogously to our description of the mesons, be centred around the concept of the constituent quark.

Nomenclature. This chapter will be solely concerned with those baryons which are made up of u-, d- and s-quarks. The baryons whose valence quarks are just u- and d-quarks are the nucleons (isospin $I = 1/2$) and the Δ particles ($I = 3/2$). Baryons containing s-quarks are collectively known as *hyperons*. These particles, the Λ, Σ, Ξ and Ω, are distinguished from each other by their isospin and the number of s-quarks they contain.

Name	N	Δ	Λ	Σ	Ξ	Ω
Isospin I	1/2	3/2	0	1	1/2	0
Strangeness S	0		-1		-2	-3
Number of s-quarks	0		1		2	3

The antihyperons have strangeness $+1$, $+2$ or $+3$ respectively.

The discovery of baryons containing c- and b-quarks has caused this scheme to be extended. The presence of quarks heavier than the s is signified by an subscript attached to the relevant hyperon symbol: thus the Λ_c^+ corresponds to a (udc) state and the Ξ_{cc}^{++} has the valence structure (ucc). Such heavy baryons will not, however, be handled in what follows.

15.1 The Production and Detection of Baryons

Formation experiments. Baryons can be produced in many different ways in accelerators. In Sect. 7.1 we have already described how nucleon resonances may be produced in inelastic electron scattering. These excited nucleon states are also created when pions are scattered off protons.

One can then study, for example, the energy (mass) and width (lifetime) of the Δ^{++} resonance in the reaction

$$\pi^+ + p \rightarrow \Delta^{++} \rightarrow p + \pi^+$$

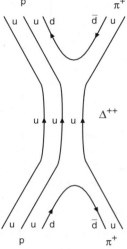

by varying the energy of the incoming pion beam and measuring the total cross section. The largest and lowest energy peak in the cross section is found at 1232 MeV. This is known as the $\Delta^{++}(1232)$. The diagram shows its creation and decay in terms of quark lines. In simple terms we may say that the energy which is released in the quark-antiquark annihilation is converted into the excitation energy of the resonance and that this process is reversed in the decay of the resonance to form a new quark-antiquark pair. This short lived state decays about $0.5 \cdot 10^{-23}$ s after it is formed and it is thus only possible to detect the decay products, i.e., the proton and the π^+. Their angular distribution, however, may be used to determine the resonances's spin and parity. The result is found to be $J^P = 3/2^+$. The extremely short lifetime attests to the decay taking place through the strong interaction. At higher centre of mass energies in this reaction further resonances may be seen in the cross section. These correspond to excited Δ^{++} states where the quarks occupy higher energy levels. Strangeness may be brought into the game by replacing the pion beam by a kaon beam and one may thus generate hyperons. A possible reaction is

$$K^- + p \rightarrow \Sigma^{*0} \rightarrow p + K^- .$$

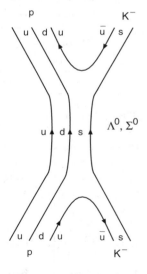

The intermediate resonance state, an excited state of the Σ^0, is, like the Δ^{++}, extremely short lived and "immediately" decays, primarily back into a proton and a negatively charged kaon. The quark line diagram offers a general description of all those resonances whose quark composition is such that they may be produced in this process. Thus excited Λ^0's may also be created in the above reaction. The cross sections of the above reactions are displayed in Fig. 15.1 as functions of the centre of mass energy. The resonance structures may be easily recogni-
sed. The individual peaks, which give us the masses of the excited baryon states, are generally difficult
to separate from each other. This is because their widths are typically of the order of 100 MeV and the various peaks hence overlap. Such large widths are characteristic for particles which decay via strong processes.

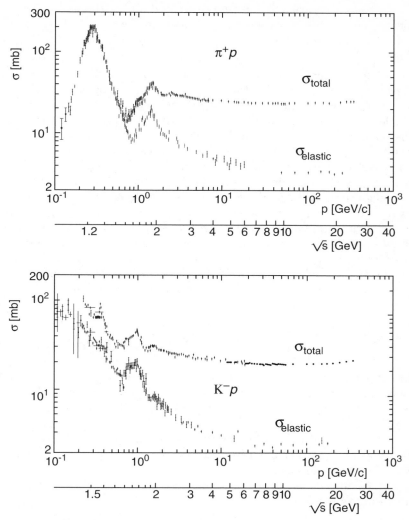

Fig. 15.1. The total and elastic cross sections for the scattering of π^+ mesons off protons (*top*) and of K^- mesons off protons (*bottom*) as a function of the mesonic beam energy (or centre of mass energy) [PD98]. The peaks are associated with short lived states, and since the total initial charge in π^+p scattering is $+2e$ the relevant peaks must correspond to the Δ^{++} particle. The strongest peak, at a beam energy of around 300 MeV/c is due to the ground state of the Δ^{++} which has a mass of 1232 MeV/c^2. The resonances that show up as peaks in the K^-p cross section are excited, neutral Σ and Λ baryons. The most prominent peaks are the excited $\Sigma^0(1775)$ und $\Lambda^0(1820)$ states which overlap significantly.

In *formation experiments*, like those treated above, the baryon which is formed is detected as a resonance in a cross section. Due to the limited number of particle beams available to us this method may only be used to generate nucleons and their excited states or those hyperons with strangeness $S = -1$.

Production experiments. A more general way of generating baryons is in *production experiments*. In these one fires a beam of protons, pions or kaons with as high an energy as possible at a target. The limit on the energy available for the production of new particles is the centre of mass energy of the scattering process. As can be seen from Fig. 15.1, for centre of mass energies greater than 3 GeV no further resonances can be recognised and the elastic cross section is thereafter only a minor part of the total cross section. This energy range is dominated by inelastic particle production.

In such production experiments one does not look for resonances in the cross section but rather studies the particles which are created, generally in generous quantities, in the reactions. If these particles are short lived, then it is only possible to actually detect their decay products. The short lived states can, however, often be reconstructed by the invariant mass method. If the momenta p_i and energies E_i of the various products can be measured, then we may use the fact that the mass M_X of the decayed particle X is given by

$$M_X^2 c^4 = p_X^2 c^2 = \left(\sum_i p_i c \right)^2 = \left(\sum_i E_i \right)^2 - \left(\sum_i \boldsymbol{p}_i c \right)^2 . \tag{15.1}$$

In practice one studies a great number of scattering events and calculates the invariant mass of some particular combination of the particles which have been detected. Short lived resonances which have decayed into these particles reveal themselves as peaks in the invariant mass spectrum. On the one hand we may identify short lived resonances that we already knew about in this way, on the other hand we can thus see if new, previously unknown particles are being formed.

As an example consider the invariant mass spectrum of the $\Lambda^0 + \pi^+$ final particles in the reaction

$$K^- + p \to \pi^+ + \pi^- + \Lambda^0 .$$

This displays a clear peak at 1385 MeV/c^2 (Fig. 15.2) which corresponds to an excited Σ^+. The Σ^{*+} baryon is therefore identified from its decay into $\Sigma^{*+} \to \pi^+ + \Lambda^0$. Since this is a strong decay all quantum numbers, e.g., strangeness and isospin, are conserved. In the above reaction it is just as likely to be the case that a Σ^{*-} state is produced. This would then decay into $\Lambda^0 + \pi^-$. Study of the invariant masses yields almost identical masses for these two baryons.[1] This may also be read off from Fig. 15.2. The somewhat

---

[1] The mass difference between the Σ^{*-} and the Σ^{*+} is roughly 4 MeV/c^2 (see Table 15.1 on p. 208).

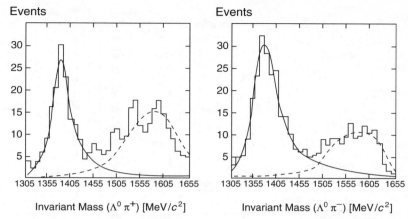

Fig. 15.2. Invariant mass spectrum of the particle combinations $\Lambda^0 + \pi^+$ (*left*) and $\Lambda^0 + \pi^-$ (*right*) in the reaction $K^- + p \rightarrow \pi^+ + \pi^- + \Lambda^0$. The momentum of the initial kaon was $1.11\,\text{GeV}/c$. The events were recorded in a bubble chamber. Both spectra display a peak around $1385\,\text{MeV}/c^2$, which correspond to Σ^{*+} and Σ^{*-} respectively. A Breit-Wigner distribution (*continuous line*) has been fitted to the peak. The mass and width of the resonance may be found in this way. The energy of the pion which is not involved in the decay is kinematically fixed for any particular beam energy. Its combination together with the Λ^0 yields a "false" peak at higher energies which does not correspond to a resonance (from [El61]).

flatter peak at higher energies visible in both spectra is a consequence of the possibility to create either of these two charged Σ resonances: the momentum and energy of the pion which is not created in the decay is fixed and so creates a „fictitious" peak in the invariant mass spectrum. This ambiguity can be resolved by carrying out the experiment at differing beam energies. There is a further small background in the invariant mass spectrum which is not correlated with the above, i.e., it does not come from $\Sigma^{*\pm}$ decay. We note that the excited Σ state was first found in 1960 using the invariant mass method [Al60].

If the baryonic state that we wish to investigate is already known, then the resonance may be investigated in individual events as well. This is, for example, important for the above identification of the Σ^{*+}, since the Λ^0 itself decays via $\Lambda^0 \rightarrow p + \pi^-$ and must first be reconstructed by the invariant mass method. The detection of the Λ^0 is rendered easier by its long lifetime of $2.6 \cdot 10^{-10}\,\text{s}$ (due to its weak decay). On average the Λ^0 transverses a distance from several centimetres to a few metres, this depends upon its energy, before it decays. From the tracks of its decay products the position of the Λ^0's decay may be localised and distinguished from that of the primary reaction.

A nice example of such a step by step reconstruction of the initially created, primary particles from a Σ^- +nucleus reaction is shown in Fig. 15.3. The method of invariant masses could be used to show a three step process of

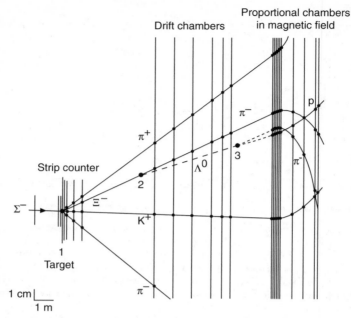

Fig. 15.3. Detection of a baryon decay cascade at the WA89 detector at the CERN hyperon beam (based upon [Tr92]). In this event a Σ^- hyperon with 370 GeV kinetic energy hits a thin carbon target. The paths of the charged particles thus produced are detected near to the target by silicon strip detectors and further away by drift and proportional chambers. Their momenta are determined by measuring the deflection of the tracks in a strong magnetic field. The tracks marked in the figure are based upon the signals from the various detectors. The baryonic decay chain is described in the text.

baryon decays. The measured reaction is

$$\Sigma^- + A \rightarrow p + K^+ + \pi^+ + \pi^- + \pi^- + \pi^- + A'.$$

The initial reaction takes place at one of the protons of a nucleus A. All of the particles in the final state were identified (except for the final nucleus A') and their momenta were measured. The tracks of a proton and a π^- could be measured in drift and proportional chambers and followed back to the point (3), where a Λ^0 decayed (as a calculation of the invariant mass of the proton and the π^- shows). Since we thus have the momentum of the Λ^0 we can extrapolate its path back to (2) where it meets the path of a π^-. The invariant mass of the Λ^0 and of this π^- is roughly 1320 MeV/c^2 which is the mass of the Ξ^- baryon. This baryon can in turn be traced to the target at (1). The analysis then shows that the Ξ^- was in fact the decay product of a primary Ξ^{*0} state which "instantaneously" decayed via the strong interaction into a Ξ^- and a π^+. The complete reaction in all its glory was therefore the following

$$\Sigma^- + A \;\;\rightarrow\;\; \Xi^{*0} + K^+ + \pi^- + A'$$
$$\hookrightarrow \Xi^- + \pi^+$$
$$\hookrightarrow \Lambda^0 + \pi^-$$
$$\hookrightarrow p + \pi^- \;.$$

This reaction also exemplifies the associated production of strange particles: the Σ^- from the beam had strangeness -1 and yet produces in the collision with the target a Ξ^{*0} with strangeness -2. Since the strange quantum number is conserved in strong interactions an additional K^+ with strangeness $+1$ was also created.

15.2 Baryon Multiplets

We now want to describe in somewhat more detail which baryons may be built up from the u-, d- and s-quarks. We will though limit ourselves to the lightest states, i.e., those where the quarks have relative orbital angular momentum $\ell = 0$ and are not radially excited.

The three valence quarks in the baryon must, by virtue of their fermionic character, satisfy the Pauli principle. The total baryonic wavefunction

$$\psi_{\text{total}} \;=\; \xi_{\text{spatial}} \cdot \zeta_{\text{flavour}} \cdot \chi_{\text{spin}} \cdot \phi_{\text{colour}}$$

must in other words be antisymmetric under the exchange of any two of the quarks. The total baryonic spin S results from adding the three individual quark spins ($s = 1/2$) and must be either $S = 1/2$ or $S = 3/2$. Since we demand that $\ell = 0$, the total angular momentum J of the baryon is just the total spin of the three quarks.

The baryon decuplet. Let us first investigate the $J^P = 3/2^+$ baryons. Here the three quarks have parallel spins and the spin wave function is therefore symmetric under an interchange of two of the quarks. For $\ell = 0$ states this is also true of the spatial wave function. Taking, for example, the uuu state it is obvious that the flavour wave function has to be symmetric and this then implies that the colour wave function must be totally antisymmetric in order to yield an antisymmetric total wave function and so fulfill the Pauli principle. Because baryons are colourless objects the totally antisymmetric colour wave function can be constructed as follows:

$$\phi_{\text{colour}} = \frac{1}{\sqrt{6}} \sum_{\alpha=r,g,b} \; \sum_{\beta=r,g,b} \; \sum_{\gamma=r,g,b} \varepsilon_{\alpha\beta\gamma} \, |q_\alpha q_\beta q_\gamma\rangle \;, \qquad (15.2)$$

where we sum over the three colours, here denoted by *red, green* and *blue*, and $\varepsilon_{\alpha\beta\gamma}$ is the totally antisymmetric tensor.

If we do not concern ourselves with radial excitations, we are left with ten different systems that can be built out of three quarks, are $J^P = 3/2^+$ and have totally antisymmetric wave functions. These are

$$|\Delta^{++}\rangle = |u^\uparrow u^\uparrow u^\uparrow\rangle \quad |\Delta^+\rangle = |u^\uparrow u^\uparrow d^\uparrow\rangle \quad |\Delta^0\rangle = |u^\uparrow d^\uparrow d^\uparrow\rangle \quad |\Delta^-\rangle = |d^\uparrow d^\uparrow d^\uparrow\rangle$$

$$|\Sigma^{*+}\rangle = |u^\uparrow u^\uparrow s^\uparrow\rangle \quad |\Sigma^{*0}\rangle = |u^\uparrow d^\uparrow s^\uparrow\rangle \quad |\Sigma^{*-}\rangle = |d^\uparrow d^\uparrow s^\uparrow\rangle$$

$$|\Xi^{*0}\rangle = |u^\uparrow s^\uparrow s^\uparrow\rangle \quad |\Xi^{*-}\rangle = |d^\uparrow s^\uparrow s^\uparrow\rangle$$

$$|\Omega^-\rangle = |s^\uparrow s^\uparrow s^\uparrow\rangle \,.$$

Note that we have only given the spin-flavour part of the total baryonic wave function here, and that in an abbreviated fashion. It must be symmetric under quark exchange. In the above notation this is evident for the pure uuu, ddd and sss systems. For baryons built out of more than one quark flavour the symmetrised version contains several terms. Thus then the symmetrised part of the wave function of, for example, the Δ^+ reads more fully:

$$|\Delta^+\rangle = \frac{1}{\sqrt{3}} \left\{ |u^\uparrow u^\uparrow d^\uparrow\rangle + |u^\uparrow d^\uparrow u^\uparrow\rangle + |d^\uparrow u^\uparrow u^\uparrow\rangle \right\} \,.$$

In what follows we will mostly employ the abbreviated notation for the baryonic quark wave function and quietly assume that the total wave function has in fact been correctly antisymmetrised.

If we display the states of this baryon decuplet on an I_3 vs. S plot, we obtain (Fig. 15.4) an isosceles triangle. This reflects the threefold symmetry of these three-quark systems.

The baryon octet. We are now faced with the question of bringing the nucleons into our model of the baryons. If three quarks, each with spin 1/2, are to yield a spin 1/2 baryon, then the spin of one of the quarks must be antiparallel to the other two, i.e., we must have ↑↑↓. This spin state is then neither symmetric nor antisymmetric under spin swaps, but rather has a mixed symmetry. This must then also be the case for the flavour wave function, so that their product, the total spin-flavour wave function, is purely symmetric. This is not possible for the uuu, ddd and sss quark combinations and indeed we do not find any ground state baryons of this form with $J = 1/2$. There are then only two different possible combinations of u and d quarks which can fulfill the necessary symmetry conditions on the wave function of a spin 1/2 baryon, and these are just the proton and the neutron.

This simplified treatment of the derivation of the possible baryonic states and their multiplets can be put on a firmer quantitative footing with the help of SU(6) quark symmetry, we refer here to the literature (see, e.g., [Cl79]).

The proton and neutron wave functions may be schematically written as

$$|p^\uparrow\rangle = |u^\uparrow u^\uparrow d^\downarrow\rangle \qquad |n^\uparrow\rangle = |u^\downarrow d^\uparrow d^\uparrow\rangle\,.$$

We now want to construct the symmetrised wave function. For a proton with, e.g., the z spin component $m_J = +1/2$, we may write the spin wave function as a product of the the spin wave function of one quark and that of the remaining pair:

$$\chi_\mathrm{p}(J=\tfrac{1}{2}, m_J=\tfrac{1}{2}) = \sqrt{2/3}\,\chi_\mathrm{uu}(1,1)\chi_\mathrm{d}(\tfrac{1}{2},-\tfrac{1}{2}) - \sqrt{1/3}\,\chi_\mathrm{uu}(1,0)\chi_\mathrm{d}(\tfrac{1}{2},\tfrac{1}{2})\,. \tag{15.3}$$

Here we have chosen to single out the d-quark and coupled the u-quark pair. (If we initially single out one of the u-quarks we obtain the same result, but the notation becomes much more complicated.) The factors in this equation are the Clebsch-Gordan coefficients for the coupling of spin 1 and spin 1/2. Replacing $\chi(1,0)$ by the correct spin triplet wave function $(\uparrow\downarrow + \downarrow\uparrow)/\sqrt{2}$ then yields in our spin-flavour notation

$$|p^\uparrow\rangle = \sqrt{2/3}\,|u^\uparrow u^\uparrow d^\downarrow\rangle - \sqrt{1/6}\,|u^\uparrow u^\downarrow d^\uparrow\rangle - \sqrt{1/6}\,|u^\downarrow u^\uparrow d^\uparrow\rangle\,. \tag{15.4}$$

This expression is still only symmetric in terms of the exchange of the first and second quarks, and not for two arbitrary quarks as we need. It can, however, be straightforwardly totally symmetrised by swapping the first and third as well as the second and third quarks in each term of this last equation and adding these new terms. With the correct normalisation factor the totally symmetric proton wave function is then

$$|p^\uparrow\rangle = \frac{1}{\sqrt{18}} \left\{\, 2|u^\uparrow u^\uparrow d^\downarrow\rangle + 2|u^\uparrow d^\downarrow u^\uparrow\rangle + 2|d^\downarrow u^\uparrow u^\uparrow\rangle - |u^\uparrow u^\downarrow d^\uparrow\rangle \right.$$
$$\left. -|u^\uparrow d^\uparrow u^\downarrow\rangle - |d^\uparrow u^\uparrow u^\downarrow\rangle - |u^\downarrow u^\uparrow d^\uparrow\rangle - |u^\downarrow d^\uparrow u^\uparrow\rangle - |d^\uparrow u^\downarrow u^\uparrow\rangle \,\right\}\,. \tag{15.5}$$

The neutron wave function is trivially found by exchanging the u- and d-quarks:

$$|n^\uparrow\rangle = \frac{1}{\sqrt{18}} \left\{\, 2|d^\uparrow d^\uparrow u^\downarrow\rangle + 2|d^\uparrow u^\downarrow d^\uparrow\rangle + 2|u^\downarrow d^\uparrow d^\uparrow\rangle - |d^\uparrow d^\downarrow u^\uparrow\rangle \right.$$
$$\left. -|d^\uparrow u^\uparrow d^\downarrow\rangle - |u^\uparrow d^\uparrow d^\downarrow\rangle - |d^\downarrow d^\uparrow u^\uparrow\rangle - |d^\downarrow u^\uparrow d^\uparrow\rangle - |u^\uparrow d^\downarrow d^\uparrow\rangle \,\right\}\,. \tag{15.6}$$

The nucleons have isospin 1/2 and so form an isospin doublet. A further doublet may be produced by combining two s-quarks with a light quark. This is schematically given by

$$|\Xi^{0\uparrow}\rangle = |u^\downarrow s^\uparrow s^\uparrow\rangle \qquad |\Xi^{-\uparrow}\rangle = |d^\downarrow s^\uparrow s^\uparrow\rangle\,. \tag{15.7}$$

The remaining quark combinations are an isospin triplet and a singlet:

$$|\Sigma^{+\uparrow}\rangle = |u^\uparrow u^\uparrow s^\downarrow\rangle$$
$$|\Sigma^{0\uparrow}\rangle = |u^\uparrow d^\uparrow s^\downarrow\rangle \qquad |\Lambda^{0\uparrow}\rangle = |u^\uparrow d^\downarrow s^\uparrow\rangle \tag{15.8}$$
$$|\Sigma^{-\uparrow}\rangle = |d^\uparrow d^\uparrow s^\downarrow\rangle\,.$$

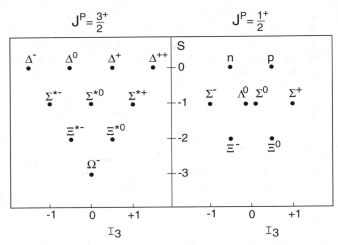

Fig. 15.4. The baryon $J^P = 3/2^+$ decuplet (*left*) and the $J^P = 1/2^+$ octet (*right*) in I_3 vs. S plots. In contradistinction to the mesonic case the baryon multiplets are solely composed of quarks. Antibaryons are purely composed of antiquarks and so form their own, equivalent antibaryon multiplets.

Note that the uds quark combination appears twice here and depending upon the relative quark spins and isospins can correspond to two different particles. If the u and d spins and isospins couple to 1, as they do for the charged Σ baryons, then the above quark combination is a Σ^0. If they couple to zero we are dealing with a Λ^0. These two hyperons have a mass difference of about 80 MeV/c^2. This is evidence that a spin-spin interaction must also play an important role in the physics of the baryon spectrum. The eight $J^P = 1/2^+$ baryons are displayed in an I_3 vs. S plot in Fig. 15.4. Note again the threefold symmetry of the states.

15.3 Baryon Masses

The mass spectrum of the baryons is plotted in Fig. 15.5 against strangeness and isospin. The lowest energy levels are the $J^P = 1/2^+$ and $J^P = 3/2^+$ multiplets, as can be clearly seen. It is also evident that the baryon masses increase with the number of strange quarks, which we can put down to the larger mass of the s-quark. Furthermore we can see that the $J^P = 3/2^+$ baryons are about 300 MeV/c^2 heavier than their $J^P = 1/2^+$ equivalents. As was the case with the mesons, this effect can be traced back to a spin-spin interaction

$$V_{\rm ss}({\rm q}_i{\rm q}_j) = \frac{4\pi}{9} \frac{\hbar^3}{c} \alpha_{\rm s} \frac{\boldsymbol{\sigma}_i \cdot \boldsymbol{\sigma}_j}{m_i m_j} \delta(\boldsymbol{x}) \,, \tag{15.9}$$

which is only important at short distances. The observant reader may notice that the 4/9 factor is only half that which we found for the quark-

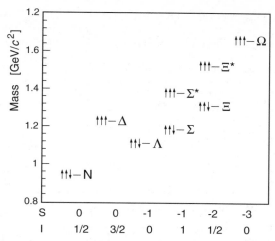

Fig. 15.5. The masses of the decuplet and octet baryons plotted against their strangeness S and isospin I. The angular momenta J of the various baryons are shown through arrows. The $J^P = 3/2^+$ decuplet baryons lie significantly above their $J^P = 1/2^+$ octet partners.

antiquark potential in the mesons (13.10), this is a result of QCD considerations. Eq. (15.9), it should be noted, describes only the interaction of two quarks with each other and so to describe the baryon mass splitting we need to sum the spin-spin interactions over all quark pairs. The easiest cases are those like the nucleons, the Δ's and the Ω where the constituent masses of all three quarks are the same. Then we just have to calculate the expectation values for the sums over $\boldsymbol{\sigma}_i \cdot \boldsymbol{\sigma}_j$. Denoting the total baryon spin by $\boldsymbol{S}$ and using the identity $\boldsymbol{S}^2 = (\boldsymbol{s}_1 + \boldsymbol{s}_2 + \boldsymbol{s}_3)^2$ we find in a similar way to (13.11):

$$\sum_{\substack{i,j=1\\i<j}}^{3} \boldsymbol{\sigma}_i \cdot \boldsymbol{\sigma}_j = \frac{4}{\hbar^2} \sum_{\substack{i,j=1\\i<j}}^{3} \boldsymbol{s}_i \cdot \boldsymbol{s}_j = \begin{cases} -3 & \text{for} \quad S = 1/2\,, \\ +3 & \text{for} \quad S = 3/2\,. \end{cases} \qquad (15.10)$$

The spin-spin energy (mass) splitting for these baryons is then just

$$\Delta M_{\mathrm{ss}} = \begin{cases} -3 \cdot \dfrac{4\,\hbar^3}{9\,c^3} \dfrac{\pi\alpha_{\mathrm{s}}}{m_{\mathrm{u,d}}^2} |\psi(0)|^2 & \text{for the nucleons}\,, \\[2ex] +3 \cdot \dfrac{4\,\hbar^3}{9\,c^3} \dfrac{\pi\alpha_{\mathrm{s}}}{m_{\mathrm{u,d}}^2} |\psi(0)|^2 & \text{for the } \Delta \text{ states}\,, \\[2ex] +3 \cdot \dfrac{4\,\hbar^3}{9\,c^3} \dfrac{\pi\alpha_{\mathrm{s}}}{m_{\mathrm{s}}^2} |\psi(0)|^2 & \text{for the } \Omega \text{ baryon}\,. \end{cases} \qquad (15.11)$$

Here $|\psi(0)|^2$ is the probability that two quarks are at the same place. Somewhat more complicated expressions may be obtained for those baryons made up of a mixture of heavier s- and lighter u- or d-quarks (see the exercises).

Table 15.1. The masses of the lightest baryons both from experiment and as fitted from (15.12). The fits were to the average values of the various multiplets and are in good agreement with the measured masses. Also included in this table are the lifetimes and most important decay channels of these baryons [PD98]. The four charged Δ resonances are not individually listed.

	S	I	Baryon	Mass [MeV/c^2] theor.	exp.	τ [s]	Primary decay channels		Decay type
Octet ($J^P = 1/2^+$)	0	1/2	p	939	938.3	stable?	—		—
			n		939.6	886.7	$\mathrm{pe^-\bar{\nu}_e}$	100 %	weak
	−1	0	Λ	1114	1115.7	$2.63 \cdot 10^{-10}$	$\mathrm{p\pi^-}$	64.1 %	weak
							$\mathrm{n\pi^0}$	35.7 %	weak
		1	Σ^+	1179	1189.4	$0.80 \cdot 10^{-10}$	$\mathrm{p\pi^0}$	51.6 %	weak
							$\mathrm{n\pi^+}$	48.3 %	weak
			Σ^0		1192.6	$7.4 \cdot 10^{-20}$	$\Lambda\gamma$	≈ 100 %	elmgn.
			Σ^-		1197.4	$1.48 \cdot 10^{-10}$	$\mathrm{n\pi^-}$	99.8 %	weak
	−2	1/2	Ξ^0	1327	1315	$2.90 \cdot 10^{-10}$	$\Lambda\pi^0$	≈ 100 %	weak
			Ξ^-		1321	$1.64 \cdot 10^{-10}$	$\Lambda\pi^-$	≈ 100 %	weak
Decuplet ($J^P = 3/2^+$)	0	3/2	Δ	1239	1232	$0.55 \cdot 10^{-23}$	$\mathrm{N}\pi$	99.4 %	strong
	−1	1	Σ^{*+}	1381	1383	$1.7 \cdot 10^{-23}$	$\Lambda\pi$	88 %	strong
			Σ^{*0}		1384		$\Sigma\pi$	12 %	strong
			Σ^{*-}		1387				
	−2	1/2	Ξ^{*0}	1529	1532	$7 \cdot 10^{-23}$	$\Xi\pi$	≈ 100 %	strong
			Ξ^{*-}		1535				
	−3	0	Ω^-	1682	1672.4	$0.82 \cdot 10^{-10}$	$\Lambda \mathrm{K}^-$	68 %	weak
							$\Xi^0\pi^-$	23 %	weak
							$\Xi^-\pi^0$	9 %	weak

With the help of this mass splitting formula a general expression for the masses of all the $\ell = 0$ baryons may be written:

$$M = \sum_i m_i + \Delta M_{\mathrm{ss}} \,. \tag{15.12}$$

The three unknowns here, i.e., $m_{\mathrm{u,d}}$, m_{s} and $\alpha_{\mathrm{s}}|\psi(0)|^2$, may be obtained by fitting to the experimental masses. As with the mesons we assume that $\alpha_{\mathrm{s}}|\psi(0)|^2$ is roughly the same for all of the baryons. We so obtain the following constituent quark masses: $m_{\mathrm{u,d}} \approx 363\,\mathrm{MeV}/c^2$, $m_{\mathrm{s}} \approx 538\,\mathrm{MeV}/c^2$ [Ga81]. The fitted baryon masses are within 1 % of their true values. (Table 15.1). The constituent quark masses obtained from such studies of baryons are a little larger than their mesonic counterparts. This is not necessarily a contradiction since constituent quark masses are generated by the dynamics of the quark-gluon interaction and the effective interactions of a three-quark system will not be identical to those of a quark-antiquark one.

15.4 Magnetic Moments

The constituent quark model is satisfyingly successfull when its predictions for baryonic magnetic moments are compared with the results of experiment. In Dirac theory the magnetic moment μ of a point particle with mass M and spin $1/2$ is

$$\mu_{\mathrm{Dirac}} = \frac{e\hbar}{2M}. \tag{15.13}$$

This relationship has been experimentally confirmed for both the electron and the muon. If the proton were an elementary particle without any substructure, then its magnetic moment should be one nuclear magneton:

$$\mu_{\mathrm{N}} = \frac{e\hbar}{2M_{\mathrm{p}}}. \tag{15.14}$$

Experimentally, however, the magnetic moment of the proton is measured to be $\mu_{\mathrm{p}} = 2.79\,\mu_{\mathrm{N}}$.

Magnetic moments in the quark model. The proton magnetic moment in the ground state, with $\ell = 0$, is a simple vectorial sum of the magnetic moments of the three quarks:

$$\boldsymbol{\mu}_{\mathrm{p}} = \boldsymbol{\mu}_{\mathrm{u}} + \boldsymbol{\mu}_{\mathrm{u}} + \boldsymbol{\mu}_{\mathrm{d}}. \tag{15.15}$$

The proton magnetic moment μ_{p} then has the expectation value

$$\mu_{\mathrm{p}} = \langle \boldsymbol{\mu}_{\mathrm{p}} \rangle = \langle \psi_{\mathrm{p}} | \boldsymbol{\mu}_{\mathrm{p}} | \psi_{\mathrm{p}} \rangle, \tag{15.16}$$

where ψ_{p} is the total antisymmetric quark wave function of the proton. To obtain μ_{p} we merely require the spin part of the wave function, χ_{p}. From (15.3) we thus deduce

$$\mu_{\mathrm{p}} = \frac{2}{3}(\mu_{\mathrm{u}} + \mu_{\mathrm{u}} - \mu_{\mathrm{d}}) + \frac{1}{3}\mu_{\mathrm{d}} = \frac{4}{3}\mu_{\mathrm{u}} - \frac{1}{3}\mu_{\mathrm{d}}, \tag{15.17}$$

where $\mu_{\mathrm{u,d}}$ are the quark magnetons:

$$\mu_{\mathrm{u,d}} = \frac{z_{\mathrm{u,d}}\,e\hbar}{2m_{\mathrm{u,d}}}. \tag{15.18}$$

The other $J^P = 1/2^+$ baryons with two identical quarks may be described by (15.17) with a suitable change of quark flavours. The neutron, for example, has a magnetic moment

$$\mu_n = \frac{4}{3}\mu_d - \frac{1}{3}\mu_u \tag{15.19}$$

and analogously for the Σ^+ we have

$$\mu_{\Sigma^+} = \frac{4}{3}\mu_u - \frac{1}{3}\mu_s \ . \tag{15.20}$$

The situation is a little different for the Λ^0. As we know this hyperon contains a u- and a d-quark whose spins are coupled to 0 and so contribute neither to the spin nor to the magnetic moment of the baryon (Sect. 15.2). Hence both the spin and the magnetic moment of the Λ^0 are determined solely by the s-quark:

$$\mu_\Lambda = \mu_s \ . \tag{15.21}$$

To the extent that the u and d constituent quark masses can be set equal to each other we have $\mu_u = -2\mu_d$ and may then write the proton and neutron magnetic moments as follows

$$\mu_p = \frac{3}{2}\mu_u \ , \qquad \mu_n = -\mu_u \ . \tag{15.22}$$

We thus obtain the following prediction for their ratio

$$\frac{\mu_n}{\mu_p} = -\frac{2}{3} \ , \tag{15.23}$$

which is in excellent agreement with the experimental result of -0.685.

The absolute magnetic moments can only be calculated if we can specify the quark masses. Let us first, however, look at this problem the other way round and use the measured value of μ_p to determine the quark masses. From

$$\mu_p = 2.79\,\mu_N = 2.79\frac{e\hbar}{2M_p} \tag{15.24}$$

and

$$\mu_p = \frac{3}{2}\mu_u = \frac{e\hbar}{2m_u} \tag{15.25}$$

we obtain

$$m_u = \frac{M_p}{2.79} = 336\ \text{MeV}/c^2 \ , \tag{15.26}$$

which is very close indeed to the mass we found in Sect. 15.3 from the study of the baryon spectrum.

Measuring the magnetic moments. The agreement between the experimental values of the hyperon magnetic moments with the predictions of the quark model is impressive (Table 15.2). Our ability to measure the magnetic moments of many of the short lived hyperons ($\tau \approx 10^{-10}$ s) is due to a combination of two circumstances: hyperons produced in nucleon-nucleon interactions are polarised and the weak interaction violates parity maximally.

Table 15.2. Experimental and theoretical values of the baryon magnetic moments [La91, PD98]. The measured values of the p, n and Λ^0 moments are used to predict those of the other baryons. The Σ^0 hyperon has a very short lifetime ($7.4 \cdot 10^{-20}$ s) and decays electromagnetically via $\Sigma^0 \to \Lambda^0 + \gamma$. For this particle the transition matrix element $\langle \Lambda^0|\mu|\Sigma^0\rangle$ is given in place of its magnetic moment.

Baryon	μ/μ_{N} (Experiment)	Quark model:	μ/μ_{N}
p	$+2.792\,847\,386 \pm 0.000\,000\,063$	$(4\mu_{\mathrm{u}} - \mu_{\mathrm{d}})/3$	—
n	$-1.913\,042\,75 \pm 0.000\,000\,45$	$(4\mu_{\mathrm{d}} - \mu_{\mathrm{u}})/3$	—
Λ^0	$-0.613 \qquad \pm 0.004$	μ_{s}	—
Σ^+	$+2.458 \qquad \pm 0.010$	$(4\mu_{\mathrm{u}} - \mu_{\mathrm{s}})/3$	$+2.67$
Σ^0		$(2\mu_{\mathrm{u}} + 2\mu_{\mathrm{d}} - \mu_{\mathrm{s}})/3$	$+0.79$
$\Sigma^0 \to \Lambda^0$	$-1.61 \qquad \pm 0.08$	$(\mu_{\mathrm{d}} - \mu_{\mathrm{u}})/\sqrt{3}$	-1.63
Σ^-	$-1.160 \qquad \pm 0.025$	$(4\mu_{\mathrm{d}} - \mu_{\mathrm{s}})/3$	-1.09
Ξ^0	$-1.250 \qquad \pm 0.014$	$(4\mu_{\mathrm{s}} - \mu_{\mathrm{u}})/3$	-1.43
Ξ^-	$-0.650\,7 \qquad \pm 0.002\,5$	$(4\mu_{\mathrm{s}} - \mu_{\mathrm{d}})/3$	-0.49
Ω^-	$-2.02 \qquad \pm 0.05$	$3\mu_{\mathrm{s}}$	-1.84

In consequence the angular distributions of their decay products are strongly dependent upon the direction of the hyperons' spins (i.e., their polarisations).

Let us clarify these remarks by studying how the magnetic moment of the Λ^0 is experimentally measured. Note that this is the most easily determined of the hyperon magnetic moments. The decay

$$\Lambda^0 \to p + \pi^-$$

is rather simple to identify and has the largest branching ratio (64 %). If the Λ^0 spin is, say, in the positive $\hat{z}$ direction, then the proton will most likely be emitted in the negative $\hat{z}$ direction, in accord with the angular distribution

$$W(\theta) \propto 1 + \alpha\cos\theta \qquad \text{where} \quad \alpha \approx 0.64\,. \qquad (15.27)$$

The angle θ is the angle between the spin of the Λ^0 and the momentum of the proton. The parameter α depends upon the strength of the interference of those terms with orbital angular momentum $\ell = 0$ and $\ell = 1$ in the p-π^- system and its size must be determined by experiment.

The asymmetry in the emitted protons then fixes the Λ^0 polarisation. Highly polarised Λ^0 particles may be obtained from the reaction

$$p + p \to K^+ + \Lambda^0 + p\,.$$

As shown in Fig. 15.6, the spin of the Λ^0 is perpendicular to the production plane defined by the path of the incoming proton and that of the Λ^0 itself.

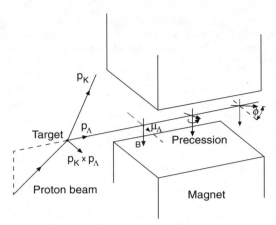

Fig. 15.6. Sketch of the measurement of the magnetic moment of the Λ^0. The hyperon is generated by the interaction of a proton coming in from the left with a proton in the target. The spin of the Λ^0 is, for reasons of parity conservation, perpendicular to the production plane. The Λ^0 then passes through a magnetic field which is orthogonal to the particle's spin. After traversing a distance d in the magnetic field the spin has precessed through an angle ϕ.

This is because only this polarisation direction conserves parity, which is conserved in the strong interaction.

If the Λ^0 baryon traverses a distance d in a magnetic field $\boldsymbol{B}$, where the field is perpendicular to the hyperon's spin, then its spin precesses with the Larmor frequency

$$\omega_{\mathrm{L}} = \frac{\mu_\Lambda \boldsymbol{B}}{\hbar} \qquad (15.28)$$

through the angle

$$\phi = \omega_{\mathrm{L}} \Delta t = \omega_{\mathrm{L}} \frac{d}{v} , \qquad (15.29)$$

where v is the speed of the Λ^0 (this may be reconstructed by measuring the momenta of its decay products, i.e., a proton and a pion). The most accurate results may be obtained by reversing the magnetic field and measuring the angle $2 \cdot \phi$ which is given by the difference between the directions of the Λ^0 spins (after crossing the various magnetic fields). This trick neatly eliminates most of the systematic errors. The magnetic moment is thus found to be [PD94]

$$\mu_\Lambda = (-0.613 \pm 0.004)\, \mu_{\mathrm{N}} . \qquad (15.30)$$

If we suppose that the s-constituent quark is a Dirac particle and that its magnetic moment obeys (15.18), then we see that this result for μ_Λ is consistent with a strange quark mass of 510 MeV/c^2.

The magnetic moments of many of the hyperons have been measured in a similar fashion to the above. There is an additional complication for the charged hyperons in that their deflection by the magnetic field must be taken into account if one wants to study spin precession effects. The best results have been obtained at Fermilab and are listed in Table 15.2. These results are compared with quark model predictions. The results for the proton, the neutron and the Λ^0 were used to fix all the unknown parameters and so predict the other magnetic moments. The results of the experiments agree with the model predictions to within a few percent.

These results support our constituent quark picture in two ways: firstly the constituent quark masses from our mass formula and those obtained from the above analysis of the magnetic moments agree well with each other and secondly the magnetic moments themselves are consistent with the quark model.

It should be noted, however, that the deviations of the experimental values from the predictions of the model show that the constituent quark magnetic moments alone do not suffice to describe the magnetic moments of the hyperons exactly. Further effects, such as relativistic ones and those due to the quark orbital angular momenta, must be taken into account.

15.5 Semileptonic Baryon Decays

The weak decays of the baryons all follow the same pattern. A quark emits a virtual $W^\pm$ boson and so changes its weak isospin and turns into a lighter quark. The $W^\pm$ decays into a lepton-antilepton pair or, if its energy suffices, a quark-antiquark pair. In the decays into a quark-antiquark pair we actually measure one or more mesons in the final state. These decays cannot be exactly calculated because of the strong interaction's complications. Matters are simpler for semileptonic decays. The rich data available to us from semileptonic baryon decays have made a decisive contribution to our current understanding of the weak interaction as formulated in the generalised Cabibbo theory.

We now want to attempt to describe the weak decays of the baryons using our knowledge of the weak interaction from Chap. 10. The weak decays take place essentially at the quark level, but free quarks do not exist and experiments always see hadrons. We must therefore try to interpret hadronic observables within the framework of the fundamental theory of the weak interaction. We will start by considering the β-decay of the neutron, since this has been thoroughly investigated in various experiments. It will then be only a minor matter to extend the formalism to the semileptonic decays of the hyperons and to nuclear β-decays.

We have seen from leptonic decays such as $\mu^- \to e^- + \overline{\nu}_e + \nu_\mu$ that the weak interaction violates parity conservation maximally, which must mean that the coupling constants for the vector and axial vector terms are of the same size. Since neutrinos are left handed and antineutrinos are right handed the coupling constants must have opposite signs (V$-$A theory). The weak decay of a hadron really means that a confined quark has decayed. It is therefore essential to take the quark wave function of the hadron into account. Furthermore strong interaction effects of virtual particles cannot be neglected: although the effective electromagnetic coupling constant is for reasons of charge conservation not altered by the cloud of sea quarks and gluons, the weak coupling is indeed so changed. In what follows we will initially take the internal structure of the hadrons into account and then discuss the coupling constants.

β-decay of the neutron. The β-decay of a free neutron

$$n \rightarrow p + e^- + \overline{\nu}_e \tag{15.31}$$

(maximum electron energy $E_0 = 782$ keV, lifetime 15 minutes) is a rich source of precise data about the low energy behaviour of the weak interaction.

To find the form of the β-spectrum and the coupling constants of neutron β-decay we consider the decay probability. This may be calculated from the golden rule in the usual fashion. If the electron has energy E_e, then the decay rate is

$$dW(E_e) = \frac{2\pi}{\hbar} |\mathcal{M}_{fi}|^2 \frac{d\varrho_f(E_0, E_e)}{dE_e} dE_e , \tag{15.32}$$

where $d\varrho_f(E_0, E_e)/dE_e$ is the density of antineutrino-electron final states with total energy E_0 and the electron having energy E_e and $\mathcal{M}_{fi}$ is the matrix element for the β-decay.

Vector transitions. A β-decay which takes place through a vector coupling is called a *Fermi transition*. The direction of the quark's spin is unaltered in these decays. The change of a d- into a u-quark is described by the ladder operator of weak isospin T_+ which changes a state with $T = -1/2$ into one with $T = +1/2$.

The matrix element for neutron β-decay has a leptonic and a quark part. Conservation of angular momentum prevents any interference between vector and axial vector transitions, i.e., a quark vector transition necessarily implies a leptonic vector transition. Since we already have $c_V = -c_A = 1$ for leptons, we do not need to worry further about their part of the matrix element.

The matrix element for Fermi decays may then be written as

$$|\mathcal{M}_{fi}|_F = \frac{G_F}{V} c_V |\langle uud | \sum_{i=1}^{3} T_{i,+} | udd \rangle| \tag{15.33}$$

where the sum is over the three quarks. According to the definition (10.4) the Fermi constant G_F includes the propagator term and the coupling to the leptons. The initial neutron state has the wave function $|udd\rangle$ and the final state is described by the quark combination $|uud\rangle$. The wave functions of the electron and the antineutrino can each be replaced by $1/\sqrt{V}$, since we have $pR/\hbar \ll 1$.

The u- and d-quarks in the proton and neutron wave functions are eigen-states of strong isospin. In β-decay we need to consider the eigenstates of the weak interaction. We therefore recall that while the ladder operators $I_\pm$ of the strong force map $|u\rangle$ and $|d\rangle$ onto each other, the $T_\pm$ operators connect the $|u\rangle$ and $|d'\rangle$ quark states. The overlap between $|d\rangle$ and $|d'\rangle$ is, according to (10.16), fixed by the cosine of the Cabibbo angle. Hence

$$\langle u|T_+|d\rangle = \langle u|I_+|d\rangle \cdot \cos\theta_C \qquad \text{where } \cos\theta_C \approx 0.98 . \qquad (15.34)$$

The vector component of the matrix element is then

$$\mathcal{M}_{fi} = \frac{G_F}{V} \cos\theta_C \cdot c_V \left\langle \text{uud} \,\Big|\, \sum_{i=1}^{3} I_{i,+} \,\Big|\, \text{udd} \right\rangle = \frac{G_F}{V} \cos\theta_C \cdot c_V \cdot 1 . \qquad (15.35)$$

Here we have employed the fact that the sum $\langle \text{uud}| \sum_i I_{i,+}|\text{udd}\rangle$ must be unity since the operator $\sum_i I_{i,+}$ applied to the quark wave function of the neutron just gives the quark wave function of the proton. This follows from isospin conservation in the strong interaction and may be straightforwardly verified with the help of (15.5) and (15.6). We thus see that the Fermi matrix element is independent of the internal structure of the nucleon.

Axial transitions. Those β-decays that take place as a result of an axial vector coupling are called *Gamow-Teller transitions*. In such cases the direction of the fermion spin flips over. The matrix element depends upon the overlap of the spin densities of the particles carrying the weak charge in the initial and final states. The transition operator is then $c_A T_+ \boldsymbol{\sigma}$.

The universality of the weak interaction means that this result should also hold for free point quarks. Since quarks are always trapped inside hadrons, we need to consider the internal structure of the nucleon if we want to calculate such matrix elements. From the constituent quark model we have

$$|\mathcal{M}_{fi}|_{GT} = \frac{G_F}{V} c_A \left| \left\langle \text{uud} \,\Big|\, \sum_{i=1}^{3} T_{i,+} \boldsymbol{\sigma} \,\Big|\, \text{udd} \right\rangle \right| . \qquad (15.36)$$

Since the squares of the expectation values of the components of $\boldsymbol{\sigma}$ are equal to each other, $\langle \sum_i \sigma_{i,x}\rangle^2 = \langle \sum_i \sigma_{i,y}\rangle^2 = \langle \sum_i \sigma_{i,z}\rangle^2$, it is sufficient to calculate the expectation value of $\sigma_z = \langle \text{uud}| \sum_i I_{i,+}\sigma_{i,z}|\text{udd}\rangle$. One finds from (15.5), (15.6) and some tedious arithmetic that

$$\left\langle \text{uud} \,\Big|\, \sum_i I_{i,+}\sigma_{i,z} \,\Big|\, \text{udd} \right\rangle = \frac{5}{3} . \qquad (15.37)$$

The total matrix element. In experiments we measure the properties of the nucleon, such as its spin, and not those of the quarks. To compare theory with experiment we must therefore reformulate the matrix element so that all operators act upon the *nucleon* wave function. The square of the neutron decay matrix element may be written as

$$|\mathcal{M}_{fi}|^2 = \frac{g_V^2}{V^2}|\langle \mathrm{p}|I_+|\mathrm{n}\rangle|^2 + \frac{g_A^2}{V^2}|\langle \mathrm{p}|I_+\boldsymbol{\sigma}|\mathrm{n}\rangle|^2 . \tag{15.38}$$

We stress that I_+ and $\boldsymbol{\sigma}$ now act upon the wave function of the nucleon. The quantities g_V and g_A are those which are measured in neutron β-decay and describe the absolute strengths of the vector and axial vector contributions. They contain the product of the weak charges at the leptonic and hadronic vertices.

Since the proton and the neutron form an isospin doublet, (15.38) may be written as

$$|\mathcal{M}_{fi}|^2 = (g_V^2 + 3g_A^2)/V^2 . \tag{15.39}$$

We note that the factor of 3 in the axial vector part is due to the expectation value of the spin operator $\boldsymbol{\sigma}^2 = \sigma_x^2 + \sigma_y^2 + \sigma_z^2$.

In the constituent quark model g_V and g_A are related to the quark dependent coupling constants c_V and c_A as follows:

$$g_V = G_F \cos\theta_C\, c_V , \tag{15.40}$$

$$g_A \approx G_F \cos\theta_C \frac{5}{3} c_A . \tag{15.41}$$

The Fermi matrix element (15.35) is independent of the internal structure of the neutron and (15.40) is as exact as the isospin symmetry of the proton and the neutron. The axial vector coupling, on the other hand, does depend upon the structure of the nucleon. In the constituent quark model it is given by (15.41). It is important to understand that the factor of 5/3 is merely an estimate, since the constituent quark model only gives us an approximation of the nucleon wave function.

The neutron lifetime. The lifetime is given by the inverse of the total decay probability per unit time:

$$\frac{1}{\tau} = \int_{m_ec^2}^{E_0} \frac{\mathrm{d}W}{\mathrm{d}E_e}\mathrm{d}E_e = \int_{m_ec^2}^{E_0} \frac{2\pi}{\hbar}|\mathcal{M}_{fi}|^2 \frac{\mathrm{d}\varrho_f(E_0, E_e)}{\mathrm{d}E_e}\,\mathrm{d}E_e . \tag{15.42}$$

Assuming that the matrix element is independent of the energy, we can pull it outside the integral. The state density $\varrho_f(E_0, E_e)$ may, in analogy to (4.18) and (5.21), be written as

$$\mathrm{d}\varrho_f(E_0, E_e) = \frac{(4\pi)^2}{(2\pi\hbar)^6}\, p_e^2\, \frac{\mathrm{d}p_e}{\mathrm{d}E_e}\, p_\nu^2\, \frac{\mathrm{d}p_\nu}{\mathrm{d}E_0}\, V^2\, \mathrm{d}E_e , \tag{15.43}$$

where we have taken into account that we here have an electron and a neutrino and hence a 2-particle state density and V is the volume in which the wave functions of the electron and of the neutrino are normalised. Since this normalisation enters the matrix element (15.39) via a $1/V^2$ factor, the decay probability is independent of V.

In (15.42) we only integrate over the electron spectrum and so we need the density of states for a total energy E_0 with a fixed electron energy E_e. Neglecting recoil effects we have $E_0 = E_e + E_\nu$ and hence $dE_0 = dE_\nu$. Using the relativistic energy-momentum relation $E^2 = p^2 c^2 + m^2 c^4$ we thus find

$$p_e^2 dp_e \;=\; \frac{1}{c^2} p_e E_e \, dE_e \;=\; \frac{1}{c^3} E_e \sqrt{E_e^2 - m_e^2 c^4} \, dE_e \qquad (15.44)$$

and an analogous relation for the neutrino. Assuming that the neutrino is massless we obtain

$$d\varrho_f(E_0, E_e) = (4\pi)^2 \, V^2 \, \frac{E_e \sqrt{E_e^2 - m_e^2 c^4} \cdot (E_0 - E_e)^2}{(2\pi\hbar c)^6} \, dE_e \; . \qquad (15.45)$$

To find the lifetime τ we now need to carry out the integral (15.42). It is usual to normalise the energies in terms of the electron rest mass and so define

$$f(E_0) \;=\; \int_1^{\mathcal{E}_0} \mathcal{E}_e \sqrt{\mathcal{E}_e^2 - 1} \cdot (\mathcal{E}_0 - \mathcal{E}_e)^2 \, d\mathcal{E}_e \qquad \text{where} \quad \mathcal{E} = E/m_e c^2. \quad (15.46)$$

Together with (15.39) this leads to

$$\frac{1}{\tau} \;=\; \frac{m_e^5 c^4}{2\pi^3 \hbar^7} \cdot (g_V^2 + 3g_A^2) \cdot f(E_0) \; . \qquad (15.47)$$

For $(E_0 \gg m_e c^2)$ we have

$$f(E_0) \;\approx\; \frac{\mathcal{E}_0^5}{30} \qquad (15.48)$$

and so

$$\frac{1}{\tau} \;\approx\; \frac{1}{\hbar^7 c^6} \cdot (g_V^2 + 3g_A^2) \cdot \frac{E_0^5}{60\pi^3}. \qquad (15.49)$$

This decrease of the lifetime as the fifth power of E_0 is called *Sargent's rule*.

In neutron decays E_0 is roughly comparable to $m_e c^2$ and the approximation (15.48) is not applicable. The decay probability is roughly half the size of (15.49):

$$\frac{1}{\tau_n} \;\approx\; \frac{1}{\hbar^7 c^6} \cdot (g_V^2 + 3g_A^2) \cdot \frac{E_0^5}{60\pi^3} \cdot 0.47 \; . \qquad (15.50)$$

Experimental results. The neutron lifetime has been measured very precisely in recent years. The storage of ultra cold neutrons has been a valuable tool in these experiments [Ma89, Go94a]. Extremely slow neutrons can be stored between solid walls which represent a potential barrier. The neutrons are totally reflected since the refraction index in solid matter is smaller than that in air [Go79]. With such storage cells the lifetime of the neutron may be determined by measuring the number of neutrons in the cell as a function of time. To do this one opens the storage cell for a specific time to a cold neutron beam of a known, constant intensity. The cell is then closed and left undisturbed until after a certain time it is opened again and the remaining neutrons are counted with a neutron detector. The experiment is repeated for various storage times. The exponential decay in the number of neutrons in the cell (together with knowledge of the leakage rate from the cell) gives us the neutron lifetime. The average of the most recent measurements of the neutron lifetime is [PD98]

$$\tau_n = 886.7 \pm 1.9\,\text{s}\,. \tag{15.51}$$

To individually determine g_A and g_V we need to measure a second quantity. The decay asymmetry of polarised neutrons is a good candidate here. This comes from the parity violating properties of the weak interaction: the axial vector part emits electrons anisotropically while the vector contribution is spherically symmetric.[2] The number of electrons that are emitted in the direction of the neutron spin $N^{\uparrow\uparrow}$ is smaller than the number $N^{\uparrow\downarrow}$ emitted in the opposite direction. The asymmetry A is defined by

$$\frac{N^{\uparrow\uparrow} - N^{\uparrow\downarrow}}{N^{\uparrow\uparrow} + N^{\uparrow\downarrow}} = \beta \cdot A \qquad \text{where} \quad \beta = \frac{v}{c}\,. \tag{15.52}$$

This asymmetry is connected to

$$\lambda = \frac{g_A}{g_V} \tag{15.53}$$

by

$$A = -2\frac{\lambda(\lambda + 1)}{1 + 3\lambda^2}\,. \tag{15.54}$$

The asymmetry experiments are also best performed with ultra low energy neutrons. An electron spectrometer with an extremely high spatial resolution is needed. Such measurements yield [PD98]

$$A \quad = \quad -0.1162 \pm 0.0013\,. \tag{15.55}$$

Combining this information we have

$$\begin{aligned} \lambda &= -1.267 \pm 0.004\,, \\ g_V/(\hbar c)^3 &= +1.153 \cdot 10^{-5}\,\text{GeV}^{-2}\,, \\ g_A/(\hbar c)^3 &= -1.454 \cdot 10^{-5}\,\text{GeV}^{-2}\,. \end{aligned} \tag{15.56}$$

A comparison with (15.40) yields very exactly $c_V = 1$, which is the value we would expect for a point-like quark or lepton. The vector part of the

---

[2] The discovery of parity violation in the weak interaction was through the anisotropic emission of electrons in the β-decay of atomic nuclei [Wu57].

interaction is conserved in weak baryon decays. This is known as *conservation of vector current* (CVC) and it is believed that this conservation is exact. It is considered to be as important as the conservation of electric charge in electromagnetism.

The axial vector term is on the other hand not that of a point-like Dirac particle. Rather than $\lambda = -5/3$ experiment yields $\lambda \approx -5/4$. The strong force alters the spin dependent part of the weak decay and the axial vector current is only partially conserved (PCAC = *partially conserved axial vector current*).

Semileptonic hyperon decays. The semileptonic decays of the hyperons can be calculated in a similar way to that of the neutron. Since the decay energies E_0 are typically two orders of magnitude larger than in the neutron decay, Sargent's rule (15.49) predicts that the hyperon lifetimes should be at least a factor of 10^{10} shorter. At the quark level these decays are all due to the decay $s \to u + e^- + \bar{\nu}_e$.

The two independent measurements to determine the semileptonic decay probabilities of the hyperons are their lifetimes τ and the branching ratio $V_{\text{semil.}}$ of the semileptonic channels. From

$$\frac{1}{\tau} \propto |\mathcal{M}_{fi}|^2 \quad \text{and} \quad V_{\text{semil.}} \equiv \frac{|\mathcal{M}_{fi}|^2_{\text{semil.}}}{|\mathcal{M}_{fi}|^2}$$

we have the relationship

$$\frac{V_{\text{semil.}}}{\tau} \propto |\mathcal{M}_{fi}|^2_{\text{semil.}} \,. \tag{15.57}$$

The lifetime may most easily be measured in production experiments. High energy proton or hyperon (e.g., Σ^-) beams with an energy of a few hundred GeV are fired at a fixed target and one detects the hyperons which are produced. One then calculates the average decay length of the secondary hyperons, i.e., the average distance between where they are produced (the target) and where they decay. This is done by measuring the tracks of the decay products with detectors which have a good spatial resolution and reconstructing the position where the hyperon decayed. The number of hyperons decreases exponentially with time and this is reflected in an exponential decrease in the number N of decay positions a distance l away from the target:

$$N = N_0 \, e^{-t/\tau} = N_0 \, e^{-l/L} \,. \tag{15.58}$$

The method of invariant masses must, of course, be used to identify which sort of hyperon has decayed. The average decay length L is then related to the lifetime τ as follows

$$L = \gamma v \tau \,, \tag{15.59}$$

where v is the velocity of the hyperon. With high beam energies the secondary hyperons can have time dilation factors $\gamma = E/mc^2$ of the order of 100. Since the hyperons typically have a lifetime of around 10^{-10} s the decay length will typically be a few metres – which may be measured to a good accuracy.

The measurement of the branching ratios is much more complicated. This is because the vast majority of decays are into hadrons (which may therefore be used to measure the decay length). The semileptonic decays are only about one thousandth of the total. This means that those few leptons must be detected with a very high efficiency and that background effects must be rigorously analysed.

The experiments are in fact sufficiently precise to put the Cabibbo theory to the test. The method is similar to that which we used in the case of the β-decay of the neutron. Using the relevant matrix element and phase space factors one calculates the decay probability of the decay under considerati- on. The calculation, which still contains c_V und c_A, is then compared with experiment.

Consider the strangeness-changing decay $\Xi^- \to \Lambda^0 + e^- + \bar{\nu}_e$. The matrix element for the Fermi decay is

$$|\mathcal{M}_{fi}|_{\mathrm{F}} = \frac{G_{\mathrm{F}}}{V} |\langle \mathrm{uds} | \sum_{i=1}^{3} T_{i,+} | \mathrm{dss} \rangle| , \qquad (15.60)$$

where we have assumed that the coupling constant $c_V = 1$ is unchanged. App- lying the operator T_+ to the flavour eigenstate $|s\rangle$ yields a linear combination of $|u\rangle$ and $|c\rangle$. Just as was the case for the β-decay of the neutron the ma- trix element thus contains a Cabibbo factor, here $\sin\theta_{\mathrm{C}}$. The Gamow-Teller matrix element is obtained from

$$|\mathcal{M}_{fi}|_{\mathrm{GT}} = \frac{g_{\mathrm{A}}}{g_{\mathrm{V}}} \frac{G_{\mathrm{F}}}{V} |\langle \mathrm{uds} | \sum_{i=1}^{3} T_{i,+} \boldsymbol{\sigma}_i | \mathrm{dss} \rangle| . \qquad (15.61)$$

Of course the evaluation of the $\boldsymbol{\sigma}$ operator depends upon the wave functions of the baryons involved in the decay.

The analysis of the data confirms the assumption that the ratio $\lambda = g_{\mathrm{A}}/g_{\mathrm{V}}$ has the same value in both hyperon and neutron decays. The axial current is hence modified in the same way for all three light quark flavours.

15.6 How Good is the Constituent Quark Concept?

We introduced the concept of constituent quarks so as to describe the baryon mass spectrum as simply as possible. We thus viewed constituent quarks as the building blocks from which the hadrons can be constructed. This means, however, that we should be able to derive all the hadronic quantum numbers from these effective constituents. Furthermore we have silently assumed that we are entitled to treat constituent quarks as elementary particles, whose magnetic moments, just like the electrons', obey a Dirac relation (15.13).

That these ideas work has been seen in the chapters treating the meson and baryon masses and the magnetic moments. Various approaches led us to constituent quark masses which were in good agreement with each other and furthermore the magnetic moments of the model were generally in very good agreement with experiment.

Constituent quarks are not, however, fundamental, elementary particles as we understand the term. This role is reserved for the "naked" valence quarks which are surrounded by a cloud of virtual gluons and quark-antiquark pairs. It is not at all obvious why constituent quarks may be treated as though they were elementary. Indeed we have seen the limitations of this approach: in all those phenomena where spin plays a part the structure of the constituent quark makes itself to some extent visible, for example in the magnetic moments of the hyperons with 2 or 3 s-quarks and also in the non-conserved axial vector current of the weak interaction. The picture of hadrons as being composed of (Dirac particle) constituent quarks is just not up to describing such matters or indeed any process with high momentum transfer.

Problems

1. Particle production and identification

A liquid hydrogen target is bombarded with a $|\boldsymbol{p}| = 12$ GeV/c proton beam. The momenta of the reaction products are measured in wire chambers inside a magnetic field. In one event six charged particle tracks are seen. Two of them go back to the interaction vertex. They belong to positively charged particles. The other tracks come from two pairs of oppositely charged particles. Each of these pairs appears "out of thin air" a few centimetres away from the interaction point. Evidently two electrically neutral, and hence unobservable, particles were created which later both decayed into a pair of charged particles.

a) Make a rough sketch of the reaction (the tracks).

b) Use Tables 14.2, 14.3 and 15.1 as well as [PD94] to discuss which mesons and baryons have lifetimes such that they could be responsible for the two observed decays. How many decay channels into two charged particles are there?

c) The measured momenta of the decay pairs were:

   1) $|\boldsymbol{p}_+| = 0.68$ GeV/c, $|\boldsymbol{p}_-| = 0.27$ GeV/c, $\sphericalangle\,(\boldsymbol{p}_+, \boldsymbol{p}_-) = 11°$;

   2) $|\boldsymbol{p}_+| = 0.25$ GeV/c, $|\boldsymbol{p}_-| = 2.16$ GeV/c, $\sphericalangle\,(\boldsymbol{p}_+, \boldsymbol{p}_-) = 16°$.

   The relative errors of these measurements are about 5 %. Use the method of invariant masses (15.1) to see which of your hypothesis from b) are compatible with these numbers.

d) Using these results and considering all applicable conservation laws produce a scheme for all the particles produced in the reaction. Is there a unique solution?

2. Baryon masses

Calculate expressions analogous to (15.11) for the mass shifts of the Σ and Σ^* baryons due to the spin-spin interaction. What value do you obtain for $\alpha_s|\psi(0)|^2$ if you use the constituent quark masses from Sec. 15.3?

3. Isospin coupling

The Λ hyperon decays almost solely into $\Lambda^0 \to p + \pi^-$ and $\Lambda^0 \to n + \pi^0$. Apply the rules for coupling angular momenta to isospin to estimate the ratio of the two decay probabilities.

4. Muon capture in nuclei

Negative muons are slowed down in a carbon target and then trapped in atomic 1s states. Their lifetime is then 2.02 μs which is less than that of the free muon (2.097 μs). Show that the difference in the lifetimes is due to the capture reaction $^{12}C + \mu^- \to\,^{12}B + \nu_\mu$. The mass difference between the ^{12}B and ^{12}C atoms is 13.37 MeV/c^2 and the lifetime of ^{12}B is 20.2 ms. ^{12}B has, in the ground state, the quantum numbers $J^P = 1^+$ and $\tau = 20.2$ ms. The rest mass of the electron and the nuclear charge may be neglected in the calculation of the matrix element.

5. Quark mixing

The branching ratios for the semileptonic decays $\Sigma^- \to n + e^- + \overline{\nu}_e$ and $\Sigma^- \to \Lambda^0 + e^- + \overline{\nu}_e$ are $1.02 \cdot 10^{-3}$ and $5.7 \cdot 10^{-5}$ respectively – a difference of more than an order of magnitude. Why is this? The decay $\Sigma^+ \to n + e^+ + \nu_e$ has not yet been observed (upper bound: $5 \cdot 10^{-6}$). How would you explain this?

6. Parity

a) The intrinsic parity of a baryon cannot be determined in an experiment; it is only possible to compare the parity of one baryon with that of another. Why is this?

b) It is conventional to ascribe a positive parity to the nucleon. What does this say about the deuteron's parity (see Sec. 16.2) and the intrinsic parities of the u- and d-quarks?

c) If one bombards liquid deuterium with negative pions, the latter are slowed down and may be captured into atomic orbits. How can one show that they cascade down into the 1s shell (K shell)?

d) A pionic deuterium atom in the ground state decays through the strong interaction via $d + \pi^- \rightarrow n + n$. In which $^{2S+1}L_J$ state may the two neutron system be? Note that the two neutrons are identical fermions and that angular momentum is conserved.

e) What parity from this for the pion? What parity would one expect from the quark model (see Chap. 14)?

f) Would it be inconsistent to assign a positive parity to the proton and a negative one to the neutron? What would then be the parities of the quarks and of the pion? Which convention is preferable? What are the parities of the Λ and the Λ_c according to the quark model?

16. The Nuclear Force

> Unfortunately, nuclear physics has not profited as
> much from analogy as has atomic physics. The
> reason seems to be that the nucleus is the domain
> of new and unfamiliar forces, for which men have
> not yet developed an intuitive feeling.
>
> *V. L. Telegdi* [Te62]

The enormous richness of complex structures that we see all around us (molecules, crystals, amorphous materials) is due to chemical interactions. The short distance forces through which electrically neutral atoms interact can and do produce large scale structures.

The interatomic potential can generally be determined from spectroscopic data about molecular excited states and from measuring the binding energies with which atoms are tied together in chemical substances. These potentials can be quantitatively explained in non-relativistic quantum mechanics. We thus nowadays have a consistent picture of chemical binding based upon atomic structure.

The nuclear force is responsible for holding the nucleus together. This is an interaction between colourless nucleons and its range is of the same order of magnitude as the nucleon diameter. The obvious analogy to the atomic force is, however, limited. In contrast to the situation in atomic physics, it is not possible to obtain detailed information about the nuclear force by studying the structure of the nucleus. The nucleons in the nucleus are in a state that may be described as a degenerate Fermi gas. To a first approximation the nucleus may be viewed as a collection of nucleons in a potential well. The behaviour of the individual nucleons is thus more or less independent of the exact character of the nucleon-nucleon force. It is therefore not possible to extract the nucleon-nucleon potential directly from the properties of the nucleus. The potential must rather be obtained by analysing two-body sysytems such as nucleon-nucleon scattering and the proton-neutron bound state, i.e., the deuteron.

There are also considerably greater theoretical difficulties in elucidating the connection between the nuclear forces and the structure of the nucleon than for the atomic case. This is primarily a consequence of the strong coupling constant α_s being two orders of magnitude larger than α, its electromagnetic equivalent. We will therefore content ourselves with an essentially qualitative explanation of the nuclear force.

16.1 Nucleon–Nucleon Scattering

Nucleon-nucleon scattering at low energies, below the pion production threshold, is purely elastic. At such energies the scattering may be described by non-relativistic quantum mechanics. The nucleons are then understood as point-like structureless objects that nontheless possess spin and isospin. The physics of the interaction can then be understood in terms of a potential. It is found that the nuclear force depends upon the total spin and isospin of the two nucleons. A thorough understanding therefore requires experiments with polarised beams and targets, so that the spins of the particles involved in the reaction can be specified, and both protons and neutrons must be employed.

If we consider nucleon-nucleon scattering and perform measurements for both parallel and antiparallel spins perpendicular to the scattering plane, then we can single out the spin triplet and singlet parts of the interaction. If the nucleon spins are parallel, then the total spin must be 1, while for opposite spins there are equally large (total) spin 0 and 1 components.

The algebra of angular momentum can also be applied to isospin. In proton-proton scattering we always have a state with isospin 1 (an isospin triplet) since the proton has $I_3 = +1/2$. In proton-neutron scattering there are both isospin singlet and triplet contributions.

Scattering phases. Consider a nucleon coming in "from infinity" with kinetic energy E and momentum $\boldsymbol{p}$ which scatters off the potential of another nucleon. The incoming nucleon may be described by a plane wave and the outgoing nucleon as a spherical wave. The cross section depends upon the phase shift between these two waves.

For states with well defined spin and isospin the cross section of nucleon-nucleon scattering into a solid angle element $\mathrm{d}\Omega$ is given by the scattering amplitude $f(\theta)$ of the reaction

$$\frac{\mathrm{d}\sigma}{\mathrm{d}\Omega} = |f(\theta)|^2 \,. \tag{16.1}$$

For scattering off a short ranged potential a *partial wave decomposition* is used to describe the scattering amplitude. The scattered waves are expanded in terms with fixed angular momentum ℓ. In the case of elastic scattering the following relation holds at large distances r from the centre of the scattering:

$$f(\theta) = \frac{1}{k} \sum_{\ell=0}^{\infty} (2\ell + 1) \, \mathrm{e}^{i\delta_\ell} \, \sin \delta_\ell \, P_\ell(\cos \theta) \,, \tag{16.2}$$

where

$$k = \frac{1}{\lambdabar} = \frac{|\boldsymbol{p}|}{\hbar} = \frac{\sqrt{2ME}}{\hbar} \tag{16.3}$$

is the wave number of the scattered nucleon, δ_ℓ a phase shift angle and P_ℓ, the angular momentum eigenfunction, an ℓ-th order Legendre polynomial. The phase shifts δ_ℓ describe the phase difference between the scattered and uns-cattered waves. They contain the information about the shape and strength of the potential and the energy dependence of the cross section. The fact that δ_ℓ appears not only as a phase factor but also in the amplitude ($\sin \delta_\ell$) follows from the conservation of the particle current in elastic scattering. This is also known as *unitarity*. The partial wave decomposition is especially convenient at low energies since only a few terms enter the expansion. This is because for a potential with range a we have

$$\ell \leq \frac{|\boldsymbol{p}| \cdot a}{\hbar} \, . \tag{16.4}$$

The phase shift δ_0 of the partial waves with $\ell = 0$ (i.e., s waves) is de-cisive for nuclear binding. From (16.4) we see that the s waves dominate proton-proton scattering (potential range 2 fm) for relative momenta less than 100 MeV/c. The Legendre polynomial P_0 is just 1, i.e., independent of θ. The phase shifts δ_0 as measured in nucleon-nucleon scattering are separa-tely plotted for spin triplet and singlet states against the momentum in the centre of mass frame in Fig. 16.1. For momenta larger than 400 MeV/c δ_0 is negative, below this it is positive. We learn from this that the nuclear force has a repulsive character at short distances and an attractive nature at larger separations. This may be simply seen as follows.

Consider a, by definition, spherically symmetric s wave $\psi(\boldsymbol{x})$. We may define a new radial function $u(r)$ by $u(r) = \psi(r) \cdot r$ which obeys the Schrödinger equation

$$\frac{\mathrm{d}^2 u(r)}{\mathrm{d}r^2} + \frac{2m(E - V)}{\hbar^2} \, u(r) = 0 \, . \tag{16.5}$$

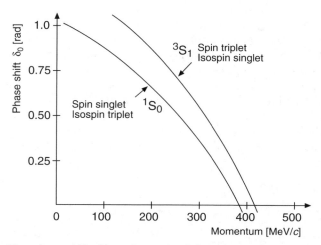

Fig. 16.1. The phase shift δ_0 as determined from experiment both for the spin triplet-isospin singlet ^{3}S$_1$ and for the spin singlet-isospin triplet ^{1}S$_0$ systems plotted against the relative momenta of the nucleons. The rapid variation of the phases at small momenta is not plotted since the scale of the diagram is too small.

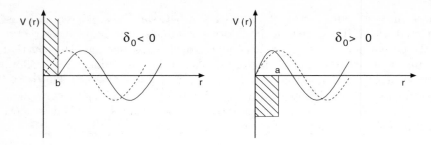

Fig. 16.2. Sketch of the scattering phase for a repulsive (*left*) and an attractive (*right*) potential. The dashed curves denote unscattered waves, the continuous ones the scattered waves.

If we now solve this equation for a repulsive rectangular potential V with radius b and $V \to \infty$ (Fig. 16.2), we find

$$\delta_0 = -kb. \tag{16.6}$$

The scattering phase is negative and proportional to the range of the potential. A negative scattering phase means that the scattered wave lags behind the unscattered one.

For an attractive potential the scattered wave runs ahead of the unscattered one and δ_0 is positive. The size of the phase shift is the difference between the phase of the wave scattered off the edge of the potential a and that of the unscattered wave:

$$\delta_0 = \arctan\left(\sqrt{\frac{E}{E+|V|}}\tan\frac{\sqrt{2mc^2(E+|V|)}\cdot a}{\hbar c}\right) - \frac{\sqrt{2mc^2 E}\cdot a}{\hbar c}. \tag{16.7}$$

The phase shift δ_0 is then positive and decreases at higher momenta. If we superimpose the phase shifts associated with a short ranged repulsive potential and a longer ranged attractive one we obtain Fig. 16.3, where the effective phase shift changes sign just as the observed one does.

The relationship between the scattering phase δ_0 and the scattering potential V is contained, in principle, in (16.6) and (16.7) since the wave number k in the region of the potential depends both upon the latter's size and shape and upon the initial energy E of the projectile. A complete scattering phase analysis leads to the nuclear potential shown in Fig. 16.4 which has – as

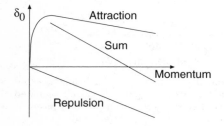

Fig. 16.3. Superposition of negative and positive scattering phases δ_0 plotted against the relative momenta of the scattered particles. The resulting effective δ_0 is generated by a short distance repulsive and a longer range attractive nucleon-nucleon potential.

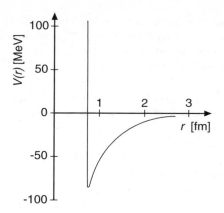

Fig. 16.4. Sketch of the radial dependence of the nucleon-nucleon potential for $\ell = 0$. Note that the spin and isospin dependence of the potential is not shown.

remarked above – a short ranged repulsive and a longer ranged attractive nature. Since the repulsive part of the potential increases rapidly at small r it is known as the *hard core*.

The nucleon–nucleon potential. We may obtain a general form of the nucleon-nucleon potential from a consideration of the relevant dynamical quantities. We will, however, neglect the internal structure of the nucleons, which means that this potential will only be valid for nucleon-nucleon bound states and low energy nucleon-nucleon scattering.

The quantities which determine the interaction are the separation of the nucleons $\boldsymbol{x}$, their relative momenta $\boldsymbol{p}$, the total orbital angular momentum $\boldsymbol{L}$ and the relative orientations of the spins of the two nucleons, $\boldsymbol{s}_1$ and $\boldsymbol{s}_2$. The potential is a scalar and must at the very least be invariant under translations and rotations. Furthermore it should be symmetric under exchange of the two nucleons. These preconditions necessarily follow from various properties, such as parity conservation, of the underlying theory of the strong force and they limit the scalars which may appear in the potential. At the end of the day the potential, for fixed isospin, has the form [Pr63]:

$$
\begin{aligned}
V(r) \quad = \quad & V_0(r) \\
& + V_{\mathrm{ss}}(r)\, \boldsymbol{s}_1 \cdot \boldsymbol{s}_2/\hbar^2 \\
& + V_{\mathrm{T}}(r)\, \left(3(\boldsymbol{s}_1 \cdot \boldsymbol{x})(\boldsymbol{s}_2 \cdot \boldsymbol{x})/r^2 - \boldsymbol{s}_1\boldsymbol{s}_2\right)/\hbar^2 \\
& + V_{\mathrm{LS}}(r)\, (\boldsymbol{s}_1 + \boldsymbol{s}_2) \cdot \boldsymbol{L}/\hbar^2 \\
& + V_{\mathrm{Ls}}(r)\, (\boldsymbol{s}_1 \cdot \boldsymbol{L})(\boldsymbol{s}_2 \cdot \boldsymbol{L})/\hbar^4 \\
& + V_{\mathrm{ps}}(r)\, (\boldsymbol{s}_2 \cdot \boldsymbol{p})(\boldsymbol{s}_1 \cdot \boldsymbol{p})/(\hbar^2 m^2 c^2)\,.
\end{aligned}
\tag{16.8}
$$

V_0 is a standard central potential. The second term describes a pure spin-spin interaction, while the third term is called the *tensor potential* and describes a non-central force. These two terms have the same spin dependence as the interaction between two magnetic dipoles in electromagnetism. The tensor term is particularly interesting, since it alone can mix orbital angular momentum states. The fourth term originates from a spin-orbit force, which is generated by the strong interaction (the analogous force in atomic physics is of magnetic origin). The final two terms in (16.8) are included on formal grounds, since symmetry arguments do not exclude them. They are, however, both quadratic in momentum and thus mostly negligible in comparison to the LS-term.

The significance of this ansatz for the potential is not that the various terms can be merely formally written down, but rather that, as we will see in Sect. 16.3, the spin and isospin dependence of the nuclear force can be explained in meson exchange models. Attempts to fit the potential terms to the experimental data have not fixed it exactly, but a general agreement exists for the first four terms. It should be also noted that many body forces need to be taken into account for conglomerations of nucleons.

The central potential for the $S = 0$ case is applicable to the low energy proton-proton and neutron-neutron interactions. The attractive part is, however, not strong enough to create a bound state. For $S = 1$ on the other hand this potential together with the tensor force and the spin-spin interaction is strong enough to present us with a bound state, the deuteron.

16.2 The Deuteron

The deuteron is the simplest of all the nucleon bound states i.e., the atomic nuclei. It is therefore particularly suitable for studying the nucleon-nucleon interaction. Experiments have yielded the following data about the deuteron ground state:

Binding energy	$B = 2.225$ MeV
Spin and parity	$J^P = \quad 1^+$
Isospin	$I = \quad 0$
Magnetic moment	$\mu = 0.857\ \mu_N$
Elec. quadrupole moment	$Q = 0.282\ e\cdot\text{fm}^2$.

The proton-neutron system is mostly made up of an $\ell=0$ state. If it were a pure $\ell=0$ state then the wave function would be spherically symmetric, the quadrupole moment would vanish and the magnetic dipole moment would be just the sum of the proton and neutron magnetic moments (supposing that the nucleonic magnetic moments are not altered by the binding interaction). This prediction for the deuteron magnetic moment

$$\mu_p + \mu_n = 2.792\,\mu_N - 1.913\,\mu_N = 0.879\,\mu_N \qquad (16.9)$$

differs slightly from the measured value of $0.857\,\mu_N$. Both the magnetic dipole moment and the electric quadrupole moment can be explained by the admixture of a state with the same J^P quantum numbers

$$|\psi_d\rangle = 0.98 \cdot |\,^3S_1\rangle + 0.20 \cdot |\,^3D_1\rangle\,. \qquad (16.10)$$

In other words there is a 4 % chance of finding the deuteron in a $^3D_1$ state. This admixture can be explained from the tensor components of the nucleon-nucleon interaction.

We now want to calculate the nucleon wave function inside a deuteron. Since the system is more or less in an $\ell = 0$ state, the wave function will be spherically symmetric. We will need the depth V of the potential well (averaged over the attractive and repulsive parts) and its range, a. The binding energy of the deuteron alone gives us one parameter – the "volume" of the potential well, i.e., Va^2. The solutions of the Schrödinger equation (16.5) are

$$
\begin{aligned}
&\text{if } r < a: \ u_I(r) = A\sin kr \quad \text{where } k = \sqrt{2m(E - V)}/\hbar, \ (V < 0), \\
&\text{if } r > a: \ u_{II}(r) = Ce^{-\kappa r} \quad \text{where } \kappa = \sqrt{-2mE}/\hbar, \qquad (E < 0),
\end{aligned}
\tag{16.11}
$$

and $m \approx M_p/2$ is the reduced mass of the proton-neutron system.

Continuity of $u(r)$ and $du(r)/dr$ at the edge of the well, i.e., $r = a$, implies that [Sc95]

$$
k \cot ka = -\kappa \qquad ak \approx \frac{\pi}{2} \tag{16.12}
$$

and

$$
Va^2 \approx Ba^2 + \frac{\pi^2}{8}\frac{(\hbar c)^2}{mc^2} \approx 100 \ \mathrm{MeV\,fm^2}\,. \tag{16.13}
$$

Current values for the range of the nuclear force, and hence the effective extension of the potential $a \approx 1.2 \cdots 1.4$ fm, imply that the depth of the potential is $V \approx 50$ MeV. This is much greater than the deuteron binding energy B (just 2.25 MeV). The tail of the wave function, which is characterised by $1/\kappa \approx 4.3$ fm, is large compared to the range of the nuclear force.

The radial probability distribution of the nucleons is sketched in Fig. 16.5 for two values of a but keeping the volume of the potential well Va^2 constant. Since deuterium is a very weakly bound system the two calculations differ only slightly, especially at larger separations.

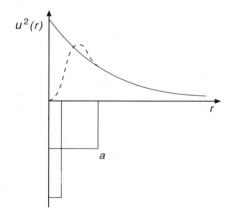

Fig. 16.5. Radial probability distribution $u^2(r) = r^2|\psi|^2$ of the nucleons in deuterium for an attractive potential with range a (*dashed curve*) and for the range $a \to 0$ with a fixed volume Va^2 for the potential well (*continuous curve*).

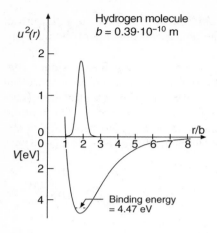

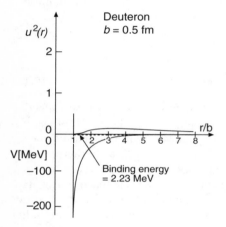

Fig. 16.6. The radial probability distribution $u^2(r)$ of the hydrogen atoms in a hydrogen molecule (*top*) [He50] and of nucleons in a deuteron (*bottom*) in units of the relevant hard cores (from [Bo69]). The covalent bond strongly localises the H atoms, since the binding energy is comparable to the depth of the potential. The weak nuclear bond, since the potential energy is comparable in size to the kinetic energy, means that the nucleons are delocalized.

A more detailed calculation which takes the repulsive part of the potential into account only changes the above wave function at separations smaller than 1 fm (cf. Fig. 16.5). In Fig. 16.6 the probability distribution of nucleons in deuterium and of hydrogen atoms in a hydrogen molecule are given for comparison. The separations are in both cases plotted in units of the spatial extension of the relevant hard core. The hard core sizes are about $0.4 \cdot 10^{-10}$ m for the hydrogen molecule and roughly $0.5 \cdot 10^{-15}$ m for the deuteron. The atoms in the molecule are well localised – the uncertainty in their separation ΔR is only about 10 % of the separation (cf. Fig. 16.6). The nuclear binding in deuterium is relatively "weak" and the bound state is much more spread out. This means that *the average kinetic energy is comparable to the average depth of the potential* and so the binding energy, which is just the sum of the kinetic and potential energies, must be very small.

The binding energy of the nucleons in larger nuclei are somewhat greater than that in deuterium and the density is accordingly larger. Qualitatively we still have the same situation: a relatively weak effective force is just strong enough to hold nuclei together. The properties of the nuclei bear witness to this fact: it is a precondition both for the description of the nucleus as a degenerate Fermi gas and for the great mobility of the nucleons in nuclear matter.

16.3 Nature of the Nuclear Force

We now turn to the task of understanding the strength and the form of the nuclear force from the structure of the nucleons and the strong interaction of the quarks inside the nucleons. In the following discussion we will employ qualitative arguments. The structure of the nucleon will be approached via the nonrelativistic quark model where the nucleons are built out of three constituent quarks. The nuclear force is primarily transmitted by quark-antiquark pairs, which we can only introduce ad hoc through plausibility arguments. A consistent theory of the nuclear force, based upon the interaction of quarks and gluons, does not yet exist.

Short distance repulsion. Let us begin with the short distance repulsive part of the nuclear force and try to construct some analogies to better understood phenomena. That atoms repel each other at short distances is a consequence of the Pauli principle. The electron clouds of both atoms occupy the lowest possible energy levels and if the clouds overlap then some electrons must be elevated into excited states using the kinetic energy of the colliding atoms. Hence we observe a repulsive force at short distances.

The quarks in a system of two nucleons also obey the Pauli principle, i.e., the 6 quark wave function must be totally antisymmetric. It is, however, possible to put as many as 12 quarks into the lowest $\ell = 0$ state without violating the Pauli principle, since the quarks come in three colours and have two possible spin ($\uparrow$, $\downarrow$) and isospin (u-quark, d-quark) directions. The spin-isospin part of the complete wave function must be symmetric since the colour part is antisymmetric and, for $\ell = 0$, the spatial part is symmetric. We thus see that the Pauli principle does not limit the occupation of the lowest quark energy levels in the spatial wave function, and so the fundamental reason for the repulsive core must be sought elsewhere.

The real reason is the spin-spin interaction between the quarks [Fa88]. We have already seen how this makes itself noticeable in the baryon spectrum: the Δ baryon, where the three quark spins are parallel to one another, is about 350 MeV/c^2 heavier than the nucleon. The potential energy then increases if two nucleons overlap and all 6 quarks remain in the $\ell = 0$ state since the number of quark pairs with parallel spins is greater than for separated nucleons. For each and every quark pair with parallel spins the potential energy increases by half the Δ-nucleon energy difference (15.11).

Of course the nucleon-nucleon system tries to minimise its "chromomagnetic" energy by maximising the number of antiparallel quark spin pairs. But this is incompatible with remaining in an $\ell = 0$ state since the spin-flavour part of the wave function must be completely symmetric. The colourmagnetic energy can be reduced if at least two quarks are put into the $\ell = 1$ state. The necessary excitation energy is comparable to the decrease in the chromomagnetic energy, so the total energy will in any case increase if the nucleons strongly overlap. Hence the effective repulsion at short distances is in equal parts a consequence of an increase in the chromomagnetic and the excitation energies (Fig. 16.7). If the nucleons approach each other very closely ($r = 0$) one finds in a non-adiabatic approximation that there is an 8/9 probability of two of the quarks being in a p state [Fa82, St88]. This configuration expresses itself in the relative wave function of the nucleons through a node at 0.4 fm. This together with the chromomagnetic energy causes a strong, short range repulsion. The nuclear force may be described by a nucleon-nucleon potential which rises sharply at separations less than 0.8 fm.

Attraction. Let us now turn to the attractive part of the nuclear force. Again we will pursue analogies from atomic physics. As we know the bonds between atoms are connected to a change in their internal structure and we expect something similar from the nucleons bound in the nucleus. Indeed a change in the quark structure of bound nucleons compared to that of their free brethren has been observed in deep inelastic scattering off nuclei (EMC effect, see Sect. 7.4).

It is clear upon a moments reflection that the nuclear force is not going to be well described by an *ionic bond*: the confining forces are so strong that it is not possible to lend a quark from one nucleon to another.

A *Van der Waals force*, where the atoms polarise each other and then stick to each other via the resulting dipole-dipole interaction can also not serve us as a paradigm. A Van der Waals force transmitted by the exchange of two gluons (in analogy to two photon exchange in the atomic case) would be too weak to explain the nuclear force at distances where the nucleons overlap and confinement does not forbid gluon exchange. At greater separations gluons

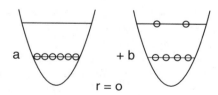

Fig. 16.7. a,b. The quark state for overlapping nucleons. This is composed of (**a**) a configuration with 6 quarks in the $\ell=0$ state and (**b**) a configuration with 2 quarks in the $\ell=1$ state. In a non-adiabatic approximation it is found that the state (**b**) dominates at separation $r = 0$ (probability 8/9) [Fa82, St88]. For larger distances this state becomes less important and disappears as $r \to \infty$.

cannot be exchanged because of confinement. Although colour neutral gluonic states (glueballs) could still be exchanged, none which are light enough have ever been experimentally observed.

The only analogy left to us to explain the nuclear force is a *covalent bond*, such as that which is, e.g., responsible for holding the H_2 molecule together. Here the electrons of the two H atoms are continually swapped around and can be ascribed to both atoms. The attractive part of the nuclear force is strongest at distances of around 1 fm and indeed reminds us of the atomic covalent bond. To simplify what follows, let us assume that the nucleon is made up of a two quark system (diquark) and a quark (see Fig. 16.8). Such a description has proven to be very successful in describing many phenomena. The most energetically favourable configuration is that where a u- and a d-quark combine to form a diquark with spin 0 and isospin 0. The alternative spin 1 and isospin 1 diquark is not favoured. The covalent bond is then expressed by the exchange of the "single" quarks, as sketched in Fig. 16.9. To push home the analogy we also show the equivalent covalent binding of the hydrogen molecule.

Since the nuclear attraction is strongest at distances of the order of 1 fm we do not need to worry about confinement effects. The covalent bond contribution to this force can be worked out analogously to the molecular case. However, the depth of the potential that is found in this way is only about one third of the experimental value [Ro94]. In fact quark exchange is less effective than its atomic counterpart of electron exchange. This is partly because to be exchanged the quarks must have the same colour, and there is only a 1/3 probability of this. The contribution of direct quark exchange sinks still further if one takes the part of the nucleon wave function into account where the diquarks have spin 1 and isospin 1. Thus the covalent bond concept, if it is directly transferred from molecules to nuclei, does not give us a good quantitative description of what is going on in nuclei. It should be noted that this is not a consequence of confinement, but rather of direct quark exchange being suppressed as a result of the quarks having three different colour charges.

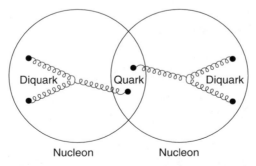

Fig. 16.8. Quark configurations in a covalent bond picture. At large separations, when the nucleons just overlap, we may understand them as each being diquark-quark systems.

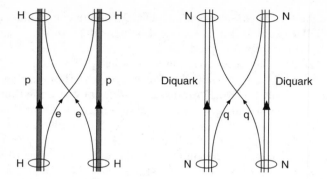

Fig. 16.9. Symbolic representation of the covalent bonds in a hydrogen molecule (*left*) and in a two nucleon system (*right*). The time axis runs vertically upwards. The electron exchange of the hydrogen molecule is replaced by quark exchange in the nucleonic system.

Meson exchange. Up to now we have neglected the fact that as well as the three constituent quarks in the nucleon there are additional quark-antiquark pairs (sea quarks) which are continually being created from gluons and annihilated back into them again. We may interpret this admixture of quark-antiquark pairs as a relativistic effect, which, due to the size of the strong coupling constant α_s, we would be wrong to neglect. An effective quark-quark exchange may be produced by colour neutral quark-antiquark pairs, as is shown in Fig. 16.10a.

This quark-antiquark exchange actually plays a larger role in the nucleon-nucleon interaction than does the simple swapping of two quarks. It must be stressed that this exchange of colour neutral quark-antiquark pairs does

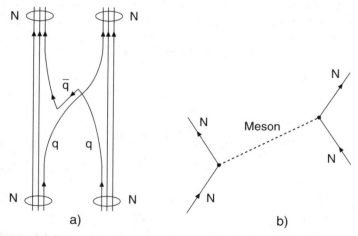

Fig. 16.10. (a) Representation of quark exchange between nucleons via the exchange of a quark-antiquark pair. Antiquarks are here depicted as quarks moving backwards in time. (b) The exchange of a meson is rather similar to this.

not only dominate at great separations where confinement only allows the exchange of colour neutral objects but also at relatively short distances. One may thus understand the nuclear force as a relativistic generalisation of the covalent strong force via which the nucleons finally exchange quarks.

Ever since Yukawa in 1935 first postulated the existence of the pion [Yu35, Br65], there have been attempts to describe the inter-nuclear forces in terms of mesonic exchange. The exchange of mesons with mass m leads to a potential of the form

$$V = g \cdot \frac{\mathrm{e}^{-\frac{mc}{\hbar}r}}{r} \,, \tag{16.14}$$

where g is a charge-like constant. This is known as the *Yukawa potential*.

■ To derive the Yukawa potential we first assume that the nucleon acts as a source of virtual mesons in the same way as an electric charge may be viewed as a source of virtual photons.

We start with the wave equation of a free, relativistic particle with mass m. If we replace the energy E and momentum $\boldsymbol{p}$ in the energy momentum relationship $E^2 = \boldsymbol{p}^2 c^2 + m^2 c^4$ by the operators $i\hbar\partial/\partial t$ and $-i\hbar\boldsymbol{\nabla}$, as is done in the Schrödinger equation, we obtain the *Klein-Gordon equation*:

$$\frac{1}{c^2}\frac{\partial^2}{\partial t^2}\Psi(\boldsymbol{x},t) = \left(\boldsymbol{\nabla}^2 - \mu^2\right)\Psi(\boldsymbol{x},t) \qquad \text{where} \quad \mu = \frac{mc}{\hbar}\,. \tag{16.15}$$

For a massless particle ($\mu = 0$) this equation describes a wave travelling at the speed of light. If we replace Ψ by the electromagnetic four-potential $A = (\phi/c, \boldsymbol{A})$ we obtain the equation for electromagnetic waves in vacuo at a great distance from the source. One may thus interpret $\Psi(\boldsymbol{x},t)$ as the wave function of the photon.

Consider now the static field limit where (16.15) reduces to

$$\left(\boldsymbol{\nabla}^2 - \mu^2\right)\psi(\boldsymbol{x}) = 0\,. \tag{16.16}$$

If we demand a spherically symmetric solution, i.e., one that solely depends upon $r = |\boldsymbol{x}|$ we find

$$\frac{1}{r^2}\frac{\mathrm{d}}{\mathrm{d}r}\left(r^2\frac{\mathrm{d}\psi(r)}{\mathrm{d}r}\right) - \mu^2\psi(r) = 0\,. \tag{16.17}$$

A particularly simple ansatz for the potential V that results from exchanging the particle is $V(r) = g \cdot \psi(r)$, where g is an arbitrary constant. It is clear that this ansatz can make sense if we consider the electromagnetic case: in the limit $\mu \to 0$ we obtain the Poisson equation for a space without charges from (16.16) and we obtain from (16.17) the Coulomb potential $V_C \propto 1/r$, i.e., the potential of a charged particle at a great separation where the charge density is zero. If we now solve (16.17) for the massive case, we obtain the Yukawa potential (16.14). This potential initially decreases roughly as $1/r$ and then much more rapidly. The range is of the order of $1/\mu = \hbar/mc$, which is also what one would expect from the uncertainty relation [Wi38]. The interaction due to pion exchange has a range of about 1.4 fm.

The above remarks are somewhat naive and not an exact derivation. We have ignored the spin of the particle: the Klein-Gordon equation holds for spinless particles (luckily this is true of the pion). Additionally a virtual meson does not automatically have the rest mass of a free particle. Furthermore these interactions take place in the immediate vicinity of the nucleons and the mesons can strongly interact with them. The wave equation of a free particle can at best be an approximation.

Since the range of this potential decreases as the meson mass m increases, the most important exchange particles apart from the pion itself are the lightest vector mesons, the ϱ and the ω. The central potential of the nuclear force can be understood in this framework as a consequence of two pion exchange, where the pions combine to $J^P(I) = 0^+(0)$. The spin and isospin dependence of the nuclear force comes from 1 meson exchange and in particular because both pseudoscalar and vector mesons are exchanged. The trading of pions between the nucleons is especially important since the pion mass is so small that they can be exchanged at relatively large distances (> 2 fm). In these models one neglects the internal structure of nucleons and mesons and assumes that they are point particles. The meson-nucleon coupling constants that emerge from experiment must be slightly adapted to take this into account.

Since mesons are really colour neutral quark-antiquark pairs their exchange and that of colour neutral q$\overline{\text{q}}$ pairs give us, in principle, two equivalent ways of describing the nucleon-nucleon interaction (Fig. 16.10b). At shorter distances, where the structure of the nucleons must definitely play a part, a description in terms of meson exchange is inadequate. The coupling constant for the exchange of ω mesons, which is responsible for the repulsive part of the potential, has to be given an unrealistically high value – about two or three times the size one would accept from a comparison with the other meson-nucleon couplings. The repulsive part of the potential is better described in a quark picture. On the other hand one pion exchange models give an excellent fit to the data at larger separations. At intermediate distances various parameters need to be fitted by hand in both types of model.

In this way we see that it could be possible to trace back the nuclear force to the fundamental constituents of matter. This is very satisfying for our theoretical understanding of the nuclear force, but a quantitative description of the nuclear force is not made any easier by this transition from a mesonic to a quark picture. To describe the forces emanating from meson exchange inside a quark picture we would need to know the probability with which the quark-antiquark pairs in the nucleus can turn into mesons. These calculations are intractable since the strong coupling constant α_s is very large at small momenta. For this reason phenomenological meson exchange models are still today the best way to quantitatively describe the nuclear force.

_____ **Problems**

1. **The nuclear force**
   The nuclear force is transmitted by exchanging mesons. What are the ranges of
   the forces generated by exchanging the following: a π, two π's, a ϱ, an ω? Which
   properties of the nuclear force are determined by the exchange particles?

2. **Neutron-proton scattering**
   How large would the total cross-section for neutron-proton scattering be if only
   the short range repulsion (range, $b = 0.7\,\mathrm{fm}$) contributed? Consider the energy
   regime in which $\ell = 0$ dominates.

17. The Structure of Nuclei

Nuclei that are in their ground state or are only slightly excited are examples of degenerate Fermi gases. The nuclear density is determined by the nucleon-nucleon interaction – essentially by the strong repulsion at short distances and the weak attraction between nucleons that are further apart. We have already seen in Sect. 6.2 that nucleons are not localised in the nuclei but rather move around with rather large momenta of the order of 250 MeV/c. This mobility on the part of the nucleons is a consequence of the fact that, as we have seen for the deuteron, the bonds between nucleons in the nucleus are "weak". The average distance between the nucleons is much larger than the radius of the nucleon hard core.

The fact that nucleons actually move freely inside the nucleus is not at all obvious and of such great conceptual importance that we shall demonstrate it by considering hypernuclei, i.e., those nuclei containing a hyperon as well as the usual nucleons. We will see that a Λ particle moves inside such nuclei like a free particle inside a potential whose depth is independent of the nucleus under consideration and whose range is the nuclear radius.

The shell model is an improvement upon the Fermi gas model in that it has a more realistic potential and the spin-orbit interaction is now taken into consideration. Not only the nuclear density but also the shapes of the nuclei are fixed by the nucleon-nucleon interaction. A nucleus in equilibrium is not always a sphere; it may be ellipsoidal or even more deformed.

17.1 The Fermi Gas Model

We wish to show in this chapter that both the nucleonic momentum distribution that we encountered in quasi-elastic electron-nucleus scattering (Sect. 6.2) and the nucleon binding energies can be understood in terms of the Fermi gas model and that, furthermore, the principal terms of the semi-empirical mass formula (2.8) necessarily emerge from this model. The protons and neutrons that together build up the nucleus are viewed in the Fermi gas model as comprising two independent systems of nucleons. As spin 1/2 particles they naturally obey Fermi-Dirac statistics. It is assumed that the nucleons, inside those constraints imposed by the Pauli principle, can move freely inside the entire nuclear volume.

The potential that every nucleon feels is a superposition of the potentials of the other nucleons. We now assume in our model that this potential has the shape of a well, i.e., that it is constant inside the nucleus and stops sharply at its edge (Fig. 17.1).

The number of possible states available to a nucleon inside a volume V and a momentum region dp is given by

$$dn = \frac{4\pi p^2 dp}{(2\pi\hbar)^3} \cdot V \,. \tag{17.1}$$

At zero temperature, i.e., in the nuclear ground state, the lowest states will all be occupied up to some maximal momentum which we call the *Fermi momentum* p_F. The number of such states may be found by integrating over (17.1)

$$n = \frac{V p_F^3}{6\pi^2 \hbar^3} \,. \tag{17.2}$$

Since every state can contain two fermions of the same species, we can have

$$N = \frac{V(p_F^n)^3}{3\pi^2 \hbar^3} \quad \text{and} \quad Z = \frac{V(p_F^p)^3}{3\pi^2 \hbar^3} \tag{17.3}$$

neutrons and protons respectively (p_F^n and p_F^p are the Fermi momenta for the neutrons and protons). With a nuclear volume

$$V = \frac{4}{3}\pi R^3 = \frac{4}{3}\pi R_0^3 A \tag{17.4}$$

and the experimental value $R_0 = 1.21$ fm (5.56), which is obtained from electron scattering, and after assuming that the proton and neutron potential wells have the same radius, we find for a nucleus with $Z = N = A/2$ the Fermi momentum

$$p_F = p_F^n = p_F^p = \frac{\hbar}{R_0}\left(\frac{9\pi}{8}\right)^{1/3} \approx 250 \text{ MeV}/c \,. \tag{17.5}$$

The nucleons it seems move freely inside the nucleus with large momenta.

Quasi-elastic electron-nucleus scattering yields a value for the Fermi momentum (6.22) which agrees well with this prediction. For lighter nuclei p_F tends to be somewhat smaller (Table 6.1, page 80) and the Fermi gas model is not so good in such cases.

The energy of the highest occupied state, the *Fermi energy* E_F, is

$$E_F = \frac{p_F^2}{2M} \approx 33 \text{ MeV} \,, \tag{17.6}$$

where M is the nucleon mass. The difference B' between the top of the well and the Fermi level is constant for most nuclei and is just the average binding energy per nucleon $B/A = 7-8$ MeV. The depth of the potential and the Fermi energy are to a good extent independent of the mass number A:

$$V_0 = E_F + B' \approx 40 \text{ MeV} \,. \tag{17.7}$$

Similarly to the case of a free electron gas in metals, the kinetic energy of the nucleon gas in nuclear matter is comparable to the depth of the potential. This is further evidence that nuclei are rather weakly bound systems.

Generally speaking heavy nuclei have a surplus of neutrons. Since the Fermi level of the protons and neutrons in a stable nucleus have to be equal (otherwise the nucleus would enter a more energetically favourable state through β-decay) this implies that the depth of the potential well as it is experienced by the neutron gas has to be greater than of the proton gas (Fig. 17.1). Protons are therefore on average less strongly bound in nuclei than are neutrons. This may be understood as a consequence of the Coulombic repulsion of the charged protons and leads to an extra term in the potential

$$V_{\mathrm{C}} = (Z-1)\frac{\alpha \cdot \hbar c}{R}\,. \tag{17.8}$$

The dependence of the binding energy upon the surplus of neutrons may also be calculated inside the Fermi gas model. First we find the average kinetic energy per nucleon

$$\langle E_{\mathrm{kin}} \rangle = \frac{\int_0^{p_F} E_{\mathrm{kin}}\, p^2 \mathrm{d}p}{\int_0^{p_F} p^2 \mathrm{d}p} = \frac{3}{5}\cdot\frac{p_F^2}{2M} \approx 20\,\mathrm{MeV}\,. \tag{17.9}$$

The total kinetic energy of the nucleus is therefore

$$E_{\mathrm{kin}}(N,Z) = N\langle E_{\mathrm{n}} \rangle + Z\langle E_{\mathrm{p}} \rangle = \frac{3}{10M}\left(N\cdot (p_F^{\mathrm{n}})^2 + Z\cdot (p_F^{\mathrm{p}})^2\right) \tag{17.10}$$

which may be reexpressed with the help of (17.3) and (17.4) as

$$E_{\mathrm{kin}}(N,Z) = \frac{3}{10M}\frac{\hbar^2}{R_0^2}\left(\frac{9\pi}{4}\right)^{2/3}\frac{N^{5/3}+Z^{5/3}}{A^{2/3}}\,. \tag{17.11}$$

Note that we have again assumed that the radii of the proton and neutron potential wells are the same. This average kinetic energy has for fixed mass number A but varying N or, equivalently, Z a minimum at $N = Z$. Hence

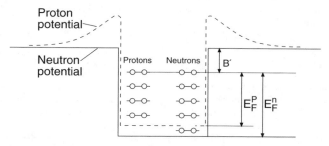

Fig. 17.1. Sketch of the proton and neutron potentials and states in the Fermi gas model.

the binding energy shrinks for $N \neq Z$. If we expand (17.11) in the difference $N - Z$ we obtain

$$E_{\text{kin}}(N, Z) = \frac{3}{10M} \frac{\hbar^2}{R_0^2} \left(\frac{9\pi}{8} \right)^{2/3} \left(A + \frac{5}{9} \frac{(N - Z)^2}{A} + \cdots \right) \qquad (17.12)$$

which gives us the functional dependence upon the neutron surplus. The first term contributes to the volume term in the mass formula while the second describes the correction which results from having $N \neq Z$. This so-called *asymmetry energy* grows as the square of the neutron surplus and the binding energy shrinks accordingly. To reproduce the asymmetry term in (2.8) to a reasonable accuracy it is necessary to take the change in the potential for $N \neq Z$ into account. This additional correction is as important as the change in the kinetic energy.

We thus see that the simple Fermi gas model, where nucleons move freely in an averaged out potential, can already render the volume and asymmetry terms in the semiempirical mass formula plausible.

■ The Fermi gas model may also usefully be applied to a very different form of nuclear matter — *neutron stars*. For these no Coulomb energy has to be considered. As well as the attractive nuclear force, which would lead to a density ϱ_0, we also have the gravitational force and the resulting density can be up to ten times larger.

Neutron stars are produced in supernova explosions. The burnt out centre of the star, which is primarily made of iron and whose mass is between one and two solar masses, collapses under the gravitational force. The high density increases the Fermi energy of the electrons so much that the inverse β-decay $p + e^- \rightarrow n + \nu_e$ takes place, while $n \rightarrow p + e^- + \bar{\nu}_e$ is forbidden by the Pauli principle. All the protons in the atomic nuclei are step by step converted into neutrons. The Coulomb barrier is thus removed, the nuclei lose their identity and the interior of the star is eventually solely composed of neutrons:

$$^{56}_{26}\text{Fe} + 26e^- \rightarrow 56n + 26\nu_e.$$

The implosion is only stopped by the Fermi pressure of the neutrons at a density of $10^{18} \, \text{kg/m}^3$. If the mass of the central core is greater than two solar masses, the Fermi pressure cannot withstand the gravitational force and the star ends up as a black hole.

The best known neutron stars have masses between 1.3 and 1.5 solar masses. The mass of a neutron star which is part of a binary system may be read off from its motion. The radius R can be measured if enough emission lines can still be measured and a gravitational Doppler shift is observable. This is proportional to M/R. Typically one finds values like 10 km for the radius.

We only have theoretical information about the internal structure of neutron stars. In the simplest model the innermost core is composed of a degenerate neutron liquid with a constant density. The roughly 1 km thick crust is made out of atoms which despite the high temperature are bound by a strong gravitational pressure in a solid state. It is therefore a good approximation to treat the neutron star as a gigantic nucleus held together by its own gravitational force.

We will assume that the density of the star is constant in the following estimate of the size of a neutron star. We may then neglect any radial dependence of the gravitational pressure and employ an average pressure. Let us consider a typical neutron star with a mass $M = 3 \cdot 10^{30}$ kg, which is about 1.5 solar masses and corresponds to a neutron number of $N = 1.8 \cdot 10^{57}$. If we view the neutron star as a cold neutron gas, the Fermi momentum is from (17.5)

$$p_{\mathrm{F}} = \left(\frac{9\pi N}{4}\right)^{1/3} \frac{\hbar}{R}. \tag{17.13}$$

The average kinetic energy per neutron is from (17.9)

$$\langle E_{\mathrm{kin}}/N\rangle = \frac{3}{5} \cdot \frac{p_{\mathrm{F}}^2}{2M_{\mathrm{n}}} = \frac{C}{R^2} \quad \text{where} \quad C = \frac{3\hbar^2}{10M_{\mathrm{n}}}\left(\frac{9\pi N}{4}\right)^{2/3}. \tag{17.14}$$

The gravitational energy of a star with constant density implies that the average potential energy per neutron is

$$\langle E_{\mathrm{pot}}/N\rangle = -\frac{3}{5}\frac{GNM_{\mathrm{n}}^2}{R}, \tag{17.15}$$

where M_{n} is the mass of the neutron and G is the gravitational constant. The star is in equilibrium if the total energy per nucleon is minimised:

$$\frac{\mathrm{d}}{\mathrm{d}R}\langle E/N\rangle = \frac{\mathrm{d}}{\mathrm{d}R}\left[\langle E_{\mathrm{kin}}/N\rangle + \langle E_{\mathrm{pot}}/N\rangle\right] = 0. \tag{17.16}$$

and so

$$R = \frac{\hbar^2 (9\pi/4)^{2/3}}{GM_{\mathrm{n}}^3 N^{1/3}}. \tag{17.17}$$

One so finds a radius of about 12 km for such a neutron star, which is very close to the experimental value, and an average neutron density of 0.25 nucleons/fm^3, which is about 1.5 times the density $\varrho_0 = 0.17$ nucleons/fm^3 inside an atomic nucleus (5.59).

This good agreement between the predicted and measured values is, however, rather coincidental. In a more exact calculation one must take into account the fact that the density inside a neutron star grows up to $10\,\varrho_0$ and one then would obtain radii which are much smaller than those measured. On the other hand at a density of $10\,\varrho_0$, the inter-neutron separations are only about 0.8 fm, this means that the hard cores touch and a strong repulsion takes place. Taking this into account we can conclude that the gravitational pressure is in equal measure compensated by the Fermi pressure and by nucleon-nucleon repulsion.

We can also expect an admixture of hyperons in equilibrium with the neutrons for such high densities as are found at the centre of neutron stars. It may also be that the overlap of the neutrons, which is largest at the centre of the star, means that the quarks are no longer confined in the individual neutrons. Neutron stars could be also partially composed of *quark matter*.

17.2 Hypernuclei

The Fermi gas model is generally employed to describe large scale systems (conduction electrons in metals, nucleons in neutron stars, electrons in a white dwarf, etc.) where the quantisation of angular momentum may be neglected. The system of nucleons inside a nucleus is, by contrast, so small that it possesses discrete energy levels with distinct angular momenta. If one calculates the energy levels in a spherically symmetric potential, one finds states with orbital angular momentum $\ell = 0, 1, 2, \ldots$

At zero temperature the lowest lying states are without exception occupied. The interaction between the nucleons can thus merely cause the individual nucleons to swap their places in the energy level spectrum. As this does not change the total energy of the nucleon it is unobservable. This is why we may talk as though each individual nucleon in the nucleus is in a definite energy and angular momentum state. The wave function that describes such a state is the one-particle wave function. The nuclear wave function is just the product of all the one-particle wave functions.

It would be nice, in order to investigate the energy levels of the individual nucleons, if we could somehow "mark" them. An elegant way to more or less do this in an experiment is to introduce a hyperon into the nucleus, ideally a Λ particle, as a probe. A The resulting nucleus is known as a *hypernucleus*.

A Λ particle in the nucleus cannot decay strongly, since strangeness is preserved in that interaction. Its lifetime is therefore roughly that of a free Λ particle, in other words about 10^{-10} s. This is a long enough time to perform a spectroscopic analysis and investigate the properties of hypernuclei.

Hypernuclei are most efficiently produced in the strangeness exchange reaction

$$\mathrm{K}^- + \mathrm{A} \rightarrow {}_\Lambda\mathrm{A} + \pi^- \tag{17.18}$$

where the index shows that a neutron in the nucleus is transformed into a Λ by the reaction

$$\mathrm{K}^- + \mathrm{n} \rightarrow \Lambda + \pi^- . \tag{17.19}$$

Figure 17.2 shows an apparatus that was used at CERN in the 1970's to generate and detect hypernuclei. The kinematics are particularly convenient if the incoming kaon momentum is 530 MeV/c and the final state pions are observed at an angle of $\theta = 0°$ since in this case no momentum is transferred to the scattered nucleus. In practice one uses kaon beams with momenta between 300 and 1000 MeV/c. The transferred momentum is then still small compared to the Fermi momentum of the nucleons in the nucleus, which can then be to a certain extent considered as undisturbed.

The energy balance of the reaction (17.19) with a free neutron just depends upon the masses of the particles involved. If, however, the neutron is bound inside a nucleus and the Λ also remains inside the nucleus then the energy difference between the K^- and the π^- yields the difference between the binding energies of the neutron and the Λ:

$$B_\Lambda = B_\mathrm{n} + E_\pi - E_\mathrm{K} + (M_\Lambda - M_\mathrm{n}) \cdot c^2 + \text{recoil} . \tag{17.20}$$

Figure 17.3 shows such a pion spectrum for this reaction for a ${}^{12}\mathrm{C}$ nucleus as a function of the Λ binding energy, B_Λ. The experimental value for the neutron separation energy in ${}^{12}\mathrm{C}$, i.e., that needed to pull a neutron out of the nucleus, was taken for B_n. As well as a clear peak around $B_\Lambda = 0$ a second, smaller maximum at 11 MeV is observed. This may be interpreted

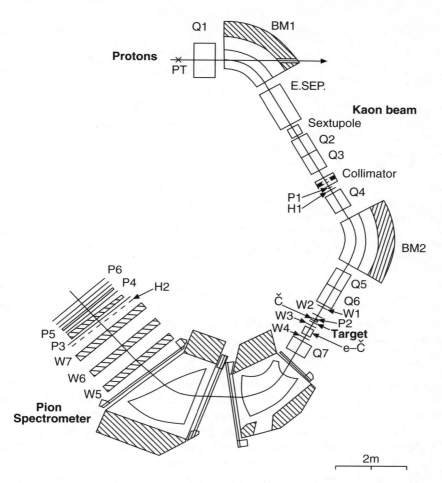

Fig. 17.2. Experimental apparatus for creating and detecting hypernuclei (from [Po81]). A beam of K^- particles hits a 1 cm thick carbon target, generating hypernuclei which, as they are produced, emit π^- mesons. The spectrometer has two stages: initially the momenta of the kaons are measured, then that of the produced pions. The particles are detected and identified with the help of scintillation counters (P), wire chambers (W) and Cherenkov counters (Č). The momenta are measured with dipole magnets (BM) while quadrupole lenses (Q) are responsible for the focusing. The excitation energies of the hypernuclei may be read off from the difference in the kaon and pion energies.

as follows: the transformation of a neutron into a Λ sets free some additional energy which is given to the pion. This energy can only come from the nuclear binding.

We have the following explanation for this. The Pauli principle prevents a proton or a neutron in the nucleus from occupying a lower energy level that is already "taken" – the states in the nucleus get filled "from the bottom up". If

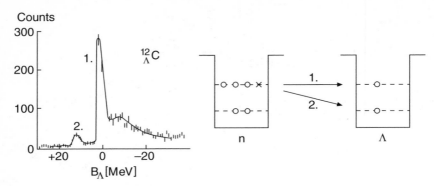

Fig. 17.3. The pion spectrum from the reaction $K^- + {}^{12}C \to \pi^- + {}^{12}_\Lambda C$ for a kaon momentum of 720 MeV/c [Po81]. The pion counting rate at $0°$ is plotted as a function of the transferred energy B_Λ, which may be interpreted as the binding energy of the Λ in the nucleus. Peak no. 1 corresponds to binding energy $B_\Lambda = 0$ and peak no. 2, which is the ${}^{12}_\Lambda C$ ground state, has a binding energy of 11 MeV.

we, however, change a neutron into a Λ particle, then this can occupy any of the states in the nucleus. The Λ does not experience the individual presence of the nucleons, but rather just the potential that they create. This potential is, it should be noted, shallower than that which the nucleons experience. This is because the Λ-nucleon interaction is weaker than that between the nucleons themselves. That this is the case may also be seen from the lack of any bound state formed from a Λ and a single nucleon.

The spectrum of Fig. 17.3 now makes sense: the protons and neutrons in the ${}^{12}C$ nucleus occupy 1s and 1p energy levels and should one of the neutrons in a 1p state be transformed into a Λ, then this can also take up a 1p state. In this case the binding energy of the Λ is close to zero. Alternatively it can land in a 1s state and it then has a binding energy of about $B_\Lambda \approx 11$ MeV.

The smeared out peak with $B_\Lambda < 0$ can be interpreted as arising from the transformation not of weakly bound neutrons near the Fermi level, but rather of deeper lying neutrons.

The Λ one-particle states may be seen even more clearly in heavier nuclei. Systematic investigations, based upon the reaction

$$\pi^+ + A \to {}_\Lambda A + K^+, \tag{17.21}$$

have yielded the binding energies of the 1s states and, furthermore, those of the excited p, d and f states for various nuclei as shown in Fig. 17.4. This shows the dependence of these binding energies upon the mass number A in the nuclei concerned.

In this way it is seen that the Λ hyperons occupy *discrete energy levels*, whose binding energies increase with the mass number. The curves shown are the results of calculations assuming both a potential with uniform depth $V_0 \approx 30$ MeV and that the nuclear radius increases as $R = R_0 A^{1/3}$ [Po81, Ch89].

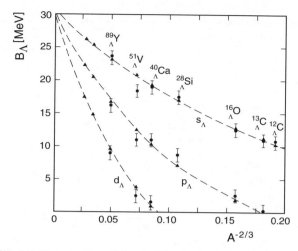

Fig. 17.4. The binding energy of Λ particles in hypernuclei as a function of the mass number A [Ch89]. The symbols s_Λ, p_Λ and d_Λ refer to the state of the Λ in the nucleus. The triangles which are connected by the dashed lines are theoretical predictions.

The scale $A^{-2/3}$ corresponds then to R^{-2} and was chosen because $B_\Lambda R^2$ is almost constant for states with the same quantum numbers, cf. (16.13).

The agreement between the calculated binding energies of the Λ particles and the experimental results is amazing, especially if one considers how simple the potential well is. The Λ moves as a free particle in the well although the nucleus is composed of densely packed matter.

17.3 The Shell Model

The consequences that we have drawn from the spectroscopy of the hypernuclei can be directly applied to the nucleons and we may assume that each nucleon occupies a well-defined energy level.

The existence of these discrete energy levels for the nucleons in the nucleus is reminiscent of the atomic electron cloud. The electrons move in the atom in a central Coulombic potential emanating from the atomic nucleus. In the nucleon, on the other hand, the nucleons move inside a *(mean field)* potential produced by the other nucleons. In both cases discrete energy levels arise which are filled up according to the dictates of the Pauli principle.

Magic numbers. In the atomic case we can order the electrons in "shells". By a shell we mean that several energy levels lie close together clearly separated from the other states. Matters seem to be similar in nuclei.

It is an observed fact that nuclides with certain proton and/or neutron numbers are exceptionally stable (cf. Fig. 2.4) [Ha48]. These numbers (2, 8,

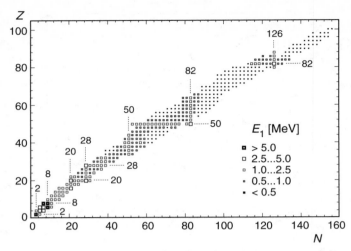

Fig. 17.5. The energy E_1 of the first excited state of even-even nuclei. Note that it is particularly big for nuclei with "magic" proton or neutron number. The excited states generally have the quantum numbers $J^P = 2^+$. The following nuclei are exceptions to this rule: $^4_2\text{He}_2$, $^{16}_8\text{O}_8$, $^{40}_{20}\text{Ca}_{20}$, $^{72}_{32}\text{Ge}_{40}$, $^{90}_{40}\text{Zr}_{50}$ (0^+), $^{132}_{50}\text{Sn}_{82}$, $^{208}_{82}\text{Pb}_{126}$ (3^-) and $^{14}_6\text{C}_8$, $^{14}_8\text{O}_6$ (1^-). E_1 is small further away from the "magic" numbers – and is generally smaller for heavier nuclei (data from [Le78]).

20, 28, 50, 82, 126) are known as *magic numbers*. Nuclei with a magic proton or neutron number possess an unusually large number of stable or very long lived nuclides (cf. Fig. 2.2). If a nucleus has a magic neutron number, then a lot of energy is needed to extract a neutron from it; while if we increase the neutron number by one then the separation energy is much smaller. The same is true of protons. It is also found that a lot of energy is needed to excite such nuclei (Fig. 17.5).

These jumps in the excitation and separation energies for individual nucleons are reminiscent of chemistry: the noble gases, i.e., those with full shells, are particularly attached to their electrons, while the alkali metals, i.e., atoms with just one electron in their outermost shell, have very small separation (ionisation) energies.

The doubly magic nuclei, those with both magic proton and magic neutron numbers, are exceptionally stable. These are the following nuclides:

$$^4_2\text{He}_2\,, \quad ^{16}_8\text{O}_8\,, \quad ^{40}_{20}\text{Ca}_{20}\,, \quad ^{48}_{20}\text{Ca}_{28}\,, \quad ^{208}_{82}\text{Pb}_{126}\,.$$

The existence of these magic numbers can be explained in terms of the socalled *shell model*. For this we need first to introduce a suitable global nuclear potential.

Eigenstates of the nuclear potential. The wave function of the particles in the nuclear potential can divided into two parts: a radial one $R_{n\ell}(r)$, which only depends upon the radius, and a part $Y_\ell^m(\theta, \varphi)$ which only depends upon the orientation (this division is possible for all spherically symmetric potentials; e.g., atoms or quarkonium). The spectroscopic nomenclature for quarkonium is also employed for the quantum numbers here (see p. 170):

$$n\ell \quad \text{with} \quad \begin{cases} n = 1, 2, 3, 4, \cdots & \text{number of nodes} + 1 \\ \ell = \text{s, p, d, f, g, h}, \cdots & \text{orbital angular momentum}. \end{cases}$$

The energy is independent of the m quantum number, which can assume any integer value between $\pm\ell$. Since nucleons also have two possible spin directions, this means that the $n\ell$ levels are in fact $2 \cdot (2\ell+1)$ times degenerate. The parity of the wave function is fixed by the spherical wave function Y_ℓ^m and is just $(-1)^\ell$.

Since the strong force is so short-ranged, the form of the potential ought to follow the density distribution of the nucleons in the nucleus. For very light nuclei ($A \lesssim 7$) this would mean a Gaussian distribution. The potential can then be approximated by that of a three dimensional harmonic oscillator. The Schrödinger equation can be solved analytically in this particularly simple case [Sc95]. The energy depends upon the sum N of the oscillating quanta in all three directions as follows

$$E_{\text{harm. osc.}} = (N + 3/2) \cdot \hbar\omega = (N_x + N_y + N_z + 3/2) \cdot \hbar\omega, \quad (17.22)$$

where N is related to n and ℓ by

$$N = 2(n-1) + \ell. \quad (17.23)$$

Hence states with even N have positive parity and those with odd N negative parity.

Woods-Saxon potential. The density distribution in heavy nuclei can be described by a Fermi distribution, cf. (5.52). The *Woods-Saxon potential* is fitted to this density distribution:

$$V_{\text{centr}}(r) = \frac{-V_0}{1 + e^{(r-R)/a}}. \quad (17.24)$$

States with the same N but different $n\ell$ values are no longer degenerate in this potential. Those states with smaller n and larger ℓ are somewhat lower. The first three magic numbers (2, 8 and 20) can then be understood as nucleon numbers for full shells:

N	0	1	2	2	3	3	4	4	4	$\cdots$
$n\ell$	1s	1p	1d	2s	1f	2p	1g	2d	3s	$\cdots$
Degeneracy	2	6	10	2	14	6	18	10	2	$\cdots$
States with $E \leq E_{n\ell}$	2	8	18	20	34	40	58	68	70	$\cdots$

This simple model does not work for the higher magic numbers. For them it is necessary to include spin-orbit coupling effects which further split the $n\ell$ shells.

Spin-orbit coupling. We may formally introduce the coupling of the spin and the orbital angular momentum (16.8) in the same manner as for the (atomic) electromagnetic interaction. We therefore describe it by an additional ℓs term in the potential:

$$V(r) = V_{\text{centr}}(r) + V_{\ell s}(r)\,\frac{\langle \ell s \rangle}{\hbar^2}\;. \tag{17.25}$$

The combination of the orbital angular momentum ℓ and the nucleon spin s leads to a total angular momenta $j\hbar = \ell\hbar \pm \hbar/2$ and hence to the expectation values

$$\frac{\langle \ell s \rangle}{\hbar^2} = \frac{j(j+1) - \ell(\ell+1) - s(s+1)}{2} = \left\{ \begin{array}{ll} \ell/2 & \text{for } j = \ell + 1/2 \\ -(\ell+1)/2 & \text{for } j = \ell - 1/2\,. \end{array} \right. \tag{17.26}$$

This leads to an energy splitting $\Delta E_{\ell s}$ which linearly increases with the angular momentum as

$$\Delta E_{\ell s} = \frac{2\ell + 1}{2} \cdot \langle V_{\ell s}(r) \rangle\;. \tag{17.27}$$

It is found experimentally that $V_{\ell s}$ is negative, which means that the $j = \ell + 1/2$ is always below the $j = \ell - 1/2$ level, in contrast to the atomic case, where the opposite occurs.

Usually the total angular momentum quantum number $j = \ell \pm 1/2$ of the nucleon is denoted by an extra index. So, for example, the 1f state is split into a $1f_{7/2}$ and a $1f_{5/2}$ state. The $n\ell_j$ level is $(2j + 1)$ times degenerate.

Figure 17.6 shows the states obtained from the potential (17.25). The spin-orbit splitting is separately fitted to the data for each $n\ell$ shell. The lowest shells, i.e., $N = 0$, $N = 1$ and $N = 2$, make up the lowest levels and are well separated from each other. This, as we would expect, corresponds to the magic numbers 2, 8 and 20. For the 1f shell, however, the spin-orbit splitting is already so large that a good sized gap appears above $1f_{7/2}$. This in turn is responsible for the magic number 28. The other magic numbers can be understood in a similar fashion.

This then is the decisive difference between the nucleus and its atomic cloud: the ℓs coupling in the atom generates the fine structure, small corrections of the order of α^2, but the spin-orbit term in the nuclear potential leads to sizeable splittings of the energy states which are indeed comparable with the gaps between the $n\ell$ shells themselves. Historically speaking, it was a great surprise that the the nuclear spin-orbit interaction had such important consequences [Ha49, Gö55].

One particle and one hole states. The shell model is very successful when it comes to explaining the magic numbers and the properties of those nuclei with "one nucleon too many" (or too few).

Those nuclei with mass number between 15 and 17 form a particularly attractive example of this. Their excited states are shown in Fig. 17.7. The

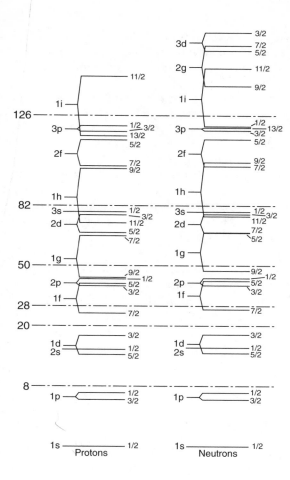

Fig. 17.6. Single particle energy levels calculated using (17.25) (from [Kl52]). Magic numbers appear when the gaps between successive energy shells are particularly large. This diagram refers to the nucleons in the outermost shells.

^{15}N and ^{15}O nuclei are so-called *mirror nuclei*, i.e., the neutron number of the one is equal to the proton number of the other and vice versa. Their spectra are exceedingly similar, both in terms of where the levels are and also in terms of their spin and parity quantum numbers. This is a consequence of the isospin independence of the nuclear force: if we swap protons and neutrons the strong force essentially does not notice it. The small differences in the spectra can be understood as electromagnetic effects. While the energy levels of ^{16}O do not resemble those of its neighbours, the ^{17}O und ^{17}F nuclei are, once again, mirror nuclei and have very similar excitation spectra. It is striking that the nuclei with mass numbers 15 and 16 require much more energy to reach their first excited states than do those with mass number 17.

These spectra can be understood inside the shell model. The ^{16}O nucleus possesses 8 protons and 8 neutrons. In the ground state the $1s_{1/2}$, $1p_{3/2}$ and $1p_{1/2}$ proton and neutron shells are fully occupied and the next highest shells, $1d_{5/2}$, are empty. Just as in atomic physics the angular momenta of the particles in a full shell add up to zero and the overall parity is positive. The ground state of ^{16}O has then the quantum numbers $J^P = 0^+$. Since the gap between the $1p_{1/2}$ and $1d_{5/2}$ energy shells is quite large (about 10 MeV) there are no easily reachable excitation levels.

The two nuclei with $A = 17$ both have a single extra nucleon in the $1d_{5/2}$ shell. The spin and parity of the nucleus are completely fixed by this one nucleon. The $2s_{1/2}$ shell happens to be just a little above the $1d_{5/2}$ shell and as small an energy as 0.5 MeV suffices to excite this single nucleon to the next shell. The nuclear quantum numbers change from $5/2^+$ to $1/2^+$ in this transition. The excited nucleon later decays, through photon emission, into the lowest possible state. Just as we talk of valence electrons in atomic physics, so these nucleons that jump between shells are known as valence nucleons. The $1d_{3/2}$ shell is about 5 MeV above the $1d_{5/2}$ one and this an amount of energy is required to reach this state.

The $A = 15$ ground states lack one nucleon in the $1p_{1/2}$ shell. One speaks of a *hole* and uses the notation $1p_{1/2}^{-1}$. The quantum numbers of the hole are those of the nucleus. Thus the ground states of these nuclei have the quantum numbers $J^P = 1/2^-$. If a nucleon from the $1p_{3/2}$ shell is excited into the vacant state in the $1p_{1/2}$, and in some sense fills the hole, a hole is

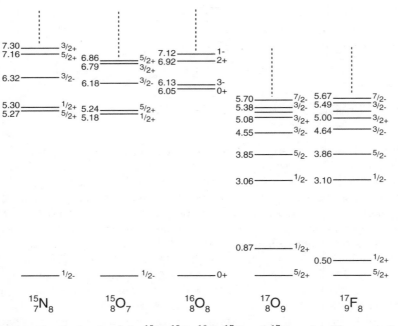

Fig. 17.7. Energy levels of the ^{15}N, ^{15}O, ^{16}O, ^{17}O and ^{17}F nuclei. The vertical axis corresponds to the excitation energy of the states with the various ground states all being set equal, i.e., the differences between the binding energies of these nuclei are not shown.

then created in the $1p_{3/2}$ shell. The new nuclear state then has the quantum numbers $J^P = 3/2^-$.

Magnetic moments from the shell model. If in the shell model we associate spin and orbital angular momentum to each individual nucleon, then we can understand the magnetic moment of the nucleus from the sum over the nucleonic magnetic moments based upon their spin and orbital angular momenta:

$$\boldsymbol{\mu}_{\text{nucleus}} = \mu_N \cdot \frac{1}{\hbar} \sum_{i=1}^{A} \{\boldsymbol{\ell}_i g_\ell + \boldsymbol{s}_i g_s\} \ . \tag{17.28}$$

Note that

$$g_\ell = \begin{cases} 1 & \text{for protons} \\ 0 & \text{for neutrons} \end{cases} \tag{17.29}$$

and (from 6.7 etc.):

$$g_s = \begin{cases} +5.58 & \text{for protons} \\ -3.83 & \text{for neutrons.} \end{cases} \tag{17.30}$$

Recall our five nuclei with mass numbers from 15 to 17. The magnetic moment of ^{16}O is zero, which makes perfect sense since in a full shell the spins and angular momenta add up to zero and so the magnetic moment must vanish.

We are in a position to make quantitative predictions for one particle and one hole states. We first assume that the nuclear magnetic moment is determined by that of the single nucleon or hole

$$\boldsymbol{\mu}_{\text{nucleus}} = \frac{1}{\hbar} \langle \psi_{\text{nucleus}} | g_\ell \boldsymbol{\ell} + g_s \boldsymbol{s} | \psi_{\text{nucleus}} \rangle \cdot \mu_N \ . \tag{17.31}$$

The Wigner-Eckart theorem tells us that the expectation value of every vector quantity is equal to its projection onto the total angular momentum, which here means the nuclear spin $\boldsymbol{J}$:

$$\boldsymbol{\mu}_{\text{nucleus}} = g_{\text{nucleus}} \cdot \mu_N \cdot \frac{\langle \boldsymbol{J} \rangle}{\hbar} \tag{17.32}$$

where

$$g_{\text{nucleus}} = \frac{\langle JM_J | g_\ell \boldsymbol{\ell} \boldsymbol{J} + g_s \boldsymbol{s} \boldsymbol{J} | JM_J \rangle}{\langle JM_J | \boldsymbol{J}^2 | JM_J \rangle} \ . \tag{17.33}$$

Since the nuclear spin $\boldsymbol{J}$ in our model is nothing but the total angular momentum of our single nucleon $\boldsymbol{j}$ and we have

$$2\boldsymbol{\ell}\boldsymbol{j} = \boldsymbol{j}^2 + \boldsymbol{\ell}^2 - \boldsymbol{s}^2 \qquad 2\boldsymbol{s}\boldsymbol{j} = \boldsymbol{j}^2 + \boldsymbol{s}^2 - \boldsymbol{\ell}^2 \tag{17.34}$$

we see that

$$g_{\text{nucleus}} = \frac{g_\ell\left\{j(j+1)+\ell(\ell+1)-s(s+1)\right\} + g_s\left\{j(j+1)+s(s+1)-\ell(\ell+1)\right\}}{2j(j+1)}.$$

(17.35)

The magnetic moment of the nucleus is defined as the value measured when the nuclear spin is maximally aligned, i.e., $|M_J| = J$. The expectation value of $\langle \boldsymbol{J} \rangle$ is then $J\hbar$ and one finds

$$\frac{|\boldsymbol{\mu}_{\text{nucleus}}|}{\mu_{\text{N}}} = g_{\text{nucleus}} \cdot J = \left(g_\ell \pm \frac{g_s - g_\ell}{2\ell+1}\right) \cdot J \qquad \text{for } J = j = \ell \pm \frac{1}{2}.$$

(17.36)

There are many different ways to measure nuclear magnetic moments, e.g., in nuclear magnetic spin resonance or from optical hyperfine structure investigations [Ko56]. The experimental values [Le78] of the magnetic moments can be compared with the predictions of (17.36).

Nucleus	State	J^P	μ/μ_{N} Model	Expt.
^{15}N	p-$1p_{1/2}^{-1}$	$1/2^-$	-0.264	-0.283
^{15}O	n-$1p_{1/2}^{-1}$	$1/2^-$	$+0.638$	$+0.719$
^{17}O	n-$1d_{5/2}$	$5/2^+$	-1.913	-1.894
^{17}F	p-$1d_{5/2}$	$5/2^+$	$+4.722$	$+4.793$

The magnetic moments of the $A = 15$ and $A = 17$ nuclei can, we see, be understood in a single particle picture. We should now perhaps admit to having chosen the example with the best agreement between the model and experiment: firstly these nuclei are, up to one single nucleon or hole, doubly magic and secondly they have a relatively small nucleon number which means that effects such as polarisation of the remainder by the valence nucleon are relatively tiny.

We assume for nuclei with odd mass number whose incomplete shells contain more than one nucleon or hole that the total nucleon magnetic moment is due to the one unpaired nucleon [Sc37]. The model then roughly reproduces the experimental trends, but disagreements as big as $\pm 1\mu_{\text{N}}$ and larger appear for many nuclei. The magnetic moment is, generally speaking, smaller than expected. The polarisation of the rest of the nucleus from the unpaired nucleon tends to explain this [Ar54].

17.4 Deformed Nuclei

The shell model approximation which assumes that nuclei are spherically symmetric objects – plus, of course, the additional spin-orbit interaction – is only good for those nuclei which are close to having doubly magic full shells. For nuclei with half full shells this is not the case. In such circumstances the nuclei are deformed and the potential is no longer spherically symmetric.

It was already realised in the 1930's, from atomic spectroscopy, that nuclei are not necessarily always spherical [Ca35, Sc35]. Deviations in the fine structure of the spectra hinted at a non-vanishing electrical quadrupole moment, i.e. that the charge distribution of the nuclei was not spherically symmetric.

Quadrupole moments. The charge distribution in the nucleus is described in terms of electric multipole moments. Since the odd moments (e.g., the dipole and octupole) have to vanish because of parity conservation, the electric quadrupole moment is the primary measure of in how far the charge distribution, and hence the nucleus, deviate from being spherical.

The classical definition of a quadrupole moment is

$$Q = \int \left(3z^2 - x^2\right) \varrho(x) \, \mathrm{d}^3 x \,. \tag{17.37}$$

An ellipsoid of diameter $2a$ in the z direction and diameter $2b$ in the other two directions (Fig. 3.9), with constant charge density $\varrho(x)$ has the following quadrupole moment:

$$Q = \frac{2}{5} Z e \left(a^2 - b^2\right) \,. \tag{17.38}$$

For small deviations from spherical symmetry, it is usual to introduce a measure for the deformation. If the average radius is $\langle R \rangle = (ab^2)^{1/3}$ and the difference is $\Delta R = a - b$ then the quadrupole moment is proportional to the *deformation parameter* [1]

$$\delta = \frac{\Delta R}{\langle R \rangle} \tag{17.39}$$

and we find

$$Q = \frac{4}{5} Z e \langle R \rangle^2 \delta \,. \tag{17.40}$$

Since the absolute value of a quadrupole moment depends upon the charge and size of the nucleus concerned, we now introduce the concept of the *reduced quadrupole moment* to facilitate the comparison of the deformations of nuclei with different mass numbers. This is a dimensionless quantity and is defined as the quadrupole moment divided by the charge Ze and the square of the average radius $\langle R \rangle$:

$$Q_{\mathrm{red}} = \frac{Q}{Ze\langle R \rangle^2} \,. \tag{17.41}$$

The experimental data for the reduced quadrupole moments are shown in Fig. 17.8. Note that no even-even nuclei are included, as quantum mechanics prevents us from measuring a static quadrupole moment for systems with angular momenta 0 or 1/2. As one sees, the reduced quadrupole moment is small around the magic number nuclei but it is large if the shells are not nearly closed – especially in the lanthanides (e.g., ^{176}Lu and ^{167}Er). If Q is positive, $a > b$, the nucleus is prolate (shaped like a cigar); if it is negative

---

[1] We skip over the exact definition of the deformation parameter here; (17.38) and (17.39) are approximations for small deformations.

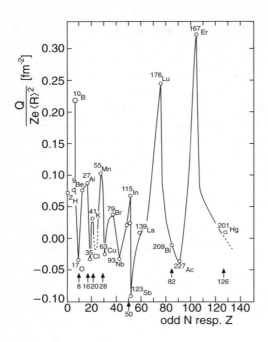

Fig. 17.8. Reduced quadrupole moments for nuclei with odd proton number Z or neutron number N plotted against this number. The quadrupole moments vanish near closed shells and reach their largest values far away from them. It is further clear that prolate nuclei ($Q > 0$) are more common than oblately deformed ones ($Q < 0$). The solid curves are based upon the quadrupole moments of very many nuclei, of which only a few are explicitly shown here.

then the nucleus is oblately deformed (shaped like a lens). The latter is the rarer case.

The electric quadrupole moments of deformed nuclei are too large to be explained solely in terms of the protons in the outermost, incomplete shell. It is rather the case that the partially occupied proton and neutron shells polarise and deform the nucleus as a whole.

Figure 17.9 shows in which nuclides such partially full shells have especially strong effects. Stable deformed nuclei are especially common among the rare earths (the lanthanides) and the transuranic elements (the actinides). The light nuclei with partially full shells are also deformed, but, due to their smaller nucleon number, their collective phenomena are less striking.

Pairing and polarisation energies. We can see why in particular nuclei with half full shells are deformed if we consider the spatial wave functions of the nucleons. Nucleons in a particular shell have a choice among various spatial and spin states. In atomic physics we have the *Hund rule*: as we fill up an $n\ell$ subshell with electrons, these initially take up the various hitherto unoccupied orbitals in position space and only when no empty orbitals are left do they start to use the space in every orbital for a further electron with opposite spin. The underlying reason is the electromagnetic repulsion of the electrons, which makes it energetically favourable to have two electrons in spatially separated orbitals rather than having two electrons with opposite spins in the same orbital. Matters are different in nuclear physics, however. The force between the nucleons is, on average, an attractive one. This has two consequences:

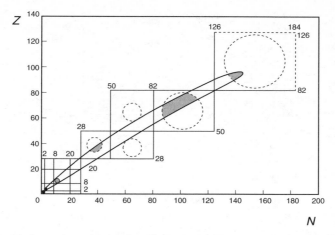

Fig. 17.9. Deformed nuclei in the N-Z plane. The horizontal and vertical lines denote the magic proton and neutron numbers respectively (i.e., they show where the closed shells are). The regions where large nuclear deformations are encountered are shaded (from [Ma63]).

– Nuclei become more stable if the nucleons are grouped in pairs with the same spatial wave function and if their angular momenta add to zero, i.e., also: $\ell_1 = \ell_2$, $m_1 = -m_2$, $\boldsymbol{j}_1 + \boldsymbol{j}_2 = \mathbf{0}$. We talk of a pairing force. Such pairs have angular momentum and parity, $J^P = 0^+$.

– Nucleon pairs prefer to occupy neighbouring orbitals (states with adjacent m values) and this leads, if the nucleus has a half full shell, to deformations. If the filled orbitals tend to be parallel to the symmetry axis (Fig. 17.10a) then the nucleus is prolately deformed and if they are perpendicular to this axis (Fig. 17.10b) the resulting nucleus is oblate.

The angular momenta and parity of nuclei are then, not only for almost magic nuclei but quite generally, fixed by individual, unpaired nucleons. Doubly even nuclei will, because of the pairing energy, always have $J^P = 0^+$ ground states, the J^P of singly odd nuclei will be determined by their one odd nucleon and, finally, the spin and parity of doubly odd nuclei will depend upon how the quantum numbers of the two unpaired nucleons combine. Experimentally determined ground state quantum numbers are in excellent agreement with these ideas.

Single particle movement of the nucleons. It is necessary, should one want to calculate the energy levels of a deformed nucleus, to recall that the nuclear potential has an ellipsoidal shape. The spin-orbit force is as strong as for the spherically symmetric potential. The one particle states of deformed nuclei may be found in a conceptionally simple way (the Nilsson model [Ni55]) but the calculations are tedious. The nucleon angular momentum is no longer a conserved quantity in a deformed potential and its place is taken by the projection of the angular momentum onto the symmetry axis of the nucleus.

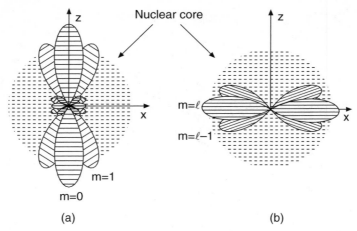

Fig. 17.10. a,b. Overlapping orbitals with adjacent m quantum numbers. If m is close to zero the orbitals are parallel to z, the symmetry axis (**a**). If $|m|$ is large they are perpendicular to this axis (**b**). The remainder of the nucleus is drawn here as a sphere. This is because nuclear deformations are primarily due to the nucleons in partially filled shells.

The Nilsson wave functions are therefore built up out of shell model wave functions with the same n but different ℓ, although their angular momentum projections m_j must be the same.

17.5 Spectroscopy Through Nuclear Reactions

Until now we have mainly concentrated upon experiments using electromagnetic probes (electrons), since the electromagnetic interaction is particularly easily described. It is, however, the case that our modern understanding of nuclear structure, and in particular the quantitative determination of the single particle properties of low lying nuclear states, comes from analysing reactions where the target and the projectile interact via the nuclear force. Our first quantitative knowledge of the various components of the wave functions goes back to studies of so-called direct reactions. The most prominent examples of these are "stripping" and "pick-up" reactions. In what follows we will restrict ourselves to a qualitative description of these two types of reactions and show how complex the problem becomes when one tries to extract quantitative information.

Stripping reactions. Stripping reactions are nuclear reactions where one or more of the nucleons from the projectile nucleus are stripped off it and transferred to the target nucleus. The simplest examples of this are the deuteron induced (d, p) and (d, n) reactions:

$$\mathrm{d} + {}^{A}Z \rightarrow \mathrm{p} + {}^{A+1}Z \quad \text{and} \quad \mathrm{d} + {}^{A}Z \rightarrow \mathrm{n} + {}^{A+1}(Z+1) \,.$$

The following shorthand notation is commonly used to denote such reactions

$$^A Z(\mathrm{d},\mathrm{p})^{A+1} Z \qquad\qquad ^A Z(\mathrm{d},\mathrm{n})^{A+1}(Z+1) \,.$$

If the incoming deuteron carries a lot of energy, compared to the binding energies of the deuteron and of a neutron in the $(A+1)$ nucleus, then a quantitative description of the stripping reaction is quite possible. The stripping reaction $^{16}\mathrm{O}(\mathrm{d},\mathrm{p})^{17}\mathrm{O}$ is depicted in Fig. 17.11.

The cross-section may be calculated from Fermi's golden rule and one finds from (5.22)

$$\frac{\mathrm{d}\sigma}{\mathrm{d}\Omega} = \frac{2\pi}{\hbar} \, |\mathcal{M}_{fi}|^2 \, \frac{p^2 \mathrm{d}p \, V^2}{(2\pi\hbar)^3 v_\mathrm{D} \mathrm{d}E} \,. \tag{17.42}$$

We write the matrix element as

$$\mathcal{M}_{fi} = \langle \psi_f | U_{\mathrm{n,p}} | \psi_i \rangle \,, \tag{17.43}$$

where ψ_i and ψ_f are the initial and final state wave functions and $U_{\mathrm{n,p}}$ is the interaction that causes the stripping reaction.

Born approximation. The physical interpretation of the stripping reaction becomes evident when we consider the matrix element in the Born approximation. We assume thereby that the interaction between the deuteron and the nucleus and also that between the proton and the nucleus are both so weak that we may describe the incoming deuteron and the outgoing proton by plane waves. In this approximation the initial state wave function is

$$\psi_i = \phi_\mathrm{A} \, \phi_\mathrm{D} \, \exp(i \boldsymbol{p}_\mathrm{D} \boldsymbol{x}_\mathrm{D} / \hbar). \tag{17.44}$$

Here ϕ_A signifies the ground state of the target nucleus and ϕ_D the internal structure of the deuteron. The incoming deuteron plane waves are contained in the function $\exp(i \boldsymbol{p}_\mathrm{D} \boldsymbol{x}_\mathrm{D} / \hbar)$. The final state wave function

$$\psi_f = \phi_{\mathrm{A}+1} \, \exp(i \boldsymbol{p}_\mathrm{p} \boldsymbol{x}_\mathrm{p} / \hbar) \tag{17.45}$$

contains the wave function of the nucleus containing the extra neutron and the outgoing proton's plane waves.

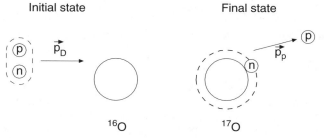

Fig. 17.11. Sketch of the stripping reaction $^{16}\mathrm{O}(\mathrm{d},\mathrm{p})^{17}\mathrm{O}$.

The only likely final states in stripping reactions are those such that the nucleon state is not too greatly changed: so we can write the final state to a good approximation as a product of the type

$$\phi_{A+1} = \phi_A \psi_n \, , \tag{17.46}$$

where ϕ_A describes the internal state of the target nucleus and ψ_n is a shell model wave function of the neutron in the potential of the nucleus A.

If the stripping process takes place via a very short ranged interaction

$$U_{n,p}(\boldsymbol{x}_n, \boldsymbol{x}_p) = U_0 \, \delta(\boldsymbol{x}_n - \boldsymbol{x}_p) \, , \tag{17.47}$$

then the matrix element has a very simple form

$$
\begin{aligned}
\langle \psi_f | U_{n,p} | \psi_i \rangle &= \int \psi_n^*(\boldsymbol{x}) \, U_0 \, \exp(i(\boldsymbol{p}_D/2 - \boldsymbol{p}_p)\boldsymbol{x}/\hbar) \, \phi_D(\boldsymbol{x} = 0) \, \mathrm{d}^3 x \\
&= U_0 \, \phi_D(\boldsymbol{x} = 0) \int \psi_n^*(\boldsymbol{x}) \, \exp(i\boldsymbol{q}\boldsymbol{x}/\hbar) \, \mathrm{d}^3 x \, . \tag{17.48}
\end{aligned}
$$

Since $\boldsymbol{p}_D/2$ is the average momentum of the proton in the deuteron before the stripping reaction, $\boldsymbol{q} = \boldsymbol{p}_D/2 - \boldsymbol{p}_p$ is just the average momentum transfer to the nucleus.

The amplitude of the stripping reaction, if we use the Born approximation and a short ranged interaction, is just the Fourier integral of the wave function of the transferred neutron. The differential cross-section of the (d, p) reaction is proportional to the square of the matrix element and hence to the square of the Fourier integral.

The most important approximation that we have made in calculating the matrix element is the assumption that the interaction which transfers the neutron from the deuteron to the nucleus leaves the motion of the proton basically unchanged. This is a good approximation for deuteron energies greater than 20 MeV or so, since the deuteron binding energy is only 2.225 MeV. The proton will remain on its course even after the neutron is detached.

Angular momentum. The orbital angular momentum transfer in the stripping reaction is just the orbital angular momentum of the transferred neutron in the state $|\psi_n\rangle$. The transfer of $\ell\hbar$ angular momentum to a nucleus with radius R requires a momentum transfer of roughly $|\boldsymbol{q}| \approx \ell\hbar/R$. This implies that the first maximum in the angular distribution $\mathrm{d}\sigma/\mathrm{d}\Omega$ of the protons will lie at an angle which corresponds to this momentum transfer. Thus the angular distribution of stripping reactions tells us the ℓ quantum number of the single particle states.

The reaction $^{16}O(d, p)^{17}O$. Figure 17.12 displays the outgoing proton spectrum as measured in the reaction $^{16}O(d, p)^{17}O$ at a scattering angle of $\theta = 45°$ and with incoming deuteron energies of 24.5 MeV. One recognises 6 peaks which all correspond to different, discrete excitation energies E_x of ^{17}O. If one measures at a smaller angle θ, and hence smaller momentum transfer,

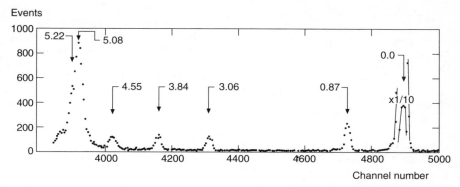

Fig. 17.12. Proton spectrum of a (d, p) reaction on ^{16}O, measured at 45° and projectile energy of 25.4 MeV (from [Co74]). The channel number is proportional to the proton energy and the excitation energies of ^{16}O are marked at the various peaks. The ground state and the excited states at 0.87 MeV and 5.08 MeV possess $J^P = 5/2^+$, $1/2^+$ and $3/2^+$ quantum numbers respectively and essentially correspond to the $(\text{n-1d}_{5/2})^1$, $(\text{n-2s}_{1/2})^1$ and $(\text{n-1d}_{3/2})^1$ single particle configurations.

three of these maxima disappear. (The mechanisms which are responsible for the population of these states are more complicated than those of the direct reactions.) The three remaining maxima correspond to the following single particle states: the $J^P = 5/2^+$ $(\text{n-1d}_{5/2})$ ground state, the $J^P = 1/2^+$ $(\text{n-2s}_{1/2})$ 0.87 MeV excited state and the $J^P = 3/2^+$ $(\text{n-1d}_{3/2})$ 5.08 MeV excited state (cf. Fig. 17.7).

The angular distributions of the protons for these three single particle states are shown in Fig. 17.13. The maximum of the data for $E_x = 0.87$ MeV is at $\theta = 0°$, i.e., at zero momentum transfer. This implies that the neutron which has been transferred to the nucleus is in a state with zero orbital angular momentum ℓ. And indeed we interpreted this state, with quantum numbers $J^P = 1/2^+$, in the shell model as an ^{16}O nucleus with an extra neutron in the $2s_{1/2}$ shell. The two other angular distributions shown have maxima at larger momentum transfers, which signify $\ell = 2$. This is also completely consistent with their quantum numbers. The relative positions of the shells can be determined from such considerations.

Limits of the Born approximation – DWBA. The results shown in Fig. 17.13 cannot be obtained using the Born approximation, since neither the deflection of the particles in the nuclear field nor absorbtion effects are taken into account in that approximation. One way to improve the approximation is to use more realistic incoming deuteron and outgoing proton wave functions, so that they describe the scattering process as exactly as possible, instead of the plane waves we have employed until now. These wave functions are produced by complicated computer analyses and the results are then compared with our experimental knowledge of elastic proton and deuteron scattering off nuclei. This calculational procedure is known as the *distorted wave Born approximation (DWBA)*. The continuous lines in Fig. 17.13

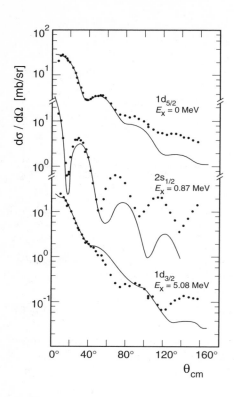

Fig. 17.13. Angular distributions from the ^{16}O(d, p)^{17}O reaction for projectile energies of 25.4 MeV (from [Co74]). The continuous curves are the results of calculations where the absorbtion of the deuteron by ^{16}O was taken into account (DWBA).

are the results of such very tedious calculations. It is obvious that even the best models are only capable of quantitatively reproducing the experimental results at small momentum transfers (small angles).

Pick up reactions. Pick up reactions are complementary to stripping reactions. A proton or neutron is carried away from the target nucleus by a projectile nucleus. Typical examples of this are the (p, d), (n, d), (d,^{3} He) and (d,^{3} H) reactions. A (p, d) reaction is shown as an example in Fig. 17.14.

The ideas we used to understand the (d, p) stripping reaction may be directly carried over to the (p, d) pick up reaction. In the Born approximation, we must only replace the wave function of the transferred neutron $|\psi_n\rangle$ in (17.48) by that of the $|\psi_n^{-1}\rangle$ hole state.

The reaction ^{16}O(d, ^{3}He)^{15}N. It may be clearly seen from Fig. 17.15 that two ^{15}N states are primarily produced in the reaction ^{16}O(d, ^{3}He)^{15}N. These two states are the $1p_{1/2}$ and $1p_{3/2}$ hole states. The other states are rather more complicated configurations (e.g., one particle and two holes) and are much less often excited.

The energy difference between the ground state ($J^P = 1/2^-$) and the $J^P = 3/2^-$ state is 6.32 MeV (cf. Fig. 17.7). This corresponds to the splitting of the 1p shell in light nuclei due to the ℓs interaction.

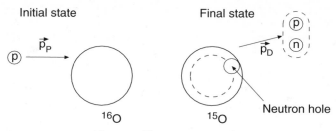

Fig. 17.14. Sketch of the $^{16}O(p,d)^{15}O$ pick up reaction.

The differential cross-sections for these states are shown in Fig. 17.16. The model calculations are based upon the simple assumption that these states are pure $p_{1/2}$ and $p_{3/2}$ hole states. They clearly reproduce the experimental data at small momentum transfers rather well. The admixture of higher configurations must then be tiny. At larger momentum transfers the reaction mechanisms become more complicated and the approximations used here are no longer good enough.

Direct reactions with heavy nuclei. Stripping and pick up reactions are well suited for the task of investigating the one particle properties of both

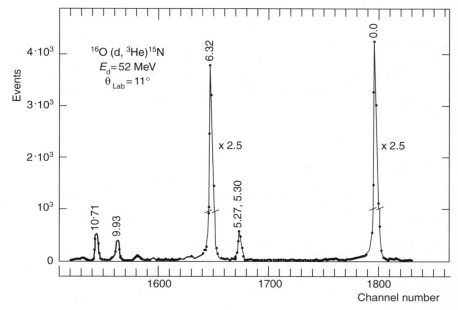

Fig. 17.15. Spectrum of ^3He nuclei detected at $11°$ when 52 MeV deuterons were scattered off ^{16}O (from [Ma73]). The cross-sections for the production of ^{15}N in the ground state and in the state with an excitation energy of 6.32 MeV are particularly large (and are scaled down in the diagram by a factor of 2.5).

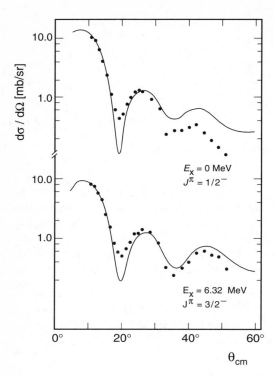

Fig. 17.16. Differential cross-section $d\sigma/d\Omega$ for the reaction $^{16}O(d, {}^{3}He)^{15}N$ (from [Be77]). See also the caption to Fig. 17.13.

spherical and deformed heavy nuclei. Valence nucleons or valence holes are again excited close to full and nearly empty shells. In those nuclei where there are half full shells, excited states cannot be described by an excited state of the shell model, rather a mixture of various shell model states must be used. The properties of the excited states are then determined by the coupling of the valence nucleons.

17.6 β-Decay of the Nucleus

β-decay provides us with another way to study nuclear structure. The β-decay of individual hadrons was treated in Sect. 15.5 where the example of free neutron decay was handled in more detail. At the quark level this transition corresponds to a d-quark changing into a u-quark. We have already seen that the axial coupling (15.38) is modified in the n → p transition by the internal hadronic structure and the influence of the strong interaction.

If the nucleon is now contained inside a nucleus, further effects need to be considered.

– The matrix element must now contain the overlap of the initial and final state nuclear wave functions. This means that the matrix element of β-decay lets us glimpse inside the nucleus containing the nucleons.

- The difference between the binding energies of the nuclei before and after the decay defines the type of decay (β^+ or β^-) and fixes the size of the phase space.
- The Coulomb interaction influences the energy spectrum of the emitted electrons or positrons, especially at small velocities, and thus also modifies the phase space.

Phase space. We calculated in (15.47) the decay rate as a function of the total energy E_0 of the electron and the neutrino. In nuclei the difference between the masses of the initial and final state nuclei yields E_0. The integral over the phase space $f(E_0)$ is now altered by the Coulomb interaction between the charge $\pm e$ of the emitted electron or positron and that $Z'e$ of the remaining nucleus. This is described by the so-called *Fermi function* $F(Z', E_e)$ which is approximately given by

$$F(Z', E_e) \approx \frac{2\pi\eta}{1 - e^{-2\pi\eta}} \qquad \text{where} \quad \eta = \mp\frac{Z'e^2}{4\pi\varepsilon_0 \hbar v_e} = \mp\frac{Z'\alpha}{v_e/c} \quad \text{for } \beta^\pm ,$$

(17.49)

where v_e is the measured final velocity of the electron or positron. The phase space function $f(E_0)$ in (15.46) is replaced by

$$f(Z', E_0) = \int_1^{\mathcal{E}_0} \mathcal{E}_e \sqrt{\mathcal{E}_e^2 - 1} \cdot (\mathcal{E}_0 - \mathcal{E}_e)^2 \cdot F(Z', \mathcal{E}_e) \, d\mathcal{E}_e$$

$$\text{where} \qquad \mathcal{E} = E/m_e c^2 ,$$

(17.50)

which can be calculated to a high precision [Be69]. The influence of the Coulomb force upon the β-spectrum is shown in Fig. 17.17.

In spectroscopy the information about the structure of the nucleus is contained in the matrix element. The product of the half life $t_{1/2}$ and $f(Z', E_0)$, which is called the *ft value*, is directly proportional to the inverse square of the matrix element. From (15.47) using $t_{1/2} = \ln 2 \cdot \tau$ one obtains:

$$f(Z', E_0) \cdot t_{1/2} = ft \text{ value} = \frac{2\pi^3 \hbar^7}{m_e^5 c^4} \cdot \ln 2 \cdot \frac{1}{V^2} \cdot \frac{1}{|\mathcal{M}_{fi}|^2}.$$

(17.51)

The *ft* values vary from as little as 10^3 s to as much as 10^{22} s. Normally therefore the base ten logarithm of its value (in seconds), the *log-ft value*, is quoted.

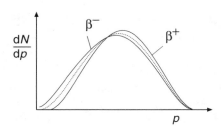

$\frac{dN}{dp}$

Fig. 17.17. Schematic appearance of the electron spectrum in β-decay. The phase space factor from (15.45) produces a spectrum with a parabolic fall off at both ends (*dotted lines*). This is modified by the interaction of the electron/positron with the Coulomb field of the final state nucleus (*continuous lines*). These latter curves were calculated from (17.49) for $Z' = 20$ and $E_0 = 1\,\mathrm{MeV}$.

The matrix element. The matrix element is influenced not only by the wave function of the nucleon in which the quark transition takes place, but also in turn by the wave function of the nucleus containing the nucleon. In both cases this depends upon how the wave functions before and after the decay overlap.

The ratio of the vector and axial vector parts is determined by the nuclear wave function. Those decays that take place through the vector part of the transition operator are called *Fermi decays*. The spin of the interacting quark does not change here and so the spin of the nucleon is unaffected. The total spin of the electron and the neutrino is thus zero. The decays due to the axial part are called *Gamow-Teller decays*. The lepton spins add up to one here. Generally both Fermi and Gamow-Teller β-decays are possible. There are, however, cases where only, or nearly only, one of the decays takes place.

Let us attempt to estimate what role is played by orbital angular momentum. The wave function of the electron and the neutrino may be written as a plane wave to a good approximation (cf. 5.18):

$$\psi(\boldsymbol{x}) = \frac{e^{i\boldsymbol{p}\boldsymbol{x}/\hbar}}{\sqrt{V}} = \frac{1}{\sqrt{V}}\left\{\ 1 + i\boldsymbol{p}\boldsymbol{x}/\hbar + \cdots\right\}\ . \tag{17.52}$$

Since $\boldsymbol{\ell} = \boldsymbol{x} \times \boldsymbol{p}$ this is an expansion in the orbital angular momentum quantum number ℓ. Since the momenta are at most of the order of a few MeV/c and the nuclear radii are a few fm, $|\boldsymbol{p}| \cdot R/\hbar$ must be of the order of 10^{-2}. The ft value contains the square of the matrix element and so we see that every extra unit of ℓ suppresses the decay by a factor of $10^{-4} - 10^{-3}$. Decays with $\ell = 0$ are called *allowed*, those with $\ell = 1$ are then *forbidden* and if $\ell = 2$ we speak of a *doubly forbidden* decay etc. If ℓ is odd the parity of the nuclear wave function changes, while if it is even parity is conserved.

The following selection rules hold for allowed decays as a result of conservation of angular momentum and parity:

$$\Delta P = 0, \quad \Delta J = 0 \qquad\qquad\qquad\quad \text{for Fermi decays,}$$
$$\Delta P = 0, \quad \Delta J = 0, \pm 1; \ (0 \rightarrow 0 \text{ forbidden}) \quad \text{for Gamow-Teller decays.}$$

Large ℓ decays only play a role if lower ℓ transitions are ruled out on grounds of angular momentum or parity conservation. Thus for example the decay of a 1^- into a 0^+ nucleus is only possible via a (once) forbidden transition and not by an allowed Gamow-Teller transition since the parity of the nucleus changes.

An example of a four times forbidden β-decay is the transition from ^{115}In ($J^P = 9/2^+$) into ^{115}Sn ($J^P = 1/2^+$). The log-ft value of this decay is 22.7 and its half life is, believe it or not, $6 \cdot 10^{14}$ years.

Super allowed decays. If the initial and final state wave functions overlap perfectly then the decay probability is particularly large. This is the case if the created proton and the decayed neutron (or the other way round) have all their quantum numbers in common, i.e., the two nuclear states are in the same isospin multiplet. Such decays are called *super allowed decays*. The ft values of such transitions are roughly that of the decay of a free neutron.

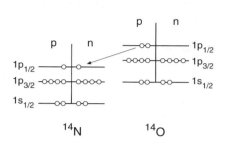

Super allowed decays are generally β^+-decays. This is because the Coulomb repulsion inside the nucleus slightly splits the states in an isospin multiplet; the excitation energy is higher for those states with more protons and fewer neutrons (cf. Fig. 2.6). Thus the protons in an isospin multiplet decay into neutrons but not the other way round. The β^--decay of ^{3}H into ^{3}He is an exception to this rule (another is free neutron decay). This is because the difference between the proton and neutron masses is larger than the decrease in the binding energy of ^{3}He from Coulomb repulsion.

An attractive example of β-decay inside an isospin triplet is provided by the process ^{14}O $\rightarrow ^{14}$N+e$^+$ + ν_e, which is a $0^+ \rightarrow 0^+$ transition (cf. Fig. 2.6) and hence purely a Fermi decay. The three lowest proton shells in the ^{14}O nucleus, i.e., the $1s_{1/2}$, $1p_{3/2}$ and $1p_{1/2}$ shells, are fully occupied as are the two lowest neutron shells, but the $1p_{1/2}$ neutron shell is empty. Thus one of the two valence nucleons (the protons in the $1p_{1/2}$ shell) can change into a neutron in the same shell and with the same wave function.

Allowed decays. Allowed decays are those with $\ell = 0$. A familiar example is the β^--decay of the nuclide ^{14}C, which is produced by cosmic rays in the upper atmosphere in the reaction ^{14}N (n, p) ^{14}C, and is used to determine the age of organic materials. The ^{14}C ground state belongs, see Fig. 2.6, to an isospin triplet which also includes the 2.31 MeV ^{14}N state and the ground state of ^{14}O.

For reasons of energy ^{14}C is only allowed to decay into the ^{14}N ground state and this can only happen if the nucleon flips its spin (a Gamow-Teller decay). The half life ($t_{1/2} = 5730$ years) and the log-ft value (9.04) are much larger than for other allowed decays. This implies that the overlap of the wave functions are extremely small – which is a stroke of luck for archaeology.

Forbidden decays. Heavy nuclei have an excess of neutrons. If a proton were to decay inside such a nucleus, it would find that the equivalent neutron shell was already full. A super allowed β^+-decay is therefore not possible in heavy nuclei. On the other hand the decay of a neutron into a proton with the same quantum numbers is possible but the resulting nucleus would be in a highly excited state and this is generally ruled out for reasons of energy.

The ^{40}K nuclide is a good example: it can turn into ^{40}Ar either through β^+-decay or by a K capture and can also β^--decay into ^{40}Ca (cf. Fig. 3.4). The ground state of ^{40}Ca is a doubly magic nucleus whose $1d_{3/2}$ (proton and neutron) shells are full while the $1f_{7/2}$ shells are empty (Fig. 17.18).

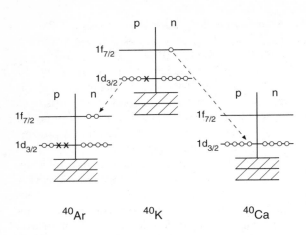

Fig. 17.18. Sketch of the β^+- and β^--decays of ^{40}K in the shell model. The energies are not to scale.

^{40}Ar ^{40}K ^{40}Ca

The ^{40}K nuclide has the configuration (p-$1d_{3/2}^{-1}$, n-$1f_{7/2}^{1}$) and ^{40}Ar has (p-$1d_{3/2}^{-2}$, n-$1f_{7/2}^{2}$). The unpaired nucleons in ^{40}K add to 4^-. Hence the decay into the ground states of ^{40}Ca and ^{40}Ar are triply forbidden. The decay into the lowest excited state of ^{40}Ar ($J^P = 2^+$) via K capture is in principle only simply forbidden, but the available phase space is very small since the energy difference is only 0.049 MeV. For these reasons ^{40}K is extremely long lived ($t_{1/2} = 1.27 \cdot 10^9$ yrs) and is still today, thousands of millions of years after the birth of the solar system, around us in substantial quantities. It is the only medium sized nuclide ($A < 200$) that gives a sizeable contribution to the natural background radioactivity.

β-decay into highly excited states. The largest excitation energy available to the daughter nucleus in a β-decay is given by the difference in the masses of the nuclei involved. We showed in Sect. 3.1 that the masses of isobars lie on a parabola. Hence the mass difference of neighbouring nuclei inside an isobar spectrum will be particularly large if their charge number Z sharply differs from that of the stable isobar. The highly neutron-rich nuclei that appear as fission products in nuclear reactions are examples of this.

A lot of energy is available to the β^--decay of such nuclei. Indeed decays into highly excited states are observed, these can in fact compete with decays into lower levels of the daughter nucleus, despite the smaller phase space available to the former. This is explained by observing that the proton in the daughter nucleus occupies a state in the same shell as the neutron did in the original nucleus. One sees here how well the shell model works even for higher nuclear excitations.

An example of this is shown in Fig. 17.19. In a few per cent of the cases the daughter ^{99}Y or ^{99}Zr nucleus is so highly excited that neutron emission is energetically allowed. Since this is a strong process it takes place "at once". One speaks of *delayed neutron emission* since it only takes place after the β-decay, typically a few seconds after the nuclear fission.

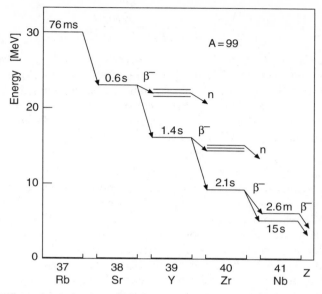

Fig. 17.19. Successive β^--decays of neutron rich isobars with $A = 99$. In a few per cent of the decays the ^{99}Sr and ^{99}Y nuclides decay into highly excited states of the daughter nuclei, from which neutrons can be emitted (from [Le78]).

■ These delayed neutrons are of great importance for reactor engineering since the chain reaction can be steered through them. A typical nuclear reactor is made up from fission material (such as ^{235}U enriched uranium) and a moderator (e.g., H_2O, D_2O or C). The absorption cross section for ^{235}U is largest for neutron energies below 1 eV. After absorbing a thermal neutron, the resulting ^{236}U nucleus divides up into two parts (fission) and emits, on average, 2 to 3 new fast neutrons whose kinetic energies are typically 0.1 to 1 MeV. These neutrons are now thermalised by the moderator and can then cause further fissions.

This cycle (neutron absorption – fission – neutron thermalisation) can lead to a self-sustaining chain reaction. Its time constant, which depends on the reactor design, is of the order of 1 ms. This time is much too short to control the chain reaction which for steady operation requires the neutron multiplication factor to be exactly equal to one. In reactor engineering therefore, the multiplication factor due to prompt neutrons is arranged to be slightly less than one. The remainder then is due to delayed neutrons whose time delay is typically of the order of seconds. This fraction, which in practice determines the multiplication rate in the reactor, can be controlled mechanically – by moving absorbing rods in and out of the reactor.

Measuring the neutrino mass. The question of whether the neutrino mass is exactly zero or in fact has a finite value is one of the most important facing contemporary physics. A direct measurement of the mass is possible from the kinematics of β-decay. The form of the β-spectrum near the end point is highly sensitive to the neutrino mass. This is best seen in a so-called *Kurie plot* where

$$K(E_e) = \sqrt{\frac{dN(E_e)/dE_e}{F(Z', E_e) \cdot E_e \cdot \sqrt{E_e^2 - m_e^2 c^4}}} \qquad (17.53)$$

is plotted against the electron energy E_e. $dN(E_e)$ is the number of electrons in the energy interval $[E_e, E_e + dE_e]$. From (15.42) and (15.45) we have that the distribution function $K(E_e)$ is a straight line which cuts the abscissa at the maximal energy E_0 – provided the neutrino is massless. If this is not the case then the curve deviates from a straight line at high E_e and crosses the axis vertically at $E_0 - m_\nu c^2$ (Fig. 17.20):

$$K(E_e) \propto \sqrt{(E_0 - E_e)\sqrt{(E_0 - E_e)^2 - m_\nu^2 c^4}} . \qquad (17.54)$$

In order to measure the neutrino mass to a good accuracy one needs nuclei where a finite neutrino mass would have a large impact, i.e., E_0 should only be a few keV. Since atomic effects must be taken into account at low energies, the initial and final atomic states should be as well understood as possible. The most suitable case is the β^--decay of tritium, $^3\text{H} \rightarrow {}^3\text{He} + e^- + \bar{\nu}_e$, where E_0 is merely 18.6 keV. The curve crosses the E axis at $E_0 - m_\nu c^2$ and E_0 is determined by linearly extrapolating the curve from lower energies.

Actually carrying out such experiments is extremely difficult since the counting rate near the maximal energy is vanishingly small. The spectrum is furthermore smeared by the limited resolution of the spectrometer, the molecular binding of the the tritium atom and the energy loss of the electrons in the source itself. It is therefore not possible to directly measure where the curve cuts the axis; rather one simulates the measured curve for various neutrino masses and looks for the best agreement.

The very best previous measurements show no deviation from zero neutrino mass. The upper bound is quoted as 15 eV/c^2 [PD98].

Measuring the neutrino helicity. The so-called *Goldhaber experiment* is an elegant method to measure the helicity of the ν_e from weak nuclear

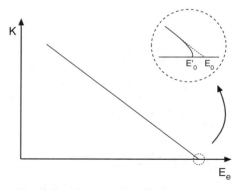

Fig. 17.20. Kurie plot of the β-spectrum. If the neutrino mass is not zero the straight line must bend near the maximum energy and cross the axis vertically at $E_0' = E_0 - m_\nu c^2$.

decays [Go58]. An isomer state of the $^{152}_{63}\text{Eu}^m$ $(J{=}0)$ nucleus can, via K capture, decay into a $J{=}1$ state of $^{152}_{62}\text{Sm}$ which has an excitation energy of 0.960 MeV. This then emits a photon to enter the $J=0$ ground state. This decay is a pure Gamow-Teller transition. Conservation of angular momentum implies that the spin of the ^{152}Sm nucleus must be parallel to that of the captured electron and antiparallel to that of the neutrino. Since the atomic recoil is opposite to the momentum of the neutrino, the helicity of the excited ^{152}Sm nucleus is equal to that of the neutrino. The emitted photon carries the angular momentum of the nucleus. Its spin must be parallel to that of the ^{152}Sm nucleus before the γ was emitted. If the photon is emitted in the recoil direction, then its helicity will be equal to that of the neutrino. To determine the neutrino's helicity one has then to measure the helicity of the photon (which corresponds to a circular polarisation) and at the same time make sure that one is only considering those photons that are emitted in the direction of the recoiling nucleus (the opposite direction to that taken by the neutrino).

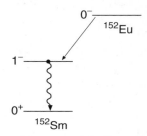

The experimental apparatus for this experiment is shown in Fig. 17.21. The photons can only reach the detector if they are resonantly scattered in a ring of Sm_2O_3. They are first absorbed and then re-emitted. Resonant

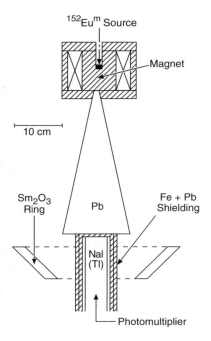

Fig. 17.21. Set up of the Goldhaber experiment (from [Go58]). Photons from the $^{152}\text{Eu}^m$ source are scattered in the Sm_2O_3 ring and detected in a NaI(Tl) scintillation detector.

absorption, i.e., the reverse of electromagnetic decay, is normally impossible in nuclear physics since the states are narrower than the shift due to the recoil. The photons from the ^{152}Eu^m source are emitted by ^{152}Sm nuclei that are already moving. If a nucleus is moving towards the Sm$_2$O$_3$ absorber before the γ emission, then the photon has a small amount of extra energy, which is sufficient to allow resonant absorption. In this way one can fix the recoil direction of the ^{152}Sm nucleus and hence that of the neutrino.

The ^{152}Eu source is inside a Fe magnet which the photons must cross to reach the ring of Sm$_2$O$_3$. Some of the photons undergo Compton scattering off the electrons in the Fe atoms. Two of the 36 electrons in the iron atom are polarised by the magnetisation. The Compton cross-section is larger if the electrons and photons are polarised in opposite directions. This permits us to determine the photon polarisation by reversing the magnetic fields and comparing the old and new counting rates.

The helicity of the neutrino was determined from this experiment as being

$$h_{\nu_e} = -1.0 \pm 0.3 \,. \tag{17.55}$$

_____ **Problems**

1. **Fermi gas model**
   Calculate the dependence of the Fermi pressure upon the nuclear density. How large is this pressure for a density $\varrho_N = 0.17$ nucleons/fm^3? What is this in macroscopic units (bar)?

2. **Shell model**
   a) In the following table we present the experimentally determined spins and parities of the ground states and first excited states of some nuclei:

	$^{7}_{3}$Li	$^{23}_{11}$Na	$^{33}_{16}$S	$^{41}_{21}$Sc	$^{83}_{36}$Kr	$^{93}_{41}$Nb
J_0^P	$3/2^-$	$3/2^+$	$3/2^+$	$7/2^-$	$9/2^+$	$9/2^+$
J_1^P	$1/2^-$	$5/2^+$	$1/2^+$	$3/2^+$	$7/2^+$	$1/2^-$

   Find the configurations of the protons and neutrons in the incomplete shells of the one-particle shell model for these nuclei and predict the quantum numbers of their ground states and first excited levels. Compare your results with the table.
   b) The spins of odd-odd nuclei are generally given by a vector addition of the total angular momenta of the two unpaired nucleons. Which possible nuclear spins and parities should $^{6}_{3}$Li and $^{40}_{19}$K have? Experimentally these nuclei have the quantum numbers 1^+ and 4^-.

3. **Shell model**
   a) Find the gap between the $1p_{1/2}$ und $1d_{5/2}$ neutron shells for nuclei with mass number $A \approx 16$ from the total binding energy of the ^{15}O (111.9556 MeV), ^{16}O (127.6193 MeV) and ^{17}O (131.7627 MeV) atoms [AM93].
   b) How does this agree with the energy of the first excited level of ^{16}O (cf. Fig. 17.7)?
   c) What information does one obtain from the energy of the corresponding state of ^{17}O?
   d) How do you interpret the difference in the total binding energies of ^{17}O und ^{17}F? Estimate the radius of these nuclei.
   e) The first excited state of ^{17}F is below the equivalent state of ^{17}O. A possible explanation of this is that the unpaired nucleon has a different spatial extension (smaller?, larger?) in the first excited state than in the ground state. What do you expect from considering the quantum numbers?

4. **Shell model**
   It is conspicuous that many of the nuclei which possess long lived isomer states have N or Z in the ranges $39 \cdots 49$ and $69 \cdots 81$ [Fe49, No49, Go52]. Why is this?

5. **Magnetic moment**
   The $^{42}_{21}$Sc nucleus has a low lying level with $J^P(I) = 7^+(0)$ and an excitation energy of 618 keV.
   a) Which shell model configuration would you assign this state to?
   b) What magnetic moment would you expect?

6. **The Goldhaber experiment**
   ^{152}Sm possesses a state with excitation energy 0.963 MeV and quantum numbers 1^- which decays via an E1 transition into the ground state.
   a) How large is the recoil energy of the nucleus?

    b) Compare this energy with the width of the state which is equivalent to an E1 one particle transition probability. Can a so-emitted photon be absorbed by another nucleus? What happens is we take the influence of thermal motion into account?

    c) Show that this energy loss is compensated if the excited ^{152}Sm nucleus was produced in an electron capture decay of ^{152}Eu and the photon was emitted in the recoil direction of the ^{152}Sm nucleus.

    The energy of the emitted neutrino is 0.950 MeV.

7. **Coupling strength of β-decay**

    A maximal energy of $E_{\text{kin}}^{\text{max}} = 1810.6 \pm 1.5$ keV is measured in the β-decay ^{14}O $\rightarrow$ ^{14}N $+ e^+ + \nu_e$ (Fig. 2.6) [EL92, Wi78]. A phase space function $f(Z', E_0)$ of 43.398 is calculated from this [Wi74]. What half-life should ^{14}O have? The experimental value is $t_{1/2} = 70\,606 \pm 18$ ms [Wi78].

18. Collective Nuclear Excitations

We showed in Sect. 17.3 that the nuclear ground states may be well described if we assume that the nucleons are in the lowest shell model orbits. The single particle picture, we further showed for the case of a *single* valence nucleon or nucleon hole, works very well if shells are nearly full or empty. Excited states are then understood as being created by a valence nucleon jumping into a higher shell model state; a direct analogy to our picture of the atom. As well as such straightforward single particle excitations, more complicated phenomena can take place in the nucleus. Collective excitations provide some of the most beautiful aspects of nuclear dynamics.

Collective excitations of many body systems can be phenomenologically understood as fluctuations around a state of equilibrium. These may be fluctuations in the density or shape. The type of collective excitation strongly depends upon the composition of the system and the manner in which its components interact with each other. We want now to show the connection between nuclear collective excitations and the forces inside and the structure of the nucleus.

Electromagnetic transitions provide us with the most elegant way to investigate collective excitations in nuclei. We will therefore first consider how electromagnetic transitions in nuclei may be determined, so that we can then say to what extent collective effects are responsible for these transitions.

The first measurements of photon absorption in nuclei led to the discovery that the lion's share of the the absorption is by a *single* state. The first description of this *giant dipole resonance* state was of an oscillation of the protons and neutrons with respect to each other. Later on it was discovered that the transition probability for electric quadrupole transitions of lower energy states was much higher than a single particle picture of the nucleus predicts. The transition probability for octupole transitions also predominantly comes from single states which we call octupole vibrations.

The single particle and collective properties of nuclei were regarded for a long time as distinct phenomena. A unified picture first appeared in the 1970's. We want to illustrate this modern framework through the example of giant dipole resonances. What we will discover can be easily extended to quadrupole and octupole oscillations.

Another important collective effect is the rotation of deformed nuclei. Such rotations form a most pleasing chapter, both didactically and aesthetically, in the story of γ spectroscopy.

18.1 Electromagnetic Transitions

Electric dipole transitions. The probability of an electric dipole transition can be somewhat simplistically derived by considering a classical Hertz dipole. The power output emitted by the dipole is proportional to ω^4. The rate of photon emission, i.e., the transition probability, may be obtained by dividing the power output by the photon energy $\hbar\omega$. One so finds

$$W_{fi} = \frac{1}{\tau} = \frac{e^2}{3\pi\varepsilon_0\hbar^4c^3}E_\gamma^3\left|\int \mathrm{d}^3x\,\psi_f^*\boldsymbol{x}\psi_i\right|^2 , \qquad (18.1)$$

where we have replaced the classical dipole $e\boldsymbol{x}$ by the matrix element. This result may also be obtained directly from quantum mechanics.

■ In the following derivation we want to treat the electromagnetic transitions semiclassically, i.e., we will not concern ourselves with quantising the radiation field or spin.

Consider first an excited nuclear state ψ_i which through γ emission enters a lower lying state, ψ_f. The golden rule says that the transition probability is

$$\mathrm{d}W = \frac{2\pi}{\hbar}\,|\langle\psi_f|\mathcal{H}_{\mathrm{int}}|\psi_i\rangle|^2\,\mathrm{d}\varrho(E) . \qquad (18.2)$$

$\mathcal{H}_{\mathrm{int}}$ describes the interaction of the moving charge with the electromagnetic field and $\varrho(E)$ is a phase space factor that describes the final state density at total energy E. For photon emission we have $E = E_\gamma$. Since γ radiation is generally not spherically symmetric, we consider the phase space in a solid angle element $\mathrm{d}\Omega$ around the momentum vector. As in (4.16) we set

$$\mathrm{d}\varrho(E) = \frac{V\,|\boldsymbol{p}|^2\,\mathrm{d}|\boldsymbol{p}|\,\mathrm{d}\Omega}{(2\pi\hbar)^3\,\mathrm{d}E} . \qquad (18.3)$$

For the photon we have $E = c\cdot|\boldsymbol{p}|$ and $\mathrm{d}E = c\cdot\mathrm{d}|\boldsymbol{p}|$, which implies

$$\mathrm{d}\varrho(E) = \frac{E_\gamma^2\,V\,\mathrm{d}\Omega}{(2\pi\hbar c)^3}. \qquad (18.4)$$

The $\mathcal{H}_{\mathrm{int}}$ operator can be obtained by considering the classical Hamiltonian for the interaction between a charge e, which emits the photon, and the electromagnetic field $A = (\phi/c, \boldsymbol{A})$ [Sc95]:

$$\mathcal{H} = \frac{1}{2m}\,(\boldsymbol{p} - e\boldsymbol{A})^2 + e\phi. \qquad (18.5)$$

Note that we have here assumed a point-like charge. The term quadratic in A is negligible and we may write

$$\mathcal{H} = \frac{\boldsymbol{p}^2}{2m} - \frac{e}{m}\boldsymbol{p}\boldsymbol{A} + e\phi . \qquad (18.6)$$

The first term corresponds to free movement of the charged particle and the last two decribe the interaction

$$\mathcal{H}_{\text{int}} = -\frac{e}{m}\boldsymbol{p}\boldsymbol{A} + e\phi \, , \tag{18.7}$$

which, for a point-like particle, is just given by the scalar product of the electric four-current

$$j = (e \cdot c, e\boldsymbol{v}) \tag{18.8}$$

and the electromagnetic field

$$A = (\phi/c, \boldsymbol{A}) \, . \tag{18.9}$$

In an electromagnetic decay $e\phi$ does not contribute to the transition probability, since real photons are transversely polarised and monopole transitions are hence forbidden.

If one replaces the momentum $\boldsymbol{p}$ by the operator $\boldsymbol{p} = -i\hbar\boldsymbol{\nabla}$ and interprets the vector $\boldsymbol{A}$ as the wave function of the photon, one obtains the matrix element

$$\langle\psi_f|\mathcal{H}_{\text{int}}|\psi_i\rangle = -\frac{ie\hbar}{m}\int \mathrm{d}^3x\, \psi_f^* \left(\boldsymbol{\nabla}\psi_i\right)\, \boldsymbol{A}\, . \tag{18.10}$$

The gradient $\boldsymbol{\nabla}$ may be replaced by the commutator of the coordinate $\boldsymbol{x}$ with the Hamilton operator, since for stationary states

$$\mathcal{H}_0 = \frac{\boldsymbol{p}^2}{2m} + V(\boldsymbol{x}) \tag{18.11}$$

we have the following relation:

$$\boldsymbol{x}\,\mathcal{H}_0 - \mathcal{H}_0\,\boldsymbol{x} = \frac{i\hbar}{m}\boldsymbol{p} = \frac{\hbar^2}{m}\boldsymbol{\nabla}. \tag{18.12}$$

In this way we have

$$-\frac{ie}{\hbar}\int \mathrm{d}^3x\, \psi_f^*\left(\boldsymbol{x}\mathcal{H}_0 - \mathcal{H}_0\boldsymbol{x}\right)\psi_i\, \boldsymbol{A} \;=\; \frac{ie}{\hbar}(E_i - E_f)\int \mathrm{d}^3x\, \psi_f^*\,\boldsymbol{x}\,\psi_i\, \boldsymbol{A}\, , \tag{18.13}$$

and the matrix element has the standard form for multipole radiation.

In the semiclassical derivation of γ emission, one writes the photon wave function as

$$A = \sqrt{\frac{\hbar}{2\varepsilon_0\omega V}}\, \boldsymbol{\varepsilon}\cos(\boldsymbol{k}\boldsymbol{x} - \omega t)\, , \tag{18.14}$$

where $\boldsymbol{\varepsilon}$ is the polarisation vector of the photon, $E_\gamma = \hbar\omega$ is its energy and $\boldsymbol{k}$ the wave vector. That this is indeed correct may be easily checked by calculating the electromagnetic radiation energy in a volume V using $\boldsymbol{A}$ from (18.14):

$$\hbar\omega = V \cdot \left(\frac{1}{2}\varepsilon_0\overline{\boldsymbol{E}^2} + \frac{1}{2}\frac{1}{\mu_0}\overline{\boldsymbol{B}^2}\right) = V\varepsilon_0\overline{\boldsymbol{E}^2} \qquad \text{with} \quad \boldsymbol{E} = -\frac{\partial\boldsymbol{A}}{\partial t}\, , \tag{18.15}$$

where the bar represents time averaging. With this result we now may write the transition probability as:

$$\begin{aligned}
\mathrm{d}W_{fi} &= \frac{2\pi}{\hbar}\frac{\hbar}{2\varepsilon_0\,\omega V}\frac{e^2 E_\gamma^2}{\hbar^2}\left|\boldsymbol{\varepsilon}\int \mathrm{d}^3x\, \psi_f^*\boldsymbol{x}\psi_i\, \mathrm{e}^{ik\boldsymbol{x}}\right|^2 \frac{E_\gamma^2 V\, \mathrm{d}\Omega}{(2\pi\hbar c)^3} \\
&= \frac{e^2}{8\pi^2\varepsilon_0\hbar^4 c^3}E_\gamma^3\left|\boldsymbol{\varepsilon}\int \mathrm{d}^3x\, \psi_f^*\boldsymbol{x}\, \mathrm{e}^{ik\boldsymbol{x}}\psi_i\right|^2 \mathrm{d}\Omega\, .
\end{aligned} \tag{18.16}$$

The wavelengths of the gamma rays are large compared to a nuclear radius. The multipole expansion

$$e^{ikx} = 1 + ikx + \cdots \qquad (18.17)$$

is very useful, since, generally speaking, only the lowest transition that the quantum numbers allow needs to be taken into account. Only very occasionally are two multipoles of equal strength in a transition. If one now sets $e^{ikx} \approx 1$, integrates (18.16) over the solid angle $\mathrm{d}\Omega$ and the polarisation one obtains (18.1).

Electric dipole (E1) transitions always connect states with different parities. The photon carries away angular momentum $|\ell| = 1\hbar$ and so the angular momenta of the initial and final states may at most differ by one unit.

Since transitions from one shell into the one immediately above play the most important role in collective excitations, we now introduce the standard notation for the wave function. A closed shell shall be denoted by the symbol $|0\rangle$ ("vacuum wave function"). If a particle in the state ϕ_{j_1} of the closed shell jumps into the state ϕ_{j_2} of the next shell a particle-hole state is created, which we symbolise by $|\phi_{j_1}^{-1}\phi_{j_2}\rangle$. The dipole matrix element

$$\langle \phi_{j_1}^{-1}\phi_{j_2}|ex|0\rangle = e \int \mathrm{d}^3x \, \phi_{j_2}^* x \phi_{j_1} \qquad (18.18)$$

describes the transition of a nucleon from the state ϕ_{j_1} to the state ϕ_{j_2}. Since $|0\rangle$ is a full shell state it must have spin and parity $J^P = 0^+$, hence the excited particle-hole state after the electric dipole transition must have the quantum numbers $J^P = 1^-$.

Magnetic dipole transitions. The transition probability of a magnetic dipole (M1) transition is obtained by replacing the electric dipole in (18.1) by a magnetic one:

$$W_{fi} = \frac{1}{\tau} = \frac{\mu_0}{3\pi\hbar^4 c^3}E_\gamma^3 \left| \int \mathrm{d}^3x \, \psi_f^* \boldsymbol{\mu} \psi_i \right|^2 \qquad \text{where} \quad \boldsymbol{\mu} = \frac{e}{2m}(\boldsymbol{L} + g\boldsymbol{s})\,. \qquad (18.19)$$

Here $\boldsymbol{L}$ is the orbital angular momentum operator and $\boldsymbol{s}$ is the spin operator.

Higher multipoles. If the electric dipole transition is forbidden, in other words if both states have the same parity or the vectorial addition of the angular momenta is inconsistent, then only higher multipole radiation can be emitted. The next highest multipoles in the transition probability hierarchy are the above magnetic dipole (M1) transition and the electric quadrupole (E2) transition [Fe53]. Both are second order in the expansion (18.17). The parity of the initial and final states must be identical in electric quadrupole transitions and the triangle inequality $|j_f - j_i| \leq 2 \leq j_f + j_i$ must be fulfilled by the angular momenta. While the transition probability for dipole radiation is, from (18.1), proportional to E_γ^3, for electric quadrupole radiation it goes as E_γ^5. This is because there is a new factor of ikx in the matrix element and $|k|$ is proportional to E_γ. The energy-independent part of the matrix element has the form $r^2 Y_2^m(\theta, \varphi)$.

18.2 Dipole Oscillations

Photon absorption in nuclei. A broad resonance, which was already known in the 1950's, dominates the absorption of gamma rays by nuclei. The experimental techniques for investigating this resonance were rather awkward since no variable energy gamma sources existed.

The method of in flight positron annihilation, which was developed in the 1960's, first permitted detailed measurements of the gamma cross-sections. Positrons, which have been produced through pair creation from a strong bremsstrahlung source, are selected according to their energy and focussed upon a target. They then partially annihilate with the target electrons and produce bremsstrahlung as an unwanted by-product (Fig. 18.1).

Such a gamma spectrum is shown in Fig. 18.2. A peak can be clearly distinguished from the bremsstrahlung at the maximal possible energy and this is presumed to come from the e^+e^- annihilation. The energy dependence of γ-induced cross-sections can be thoroughly investigated by varying the energy of the positrons. As well as the total cross-section, the cross-section for the photoproduction of neutrons (*nuclear photoeffect*)

$$^A\mathrm{X}\,(\gamma,\mathrm{n})\ ^{A-1}\mathrm{X} \tag{18.20}$$

is of especial importance. This is in fact the major part of the total cross-section. The photoproduction of protons is, by contrast, suppressed by the Coulomb barrier. In what follows we will limit ourselves to the (γ,n) reaction.

We have chosen $\sigma(\gamma,\mathrm{n})$ for neodymium isotopes as an example (Fig. 18.3). Various observations may be made.

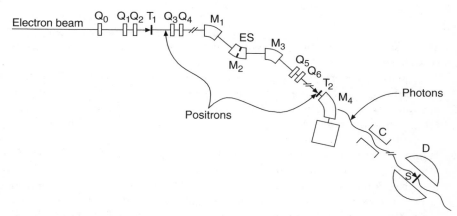

Fig. 18.1. Experimental set up for in flight positron annihilation (from [Be75]). An electron beam hits a target ($\mathrm{T_1}$). The bremsstrahlung that is produced converts into electron-positron pairs. The positrons are then selected according to their energy by three dipole magnets ($\mathrm{M_1}$, $\mathrm{M_2}$, $\mathrm{M_3}$) before hitting a second target ($\mathrm{T_2}$). Some of them annihilate in flight with target electrons. A further magnet ($\mathrm{M_4}$) deflects all charged particles and only photons arrive at the experimenter's real target (S).

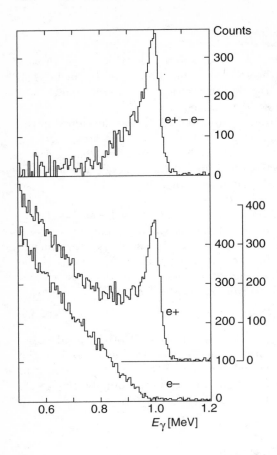

Fig. 18.2. The photon spectrum from in-flight electron positron annihilation [Be75]. This is later used for (γ, n) reactions. The background of bremsstrahlung from positrons hitting the target is determined by aiming a mono-energetic beam of electrons at the target. The cross-section for fixed photon energies is found by performing experiments with the two different photon beams and subtracting the counting rates.

- The absorption probability is centred in a resonance which we call a *giant resonance*.
- The excitation energy of the giant resonance is roughly twice the separation between neighbouring shells. This is astounding since, for reasons of parity and angular momentum conservation, many more single particle transitions are possible between one shell and the next than between a shell and the next but one.
- While a narrow resonance is observed in absorption by ^{142}Nd, this splits into two resonances as the mass number increases.
- The integrated cross-section is about as big as the sum over all expected cross-sections for the transition of a single nucleon from the last closed shell. This means that all the protons and neutrons of the outermost shell contribute coherently to this resonance.

A qualitative explanation of the giant resonances comes from the oscillation of protons and neutrons with respect to each other (Fig. 18.4). The ^{150}Nd is deformed and has a cigar-like shape. The two maxima for this nucleus cor-

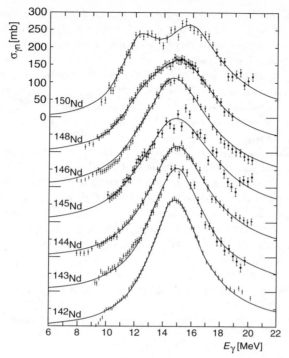

Fig. 18.3. Cross-section for γ-induced emission of neutrons in neodymium isotopes [Be75]. The curves have been shifted vertically for the sake of clarity. Neodymium isotopes progress from being spherically symmetric to being deformed nuclei. The giant resonance of the spherically symmetric ^{142}Nd nucleus is narrow, while that of the deformed ^{150}Nd nucleus shows a double peak.

respond to oscillations along the symmetry axis (lower peak) and orthogonal to it (higher peak).

We will attempt to justify this intuitive picture of giant resonances and their excitation energies in the framework of the shell model.

The giant dipole resonance. Consider once again the example of the doubly magic ^{16}O nucleus. B Let us assume that photon absorption leads

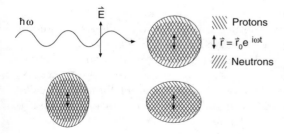

Fig. 18.4. The giant dipole resonance as oscillations of the protons and neutrons against each other. In deformed nuclei *(below)* two oscillation modes are available.

to a nucleon in the $1p_{3/2}$ or $1p_{1/2}$ shell being promoted into the $1d_{5/2}$, $1d_{3/2}$ or $2s_{1/2}$ shell. If this nucleon drops back into the 1p shell, it can pass on its excitation energy through recoils to other nucleons, which may then, for example, be themselves excited out of the 1p shell into the 1d or 2s shell. If the nuclear states that are produced by the excitation of a nucleon into a higher level were degenerate, then the probability of generating all of these states must be equal and a simple single particle picture would be doomed to failure from the start. In reality this is almost the case; the excited states are almost degenerate.

One can understand these states as a combination of a hole in the remaining nucleus and a particle in a higher shell, and the interaction between the particle and all the nucleons of the now incomplete shell may be viewed as an interaction between the particle and the hole. This interaction depends upon the spin and isospin of the particle-hole system and causes the states to mix strongly. Below we want to use a greatly simplified model to show how the transition strengths of all one particle-one hole states combine through this mixing into a single state.

We use $\mathcal{H}_0$ to denote the Hamiltonian operator of a nucleon in the central potential of the single particle shell model. In the transition of the particle from a full shell to the one above, we must also take the particle-hole interaction into account; the Hamiltonian operator must then be written as

$$\mathcal{H} = \mathcal{H}_0 + \mathcal{V}. \tag{18.21}$$

Collective excitations appear just because of the mixing generated by this particle-hole interaction, $\mathcal{V}$.

Consider now all particle-hole states with 1^- spin and parity. These can only be particle-hole combinations such that the angular momenta j_1 and j_2 add vectorially to $1\hbar$ and the sum of the orbital angular momentum quantum numbers $\ell_1 + \ell_2$ is odd (so that the parity is negative). If we restrict ourselves to the excitation of a nucleon from the 1p into the 1d or 2s shell, then we have the following possible particle-hole states:

$$\left| \phi_{1p_{3/2}}^{-1} \phi_{1d_{5/2}} \right\rangle, \quad \left| \phi_{1p_{3/2}}^{-1} \phi_{2s_{1/2}} \right\rangle, \quad \left| \phi_{1p_{3/2}}^{-1} \phi_{1d_{3/2}} \right\rangle,$$

$$\left| \phi_{1p_{1/2}}^{-1} \phi_{2s_{1/2}} \right\rangle, \quad \left| \phi_{1p_{1/2}}^{-1} \phi_{1d_{3/2}} \right\rangle.$$

Since both the proton and neutron shells are full in the ^{16}O nucleus, such states exist for both proton and neutron excitations. They have all got roughly the same energy and may be viewed as approximately degenerate.

The number of nucleons per shell is larger in heavy nuclei, and the number of nearly degenerate particle-hole $J^P = 1^-$ states is accordingly greater. N, the number of particle-hole states is between 10 to 20 for medium sized nuclei.

The connection between one particle and collective excitation can be clarified by a simple model [Br67]. We denote particle-hole states by $|\psi_i\rangle$:

$$|\psi_i\rangle = |\phi_{j_1}^{-1}\phi_{j_2}\rangle \qquad \text{where } i = 1\cdots N. \tag{18.22}$$

The $|\psi_i\rangle$ are, by definition, eigenstates of the unperturbed Hamiltonian

$$\mathcal{H}_0 |\psi_i\rangle = E_i |\psi_i\rangle. \tag{18.23}$$

The solution to the Schrödinger equation with the full Hamiltonian operator

$$\mathcal{H} |\Psi\rangle = (\mathcal{H}_0 + \mathcal{V}) |\Psi\rangle = E |\Psi\rangle, \tag{18.24}$$

is $|\Psi\rangle$. This wave function $|\Psi\rangle$ projected out upon the space spanned by $|\psi_i\rangle$ in (18.22) may be written as

$$|\Psi\rangle = \sum_{i=1}^{N} c_i |\psi_i\rangle \tag{18.25}$$

where the coefficients c_i fulfill the secular equation

$$\begin{pmatrix} E_1 + V_{11} & V_{12} & V_{13} & \cdots \\ V_{21} & E_2 + V_{22} & V_{23} & \cdots \\ V_{31} & V_{32} & E_3 + V_{33} & \cdots \\ \vdots & \vdots & \vdots & \ddots \end{pmatrix} \cdot \begin{pmatrix} c_1 \\ c_2 \\ c_3 \\ \vdots \end{pmatrix} = E \cdot \begin{pmatrix} c_1 \\ c_2 \\ c_3 \\ \vdots \end{pmatrix}. \tag{18.26}$$

We assume for simplicity that all the V_{ij} are the same

$$\langle \psi_i | V | \psi_j \rangle = V_{ij} = V_0. \tag{18.27}$$

The solution of the secular equation is then rather simple: the coefficients c_i may be written as

$$c_i = \frac{V_0}{E - E_i} \sum_{j=1}^{N} c_j, \tag{18.28}$$

where $\sum_j c_j$ is a constant. Summing over all N particle-hole states on both sides and bearing in mind that $\sum_i c_i = \sum_j c_j$, we obtain the relation

$$1 = \sum_{i=1}^{N} \frac{V_0}{E - E_i}, \tag{18.29}$$

as the solution of the secular equation.

The solutions of this equation are most easily understood graphically (Fig. 18.5). The right hand side of the equation has poles at $E = E_i$ where $i = 1\ldots N$. The solutions E_i' to (18.29) are to be found where the right hand side is unity. The new energies are marked by circles on the abscissa. $N-1$ eigenvalues (3 in the diagram) are "squeezed in" between the unperturbed energies $E_1 \ldots E_n$. The exception, denoted by E_C, is the collective state, as we will show in the following. A repulsive ($V_0 > 0$) interaction, as is assumed in the diagram, has its collective state above the particle-hole state.

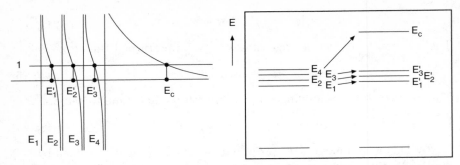

Fig. 18.5. Graphical representation of the solution to the secular equation (18.26) and a picture of how the energy levels are shifted.

To obtain a quantitative estimate of the energy shift, we now assume that $E_i = E_0$ for all i. Equation (18.29) then becomes

$$1 = \sum_{i=1}^{N} \frac{V_0}{E_C - E_i} = \frac{N V_0}{E_C - E_0} , \qquad (18.30)$$

from which

$$E_C = E_0 + N \cdot V_0 \qquad (18.31)$$

follows. The energy shift of the collective state is proportional to the number of degenerate states. From experiment we know that the energy of the giant resonance is roughly twice the separation between two shells, i.e., $N \cdot V_0 \approx E_0$. The effective interaction decreases for heavier nuclei but this is compensated by the increased number of states which can enter the collective motion.

The expansion coefficients for the collective state

$$c_i^{(C)} = \frac{V_0}{E_C - E_i} \sum_{j} c_j^{(C)} \qquad (18.32)$$

are nearly independent of i so long as the energy of the collective state E_C is well separated from the E_i. The collective state has the following configuration:

$$|\psi_C\rangle = \frac{1}{\sqrt{N}} \sum_{j_i j_k} |\phi_{j_i}^{-1} \phi_{j_k}\rangle . \qquad (18.33)$$

This state is singled out by the fact that the amplitudes of each and every particle-hole state add with the same sign (constructively), since $E_C > E_i$ for all i. For the other $N-1$ diagonal states only one of the c_j is large and the others are small and have different signs. The superposition of the amplitudes is therefore destructive. The coherent superposition of the amplitudes means that the transition probability is large for the collective case and otherwise small as we will show in what follows.

If we do not assume as in (18.27) that all the V_{ij} are equal, then the calculation becomes more tedious but the general conclusion remains the same: as long as the V_{ij} are of the same order of magnitude the highest state is shifted well above the others and manifests itself as a coherent sum of all the particle-hole states.

Estimating the transition probability. The operator for the electric dipole transition is

$$D = e \sum_{p=1}^{Z} x_p \,, \tag{18.34}$$

where x_p is the coordinate of a proton. This must be modified slightly, since it is not yet clear which coordinate system x_p refers to. The most natural coordinate system is the centre of mass system and we therefore write B

$$D = e \sum_{p=1}^{Z} (x_p - X) \qquad \text{where} \qquad X = \frac{1}{A} \left(\sum_{p=1}^{Z} x_p + \sum_{n=1}^{N} x_n \right) . \tag{18.35}$$

This may be recast as

$$D = e \frac{N}{A} \sum_{p=1}^{Z} x_p - e \frac{Z}{A} \sum_{n=1}^{N} x_n \,. \tag{18.36}$$

We interpret this expression as meaning that

$$
\begin{aligned}
e_{\mathrm{p}} &= +eN/A \quad \text{is the effective proton charge} \qquad \text{and} \\
e_{\mathrm{n}} &= -eZ/A \quad \text{is the effective neutron charge.}
\end{aligned}
\tag{18.37}
$$

A photon "pulls" the protons in one direction and the neutrons in the opposite one. The neutrons and protons always move oppositely to each other under the influence of the photon in such a way that the centre of mass stays in the same place.

If we replace ψ_i and ψ_f in (18.1) by the nucleon wave functions in the one particle shell model before and after the γ emission, we find the so-called one particle transition probabilty. This, weighted with the square of the effective charge, may be used to estimate the collective nature of transitions.

We need to use the wave function (18.33) to calculate the matrix element

$$\mathcal{M}_{fi} = \int \mathrm{d}^3 x \, \psi_f^* \, D_z \, \psi_i \tag{18.38}$$

where D_z is the z component of the dipole operator (18.34), if we want to calculate the transition probability. In our case ψ_i is just $|0\rangle$, the wave function of the ground state with closed shells and ψ_f is (18.33) the wave function of the collective excitation. Thus we have

$$\mathcal{M}_{C0} = \frac{1}{\sqrt{N}} \int d^3x \left\{ \langle \phi_{j_i}^{-1} \phi_{j_k} | + \langle \phi_{j_l}^{-1} \phi_{j_m} | + \cdots \right\} D_z \Big| 0 \Big\rangle . \qquad (18.39)$$

The matrix element between the ground state and the particle-hole excitation can be identified with the dipole transition of a particle from a closed shell into a higher one. The integrals

$$A_n = \int d^3x \, \phi_{j_k}^* D_z \phi_{j_i} \qquad (18.40)$$

represent the amplitude for the transition of a particle from the j_i shell into the j_k one. Here n is an index which denotes each of the total N particle-hole states. The phases of the transition amplitudes A_n that contribute to the collective state are the phases of the differences of the magnetic substates. In the square of the amplitudes an equal number of mixed terms with positive and negative signs occur; they therefore average out to zero. If we assume for simplicity that the moduli $|A_n|$ are also identical, then the squared matrix element becomes

$$|\mathcal{M}_{C0}|^2 = \frac{1}{N} \left| \sum_{n=1}^{N} A_n \right|^2 = \frac{N^2}{N} \cdot |A|^2 = N \cdot |M_{1\,\text{particle}}|^2 . \qquad (18.41)$$

The transition probabilities are then rearranged. Because the states mix, we no longer have N different states each excited with probability $|A|^2$, but rather the total transition probability $N|A|^2$ is taken up by the collective state.

These ideas apply equally to both protons and neutrons. But, since the proton and neutron effective charges (18.37) are of opposite signs, protons and neutrons oscillate inside the nucleus with opposite phases. This is the semiclassical interpretation of the giant dipole resonance.[1] The oscillation in deformed nuclei can take place along or orthogonal to the symmetry axis. This leads to two peaks in the excitation curve, as is seen in Fig. 18.3 for the case of ^{150}Nd.

This treatment of the collective dipole resonance in a shell model, where we limited ourselves to just a few particle-hole states and then actually only solved it schematically, explains why the dipole transition strength is essentially restricted to one state. The resonance lies above the neutron threshold, i.e., in the continuum, and primarily mixes with neutron scattering states. Thus the cross-section for photon absorption displays a broad structure instead of a narrow state.

---

[1] There is an attractive analogy to the giant dipole resonance in plasma physics: electromagnetic radiation directed at a plasma is absorbed over a broad band around the so-called plasma frequency. At this frequency the totality of the free electrons oscillate against the ions.

18.3 Shape Oscillations

Quadrupole oscillations. Other nuclear collective states have also been observed in experiments. To keep things simple, we will limit ourselves in what follows to doubly even nuclei. Their ground and first excited states always have quantum numbers $J^P = 0^+$ and $J^P = 2^+$, with the exception of doubly magic nuclei and a very few others (Figs. 17.5 and 18.6). The simplest explanation for these excited levels would be that a nucleon pair has been broken apart to produce the second lowest energy level, $J^P = 2^+$. Measurements of the lifetimes of such states show, however, that the transition probability for the electric quadrupole transition is up to two orders of magnitude more than a one particle transition would suggest. The lowest 2^+ states are in fact, for nuclei with enough particles outside closed shells, our first encounter with the ground state rotation band which we will treat in Sect. 18.4. If the configuration has only a few particles outside closed shells, then we describe these states as oscillations of the geometric shape of the nucleus around its equilibrium form, which last is approximately spherically symmetric. For such 2^+ states it seems likely that these vibrations are of the quadrupole type (Fig. 18.8a).

Near the giant dipole resonance, and so at much higher excitation energies, further collective states with $J^P = 2^+$ are observed in electron scattering. These are called giant quadrupole resonances.

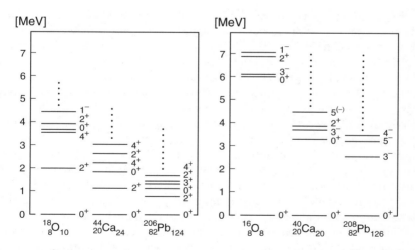

Fig. 18.6. Energy levels of three simply magic even-even nuclei, ^{18}O, ^{44}Ca and ^{206}Pb (*left*), and three doubly magic ones, ^{16}O, ^{40}Ca and ^{208}Pb (*right*). The excited states in the first case have $J^P = 2^+$. This state is lacking in the three doubly magic nuclei, which instead have a lower lying 3^- state. The transition probability into the ground state is high compared with what we would expect from a single particle excitation. These states are interpreted as collective quadrupole or octupole vibrations.

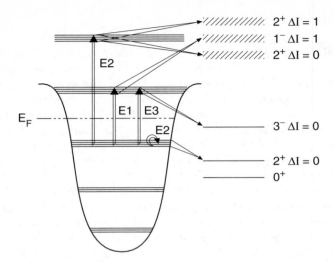

Fig. 18.7. Collective excitations in the framework of the shell model. Shape oscillations are denoted by $\Delta I = 0$. Those collective states where the protons and neutrons oscillate in phase are shifted downwards. States where they oscillate with opposite phases ($\Delta I = 1$) are pushed up to higher energies. Shells below the Fermi energy E_F are occupied by nucleons. The ground state lies at a position below the single particle excitations given by the pairing energy.

This illustrative discussion of quadrupole oscillations needs to be explained, in a similar fashion to our treatment of the giant dipole resonance, in terms of the shell model and the nature of the nuclear force. In a single particle picture collective excitations only arise if the particles in a shell are excited with correlated phases. For the giant dipole resonances we saw that this took place through coherent addition of all particle-hole excitations. To now create $J^P = 2^+$ states we need to either promote one particle into the next shell but one, or into the next level inside the same shell. This is a consequence of the spin and parity of the shell states. Shells below ^{48}Ca have alternating $+1$ and -1 parity and in heavier nuclei at least states with similar j will have opposite parities in successive shells. The particle-hole states are in this case nearly degenerate which can lead to collective states. Exciting particles inside the same shell leads to low lying quadrupole vibrations, exciting them into the next shell but one generates giant quadrupole resonances.

While the semiclassical picture of a giant dipole resonance has the protons and neutrons oscillating against each other, the protons and neutrons in nuclear quadrupole oscillations can move either with the same or opposite phase. If they move in phase the isospin is unchanged, if oppositely it is changed by unity. We will only consider the first case here. The interaction between particle-hole states which causes this in-phase motion is, obviously, of an attractive type. If we were to solve the secular equation for a collective

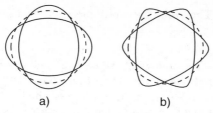

Fig. 18.8. (a) Quadrupole vibrations; (b) Octupole vibrations.

2^+ state, we would see that the attractive interaction shifts the energy levels downwards. The lowest energy state is built up out of a coherent superposition of particle-hole states with $J^P = 2^+$ and is collectively shifted down.

The various collective excitations in the framework of the shell model are depicted in Fig. 18.7. The giant quadrupole resonance splits into two parts. That with $\Delta I = 1$, which comes from proton-neutron repulsion, is, similarly to the giant dipole resonance, shifted up to higher energies. The giant quadrupole resonance which has $\Delta I = 0$ corresponds to shape oscillations and is shifted down. In both cases, however, the shift is smaller than was the case for the giant dipole resonance, which implies that the collective nature of these excitations is less pronounced. This may be explained as follows: the one particle-one hole excitations which build up the giant dipole resonance can only, for reasons of energy, enter a few other states, which themselves are one particle-one hole excitations in the same shell combinations. This state made up of single particle-hole excitations is thus long lived and displays a strongly coherent nature. This is all no longer true for excitations into the next shell but one, such as those which comprise the quadrupole resonance. The single particle-hole excitations of the next shell but one can decay into two-particle-hole states. Hence they have shorter lifetimes, are less coherent and less collective.

If the protons and neutrons move in phase this appears as a change in the shape of the nucleus. This alteration can hardly be quantitatively described in the shell model, since its particle wave functions were obtained using a spherically symmetric potential. Shape oscillations change the form of the potential and the nucleonic motion has to alter itself accordingly. Quantitative treatments of nuclei with large quadrupole oscillations are then of a hybrid form, where the total wave function has both vibrational and single particle parts.

Octupole oscillations. Nuclei with doubly closed shells, like ^{16}O, ^{40}Ca and ^{208}Pb, possess a low-lying 3^- state (Fig. 18.6) whose transition probability can be up to two orders of magnitude higher than the single particle prediction. This state can be interpreted as an octupole vibration (Fig. 18.8b). The collective 3^- states can, like the giant dipole resonance, be built up out of particle-hole excitations in neighbouring shells. Since the protons and neutrons oscillate in phase in such shape vibrations, the particle-hole interaction must be attractive. The collective octupole excitations are shifted to lower energies.

Summary. The picture of collective excitations which we have here attempted to explain is the following: since the shell energies in the nucleus are distinctly separated from each other, those particle-hole states which are created when a nucleon is excited into a higher shell are nearly degenerate. Coherent superposition of these particle-hole states then form a collective excitation. Shape oscillations can be interpreted as coherent superpositions of the movement of single particles, but a quantitative description is only possible in terms of collective variables.

18.4 Rotation States

Nuclei with sufficiently many nucleons outside of closed shells display a characteristic excitation pattern: a series of states with increasing total angular momentum, the separation between whose energies increases linearly. These excitations are interpreted as corresponding to the nucleus rotating and, in analogy to molecular physics, the series are called *rotation bands*. Electric quadrupole transitions between the states of a rotation band display a markedly collective nature. The excitation pattern, and also the collective character of the quadrupole transitions, are understood as consequences of these nuclei being highly deformed [Bo53]. Generally speaking the spin of the nuclear ground state is coupled to the angular momentum of the collective excitations. We will bypass this complication by only considering even-even nuclei, since these have spin zero in the ground state.

Rotational energy in classical mechanics depends upon the angular momentum J and the moment of inertia Θ:

$$E_{\text{rot}} = \frac{|J_{\text{rot}}|^2}{2\Theta} \ . \tag{18.42}$$

In quantum mechanics rotation is described by a Hamiltonian operator

$$\mathcal{H}_{\text{rot}} = \frac{J^2}{2\Theta} \ . \tag{18.43}$$

In such a quantum mechanical system the rotation must be perpendicular to the symmetry axis. The eigenstates of the angular momentum operator J are the spherical harmonic functions Y_J^m, which describe the angular distribution of the wave function. The associated eigenvalues are:

$$E_J = J(J+1)\frac{\hbar^2}{2\Theta} \ . \tag{18.44}$$

The gaps between successive states increase linearly because of $E_{J+1} - E_J = 2(J+1)\hbar^2/2\Theta$. This is typical of rotating states. Only even values of J are attainable, for reasons of symmetry, for those nuclei which have $J^P = 0^+$ in the ground state. The moment of inertia Θ can be found from the spins and excitation energies.

We want to discuss the experimental data through two examples which we have chosen out of the range of masses where highly deformed nuclei occur: the lanthanides and the actinides.

Coulomb excitation. Coulomb excitations in heavy ion reactions are often used to produce highly excited rotating states. To ensure that the interaction only takes place via Coulomb excitation, both partners must remain further apart than the range of the nuclear force. The projectile energy must then be so chosen that the *Coulomb threshold*

$$E_C = \frac{Z_1 Z_2 e^2}{4\pi\varepsilon_0} \frac{1}{R_1 + R_2} = \frac{Z_1 Z_2 \alpha \cdot \hbar c}{R_1 + R_2} \tag{18.45}$$

of the partners is not crossed. Larger values for the radii R_1 and R_2 of the reacting particles than in (5.56), say $R = 1.68$ fm $\cdot A^{1/3}$ are then assumed to make sure that the tails of the nuclear wave functions do not have any effects [Ch73].

Consider now the example of the Coulomb scattering of a $^{90}_{40}$Zr projectile off a $^{232}_{90}$Th target nucleus. The ^{90}Zr ion is accelerated in a tandem Van de Graaff accelerator up to a kinetic energy of $E_{Zr} = 415$ MeV. The centre of mass energy which is then available to the colliding particles is

$$E_{cm} = \frac{A_{Th}}{A_{Zr} + A_{Th}} E_{Zr} \approx 299 \text{ MeV}. \tag{18.46}$$

If we insert the charge numbers and radii of these two nuclei into (18.45), we find that $E_C \approx 300$ MeV. The centre of mass energy is, in other words, just below where the first non-electromagnetic effects would make themselves felt.

The ^{90}Zr projectile nucleus follows a hyperbolic path in the field of the target nucleus (Fig. 18.9a) and exposes the ^{232}Th nucleus to a rapidly changing electric field. The path of the ion is so sharply curved that frequencies in the time dependent electric field are generated that are high enough to produce individual excitations with energies up to about 1 MeV.

There is not just a quantitative but also a qualitative difference between Coulomb excitation and electron scattering off nuclei:

– The principal distinction is that the interaction is much stronger with a projectile charge which is Z times that of the electron. One must replace α by $Z\alpha$ in the matrix element (5.31). This means that the cross-section increases as Z^2.

– If we are not to cross the Coulomb threshold, the projectile energy must be so low that its velocity obeys $v \lesssim 0.05\,c$. Magnetic forces are hence of little importance.

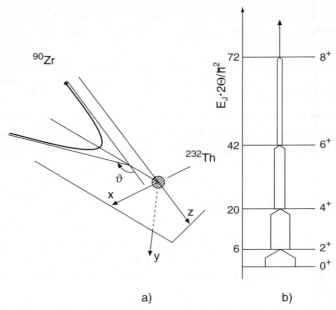

a)   b)

Fig. 18.9. (a) Kinematics of a heavy ion collision (here ^{90}Zr+^{232}Th). The projectile follows a hyperbolic orbit in the Coulomb field of the target nucleus. (b) Sketch of multiple Coulomb excitation of a rotation band. Successive quadrupole excitations lead to the 2^+, 4^+, 6^+, 8^+,... states being populated (with decreasing intensity).

– The ion orbit may be calculated classically, even for inelastic collisions. The kinetic energy of the projectile in Coulomb excitation changes by less than 1 % and thus its path is practically the same. The frequency distribution of the virtual photons is very well known and the transition amplitudes can be worked out to a high degree of accuracy.

   The large coupling strength means that successive excitation from one level to the next is now possible. This is sketched in Fig. 18.9b: the quadrupole excitation reproduces itself inside a rotation band from the 2^+ state via the 4^+ to the 6^+.

   The popularity of Coulomb excitation in gamma spectroscopy is well founded. In such reactions we primarily produce states inside rotation bands. The cross-sections into the excited states give us, through the transition probabilities, the most important information about the collective nature of the rotation bands. Measurements of the cross-sections into the various states simultaneously determine the transition probability for the electric quadrupole transition inside the rotation band.

   The introduction of germanium semiconductor detectors has marked a very significant step forward in nuclear-gamma spectroscopy. The low energy part of the gamma spectrum of Coulomb excitation of ^{232}Th from scattering with ^{90}Zr ions is shown in Fig. 18.10. This gamma spectrum was recorded

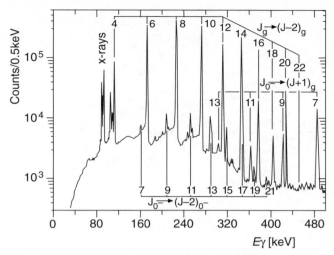

Fig. 18.10. Photon spectrum of a Coulomb excited ^{232}Th nucleus. Three series of matching lines may be seen. The strongest lines correspond to transitions in the ground state rotation band $J_g \rightarrow (J-2)_g$. The other two bands are strongly suppressed and are the results of excited states (cf. Fig. 18.12) [Ko88].

with a Ge-semiconductor counter and a coincidence condition for the backwardly scattered ^{90}Zr ions, which were measured with a Si-semiconductor detector (Fig. 18.11).

Excellent energy resolution makes it possible to see individual transitions inside rotation bands. Three series of lines can be recognised. The strongest are transitions inside the ground state rotation band ($J_g \rightarrow (J-2)_g$). According to (18.42) these lines should be equidistantly spaced out. This is only approximately the case. This may be explained by noting that the moment of inertia increases with the spin. Events with scattering angles around $180°$ are chosen because the projectile must then have got very close to the target and then at the moment of closest approach have experienced a strong acceleration. The virtual photon spectrum which the projectile emits contains high frequencies which are important for the excitation of the high spin states. The spectrum which emerges from this sort of measurement is shown in Fig. 18.12: as well as the ground state rotation band, there are other rotation bands which are built upon excited states. In this case the excitations may be understood as vibratory states.

Fusion reactions. Records in high spin excitations may be obtained with the help of fusion reactions such as

$$^{48}\text{Ca} + {}^{108}\text{Pd} \longrightarrow {}^{156}\text{Dy} \longrightarrow {}^{152}\text{Dy} + 4\text{n} \, .$$

^{48}Ca nuclei with a kinetic energy of 200 MeV can just break through the Coulomb barrier. If the fusion process takes place when the nuclei just touch,

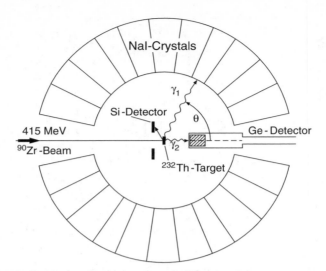

Fig. 18.11. Experimental apparatus for investigating Coulomb excitation in heavy ion collisions. In the example shown a ^{90}Zr beam hits a ^{232}Th target. The backwardly scattered Zr projectiles are detected in a silicon detector. A germanium detector, with which the γ cascades inside the rotation bands can be finely resolved, gives a precise measurement of the γ spectrum. These photons are additionally measured by a crystal ball of NaI crystals with a poorer resolution. A coincidence condition between the silicon detector and the NaI crystals can be used to single out an energy window inside which one may study the nuclear rotation states with the germanium detector (from [Ko88]).

then the ^{156}Dy fusion product receives angular momentum

$$\ell\hbar \approx (R_1 + R_2)\sqrt{2mE} \,, \tag{18.47}$$

where m is the reduced mass of the ^{48}Ca–^{108}Pd system. R_1 and R_2 are of course the correct nuclear radii from (5.56). The calculation thus yields $\ell \approx 180$. In practice the fusion reaction only takes place if the projectile and target overlap, so this number should be understood as an upper limit on the accessible angular momentum. Experimentally states up to $J^P = 60^+$ have been reached in this reaction (Fig. 18.14).

The moment of inertia. The size of the moment of inertia can with the aid of (18.44) be extracted from the measured energy levels of the rotation bands. The deformation δ can be obtained from the electric quadrupole radiation transition probability inside the rotation band. The matrix element for the quadrupole radiation is proportional to the quadrupole moment of the nucleus, which, for collective states, is given by (17.40). The observed connection between the moment of inertia and the deformation parameter is displayed in Fig. 18.13. Note that the nuclear moments of inertia are normalised to those of a rigid sphere with radius R_0

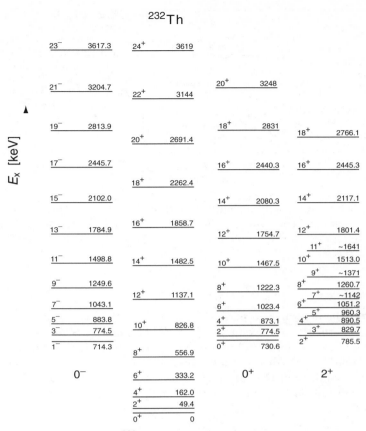

Fig. 18.12. Spectrum of the ^{232}Th nucleus. The excitation energies are in keV. As well as the ground state rotation band, which may be excited up to $J^P = 24^+$, other rotation bands have been observed which are built upon vibrational excitations (from [Ko88]). The quantum numbers of the vibrational states are given below the bands. For reasons of symmetry, the only rotation states which can be constructed upon the $J^P = 0^-$ vibrational state are those with odd angular momenta.

$$\Theta_{\text{rigid sphere}} = \frac{2}{5} M R_0^2 . \tag{18.48}$$

The moment of inertia increases with the deformation and is about half that of a rigid sphere.[2]

Two extremal models are also shown in Fig. 18.13. The moment of inertia is maximised if the deformed nucleus behaves like a rigid body. The other limit is reached if the nucleus behaves like an irrotational liquid.

Superfluid ^{4}He is an example of an ideal fluid, incompressible and frictionless. Currents in an frictionless liquid are irrotational. A massless eggshell

---

[2] The comparison with a rigid sphere is, of course, purely classical; a spherically symmetric quantum mechanical system cannot rotate.

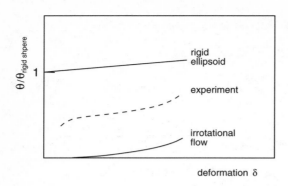

Fig. 18.13. Moments of inertia of deformed nuclei compared with a rigid sphere as a function of the deformation parameter δ. The extreme cases of a rigid ellipsoid and an irrotational liquid are given for comparison.

filled with superfluid helium would as it rotated have the moment of inertia of an irrotational current. Only the swelling out of the egg, and not the interior, would contribute to the moment of inertia. The moment of inertia for such an object is

$$\Theta = \frac{45\delta^2}{16\pi} \cdot \Theta_{\text{rigid sphere}} , \qquad (18.49)$$

where δ is the deformation parameter from (17.39).

Let us return to the example of the ^{232}Th nucleus. The transition probabilities yield a deformation parameter of $\delta = 0.25$. If the rotation of the nucleus could be described as that of an irrotational current, then its moment of inertia would, from (18.49), have to be 6 % of that of a rigid sphere. The level spacings of the ground state band yield, however

$$\frac{\Theta_{^{232}\text{Th}}}{\Theta_{\text{rigid sphere}}} \approx 0.3 . \qquad (18.50)$$

This implies that the experimentally determined moment of inertia lies between the two extremes (Fig. 18.13).

This result may be understood at a qualitative level rather easily. We mentioned in Sect. 17.4 that nuclear deformation is a consequence of an accumulation of mutually attractive orbitals either parallel to the symmetry axis (prolate shape) or perpendicular to it (oblate shape). The deformation is associated with the orbitals and one would expect deformed nuclei to rotate like rigid ellipsoids; but this clearly does not happen. This deviation from the rotation of a rigid rotator implies that nuclear matter must have a superfluid component. Indeed nuclei behave like eggshells that are filled with a mixture of a normal fluid and a superfluid.

The superfluid components of nuclear matter are presumably generated by the pairing force. Nucleons with opposite angular momenta combine to form pairs with spin zero (cf. p. 259). Such zero spin systems are spherically symmetric and cannot contribute to the rotation. The pair formation may

be understood analogously to the binding of electrons in *Cooper pairs* in superconductors [Co56b, Ba57]. The paired nucleons represent, at least as far
as rotation is concerned, the superfluid component of nuclear matter. This
means on the other hand that not all nucleons can be paired off in deformed
nuclei; the larger the deformation, the more nucleons must remain unpaired. This explains why the moment of inertia increases with the deformation
(Fig. 18.13).

A similar dependence of the moment of inertia upon the unpaired nucleons
can be seen in the rotation bands. The speed of rotation of the nucleus,
and hence the centrifugal force upon the nucleons, increases with angular
momentum. This causes nucleon pairs to break apart. Thus for large angular
momenta the moment of inertia approaches that of a rigid rotator, as one
can vividly demonstrate in ^{152}Dy.

The excitation spectrum of ^{152}Dy (Fig. 18.14) is more than a little exotic.
The ground state of ^{152}Dy is not strongly deformed, as one sees from the
fact that the levels in the ground state rotation band do not strictly follow
the $E \propto J(J+1)$ law and that transition probabilities are small. This band,
in which the 0^+ until 46^+ states have been observed, first shows a genuine
rotational character for high spins. The band which goes up to $J^P = 60^+$ is
particularly interesting [Tw86]. The moment of inertia of this band is that
of a rigid ellipsoid whose axes have the ratios $2:1:1$ [Ra86]. The transition
probabilities inside this band are of the order of 2000 single particle probabilities. Additionally to these two rotation bands, which have a prolate
character, states have been found which may be interpreted as those of an
oblately deformed nucleus. Evidently ^{152}Dy has two energy minima near to
its ground state, a prolate and an oblate shape. This example shows very
nicely that for nuclei with incomplete shells a deformed shape is more stable than a sphere. Tiny changes in the configuration of the nucleus decide
whether the prolate or oblate form is energetically favoured (Fig. 17.10).

Further excitations of deformed nuclei. We have here only treated the
collective aspects of rotation. Generally, however, excitations occur where,
as well as rotation, an oscillation around either the equilibrium shape of the
deformed nucleus or single particle excitations are seen. (The latter case may
be particularly clearly seen in odd nuclei.) The single particle excitations may
be, as described in Sect. 17.4, calculated from the movement of nucleons in
a deformed potential. Deformed nuclei may be described, similarly to their
vibrating brethren, in a hybrid model which employs collective variables for
the rotating and vibrating degrees of freedom. The single particle motion
is coupled to these collective variables. The names of Bohr and Mottelson
in particular are associated with the work that showed that a consistent
description of nuclear excitations is possible in such hybrid models.

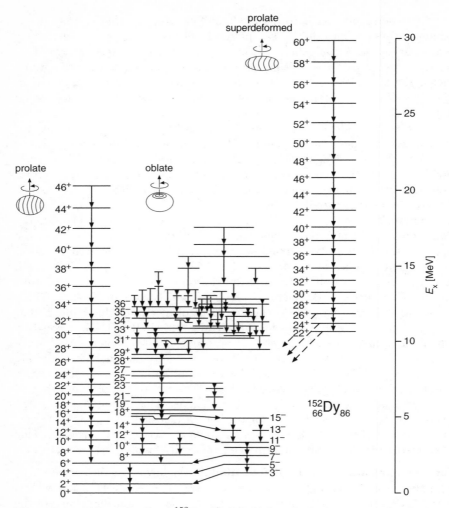

Fig. 18.14. Energy levels of ^{152}Dy [Sh90]. Although the low energy levels do not display typical rotation bands, these are seen in the higher excitations, which implies that the nucleus is then highly deformed.

_____ **Problems**

1. **The electric dipole giant resonance**
    a) How large is the average deviation between the centres of mass of the pro-
       tons and neutrons in giant dipole resonances for nuclei with $Z = N = A/2$?
       The A dependence of the resonance energy is very well described by $\hbar\omega \approx$
       $80\ \mathrm{MeV}/A^{1/3}$. Give the numerical value for $^{40}\mathrm{Ca}$.
    b) Calculate the squared matrix element for the dipole transition in this model.
    c) Calculate the matrix element for a proton or neutron dipole transition (18.36)
       in the shell model with a harmonic oscillator potential. Use the fact that sin-
       gle particle excitations are about half the size of those of the giant resonance.

2. **Deformation**
   The deformation parameter of the $^{176}_{71}\mathrm{Lu}$ nucleus is $\delta = +0.31$. Find the semi-
   axes a and b of the rotational ellipsoid, describe its shape and calculate the
   quadrupole moment of this nucleus.

3. **Rotational bands**
   The rotational band of $^{152}\mathrm{Dy}$ in Fig. 18.14 which extends up to $J^P = 60^+$
   corresponds to the rotation of an ellipsoid the ratio of whose axes is $2:1:1$.
   What would be the velocity of the nucleons at the "tip" of the ellipsoid if this
   was a rotating rigid body? Compare this velocity with the *average* speed of
   nucleons in a Fermi gas with $p = p_\mathrm{F} = 250\ \mathrm{MeV}/c$.

19. Nuclear Thermodynamics

Up to now we have concerned ourselves with the properties of nuclei in the ground state or the lower lying excited states. We have seen that the observed phenomena are characterised, on the one hand, by the properties of a degenerate fermion system and, on the other, by the limited number of the constituents. The nuclear force generates, to a good approximation, an overall mean field in which the nucleons move like free particles. In the shell model the finite size of nuclei is taken into account and the states of the individual nucleons are classified according to radial excitations and angular momenta. Thermodynamically speaking, we assign such systems zero temperature.

In the first part of this chapter we want to concern ourselves with highly excited nuclei. At high excitation energies the mean free path of the nucleon inside the nucleus is reduced; it is only about 1 fm. The nucleus is then no longer a degenerate fermionic system, but rather resembles, ever more closely for increasing excitations, the state of a normal liquid. It is natural to use statistical methods in the description of such systems. A clear description may be gained by employing thermodynamical quantities. The excitation of the nucleus is characterised by the temperature. We should not forget that strictly speaking one can only associate a temperature to large systems in thermal equilibrium and even heavy nuclei do not quite correspond to such a system. As well as this, excited nuclei are not in thermal equilibrium, but rather rapidly cool down via the emission of nucleons and photons. In any thermodynamical interpretation of experimental results we must take these deficiencies into account. In connection with nuclear thermodynamics one prefers to speak about *nuclear matter* rather than nuclei, which implies that many experimental results from nuclear physics may be extrapolated to large systems of nucleons. As an example of this we showed, when we considered the nuclear binding energy, that by taking the surface and Coulomb energies into account one can calculate the binding energy of a nucleon in nuclear matter. This is just the volume term of the mass formula, (2.8).

Heavy ion reactions have proven themselves especially useful in the investigation of the thermodynamical properties of nuclear matter. In nucleus-nucleus collisions the nuclei melt together to form for a brief time a nuclear matter system with increased density and temperature. We will try below to describe the phase diagram of nuclear matter using experimental and theoretical results about these reactions.

The results of nuclear thermodynamics are also of great importance for cosmology and astrophysics. According to our current understanding, the universe in the early stages of its existence went through phases where its temperature and density were many orders of magnitude higher than in the universe of today. These conditions cannot be reconstructed in the laboratory. Many events in the history of the universe have, however, left lasting traces. With the help of this circumstantial evidence one can try to draw up a model of the development of the universe.

19.1 Thermodynamical Description of Nuclei

We have already in Sect. 3.4 (Fig. 3.10) distinguished between three sorts of excitations in nuclei:

- The ground state and the low-lying states can be described in terms of single particle excitations or via collective motion. This was treated in Chapters 17 and 18.
- Far above the particle threshold there are no discrete states but only a continuum.
- In the transition region below and barely above the particle threshold there are lots of narrow resonances. These states do not, however, contain any information about the structure of the nucleus. The phenomena in this energy range in nuclei are widely referred to as *quantum chaos*.

In the following we shall concern ourselves with the last two of these domains. Their description involves statistical methods and so we will initially turn our attention to the concept of *nuclear temperature*.

Temperature. We want to introduce the idea of temperature in nuclear physics through the example of the spontaneous fission of ^{252}Cf. The half life of ^{252}Cf is 2.6 years and it has a 3.1 % probability of decaying via spontaneous fission. There is some friction in the separation of the fission fragments and so not all of the available energy from the fission process is converted into kinetic energy for the fragments. Rather the internal energy of the fragments is increased: the two fragments heat up.

The cooling down process undergone by the fission fragments is shown schematically in Fig. 19.1. Initially cooling down takes place via the emission of slow neutrons. Typically 4 neutrons are emitted, each of them carrying off, on average, 2.1 MeV. Once the fragments have cooled below the threshold for neutron emission, they can only cool further by photon emission.

The energy spectrum of the emitted neutrons has the form of a evaporation spectrum. It may be described by a Maxwell distribution:

$$N_{\mathrm{n}}(E_{\mathrm{n}}) \sim \sqrt{E_{\mathrm{n}}} \cdot \mathrm{e}^{-E_{\mathrm{n}}/kT} \, . \tag{19.1}$$

Figure 19.2 shows the experimental spectrum normalised by a factor of $\sqrt{E_{\mathrm{n}}}$. The exponential fall-off is characterised by the temperature T of the system.

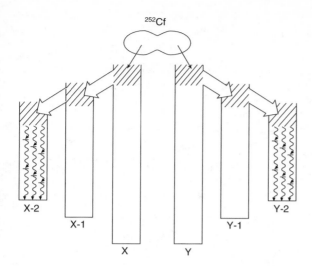

Fig. 19.1. Cooling of fission fragments (schematic). A ^{252}Cf nucleus splits into two parts with mass numbers X and Y which then cool down by emitting first neutrons and later photons.

In this case $kT = 1.41$ MeV. Fission fragments from different nuclei are found to have different temperatures. One finds, e.g., a smaller value in the fission of ^{236}U, namely $kT = 1.29$ MeV.

Figure 19.3 displays the energy spectrum of the photons emitted in the de-excitation of the produced daughter nuclei. On average about 20 photons are set free for each spontaneous fission, and 80 % of these photons have energies of less than 1 MeV. This spectrum also closely resembles an evaporation spectrum. The stronger fall-off of the photon spectrum compared to the neu-

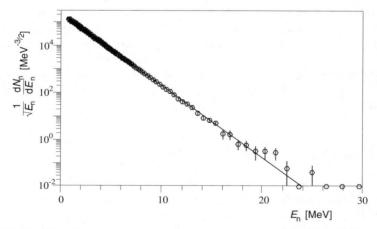

Fig. 19.2. Energy spectrum of neutrons emitted in the spontaneous fission of ^{252}Cf (from [Bu88]). The distribution is divided by $\sqrt{E_n}$ and then fitted to the exponential behaviour of a Maxwell distribution (*solid line*).

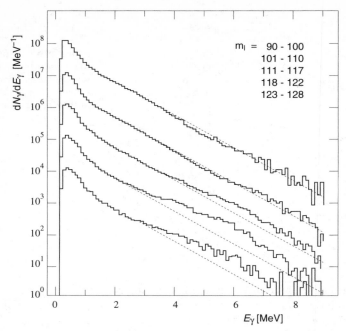

Fig. 19.3. Photon emission energy spectra in the spontaneous fission of ^{252}Cf. The various spectra correspond to different mass numbers, m_1, of the lighter fission products (*from top to bottom*). The dotted line is a common fit of an exponential function (from [Gl89]).

tron spectrum signals that the temperature in the photon emission phase, which takes place for lower nuclear excitations, is significantly lower.

Our successful statistical interpretation of these neutron and photon spectra leads to the important conclusion that the states in the neighbourhood of the particle threshold, which may be understood as a reflection of the corresponding transitions, can also be described with statistical methods. Indeed the observed form of the spectrum may be formally derived from a statistical study of the density of states of a degenerate Fermi gas.

19.2 Compound Nuclei and Quantum Chaos

Many narrow resonances may be found in the transition region below and just above the particle threshold of a heavy nucleus. The states below the particle threshold are discrete and each one of these states possesses definite quantum numbers. The same is true of the states immediately above the threshold. Decays into these states are only described statistically through the density of these states. These states therefore do not contain any specific information about the structure of the nucleus.

Compound nuclei. In neutron capture by heavy nuclei a multiplicity of resonances are observed. An example of such a measurement is seen in Fig. 19.4 where the cross-section for neutron scattering off thorium displays very many resonances. One should note that the energy scale is in eV, the separation of these resonances is thus six orders of magnitude smaller than the gaps in energy separating lower lying states. This observation was already explained in the thirties by Niels Bohr in the so-called compound nucleus model. Neutrons in the nucleus have a very short free path due to the strong interaction and they very rapidly distribute their energy among the nucleons in the nucleus. The probability that all the energy supplied is held by one single nucleon is small. The nucleons cannot therefore escape from the nucleus and this leads to a long lifetime for the compound nucleus states. This lifetime is mirrored in the narrow widths of the resonances.

This picture has been greatly refined in the intervening decades. Thus the compound nucleus state is not reached immediately, but rather the system, via successive collisions, passes through a series of intermediate states. The compound nucleus state is the limiting case in which the nucleons are in thermal equilibrium.

Quantum chaos in nuclei. In the theory of classical deterministic systems we distinguish between regular and chaotic orbits. Regular orbits are stable orbits which are not greatly affected by small external perturbations. The particles undergo periodic motion and the entire configuration of the system thus repeats itself. Chaotic orbits are very different. They are not periodic and infinitesimally small perturbations lead to big changes. While predictions for the development of regular systems may be made to an arbitrary accuracy, the uncertainties associated with predicting chaotic systems increase exponentially.

In quantum mechanics regular orbits correspond to states whose wave functions may be calculated with the help of the Schrödinger equation in some model, e.g., for nuclei the shell model. The quantum mechanical equivalent of classical chaotic motion are states which are stochastically made up of single particle wave functions. In both the classical and quantum mechanical

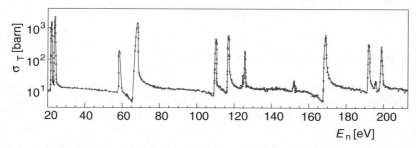

Fig. 19.4. Total cross-section for the reaction ^{232}Th+n as a function of the neutron energy. The sharp peaks correspond to resonances with orbital angular momentum $\ell = 0$ (from [Bo69]).

cases a system in a chaotic state does not contain any information about the interactions between the particles.

The stochastic composition of chaotic states can be experimentally demonstrated by measuring the energy separations between these states. For this one considers resonance spectra such as that of Fig. 19.4. In the excitation region of the compound nucleus the states are very dense, so a statistical approach is justified.

It is apparent here that states with the same spin and parity (in Fig. 19.4 all the sharp resonances) attempt to keep as far apart as possible. The most likely separation of these states is significantly greater than the most likely separation of the energy levels of states if they were, for the same state density, distributed in a statistical fashion, according to a Poisson distribution independently of each other. This behaviour of the chaotic states is just what one expects if they are made up from a mixture of single particle states with the same quantum numbers. Such quantum mechanical mixed states attempt to repel each other, i.e., their energy levels arrange themselves as far apart from each other as possible.

The existence of collective states, such as, e.g., the giant dipole resonance, for excitations above the particle threshold, i.e., in the region where the behaviour of the states is chaotic, is a very pretty example of the coexistence of regular and chaotic nuclear dynamics. Excitation of the collective state of the giant resonance takes place through photon absorption. The collective state couples to the many chaotic states via the nucleon-nucleon interaction. These partially destroy the coherence and thus reduce the lifetime of the collective state.

The continuum. The continuum is by no means flat, rather strong fluctuations are seen in the cross-section. The reason for this is that, on the one hand, at higher energies the widths of the resonances increase because more decay channels stand open to them, but on the other hand the density of states also increases. Resonances with the same quantum numbers thus interfere with each other which leads to fluctuations in the total cross-section. These fluctuations do not correspond to single resonances but to the interference of many resonances. The size of the fluctuations and their average separation can be quantitatively calculated from the known state density [Er66].

19.3 The Phases of Nuclear Matter

The liquid–gas phase transition. Peripheral heavy ion reactions have proven themselves most useful as a way to heat up nuclei in a controlled way. In a glancing collision of two nuclei (Fig. 19.5) two main fragments are produced which are heated up by friction during the reaction. In such reactions one can measure rather well both the temperature of the fragments and also the energy supplied to the system. The temperature of the fragments is found from the Maxwell distribution of the decay products, while the total energy supplied to the system is determined by detecting all of the particles produced in the final state. Since the fragment which came from the projectile moves off in the direction of the projectile, its decay products will also move in that direction and may be thus kinematically distinguished both from the decay products of the target fragments and also from the frictionally induced evaporative nucleons. The contributions from the energy supplied to the fragments and from the energy lost to friction during the glancing collision may thus be separated from one another.

Let us take as an example an experiment where gold nuclei with an energy of 600 MeV/nucleon were fired at a gold target. The reaction products were then tracked down using a detector which spanned almost the entire solid angle (a 4π detector).

The dependence of the fragments' temperature on the energy supplied to the system is shown in Fig. 19.6. For excitation energies E/A up to about 4 MeV/nucleon one observes that the temperature sharply increases. In the region $4\,\mathrm{MeV} < E/A < 10\,\mathrm{MeV}$ the temperature hardly varies at all, while at higher energies it again grows rapidly. This behaviour is reminiscent of the process of water evaporation where, around the boiling point, at the

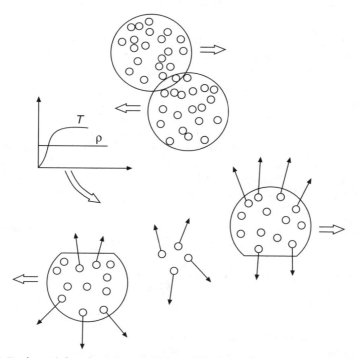

Fig. 19.5. A peripheral nuclear collision. The large fragments are heated up by friction. As well as this, individual nucleons and smaller nuclear fragments are also produced in the collision. The diagram describes the time evolution of the density ϱ and temperature T of the fragments during the collision.

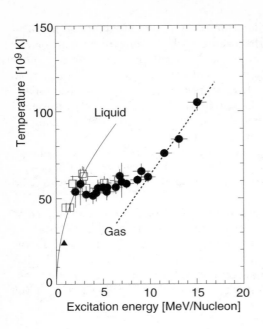

Fig. 19.6. Temperature of the fragments in a peripheral collision of two ^{197}Au nuclei as a function of the excitation energy per nucleon (from [Po95]). The behaviour of the temperature can be understood as a phase transition in nuclear matter.

phase transition from liquid into steam, the temperature remains constant, even though energy is added to the system, until the entire liquid has been converted into a gaseous state. It is therefore natural to interpret the temperature dependence described above as a nuclear matter phase transition from a liquid to a gas-like state.

The terms which we have used come from equilibrium thermodynamics. For such conditions a logical interpretation of the phase transition would be the following: at a temperature of about $kT \sim 4$ MeV a layer of nucleons in a gaseous phase forms around the nucleus. This does not evaporate away but remains in equilibrium with the liquid nucleus and exchanges nucleons with it. The nucleon gas can only be further heated up after the whole of the nucleon liquid has evaporated.

Hadronic matter. If we wish to investigate central, and not peripheral, collisions in gold-gold collisions, we have to select in the experiment those events in which many charged and neutral pions are emitted (Fig. 19.7). To keep the discussion simple, we will choose projectile energies of 10 GeV/nucleon or more for which a large number of pions is created.

At such energies the nucleonic excitation $N + N \rightarrow \Delta + N$ has a cross-section of $\sigma = 40$ mb. The corresponding path length $\lambda \approx 1/\sigma \varrho_N$ in the nucleus is of the order of 1 fm. This means that multiple collisions take place in heavy ion collisions and that for sufficiently high energies every nucleon will on average be excited once or more into a Δ baryon. In the language of thermodynamics this excitation corresponds to the opening up of a new degree of freedom.

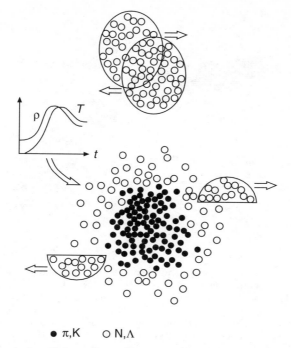

$\bullet$ π,K $\circ$ N,Λ

Fig. 19.7. Central collision of two heavy nuclei at high energies. A large number of pions are produced here. The curves show the increase of density, ϱ, and temperature, T, in the central region of the collision.

Δ baryons decay rapidly but they are continually being reformed through the inverse reaction $\pi N \rightarrow \Delta$. Creation and decay via $\pi N \leftrightarrow \Delta$ thus stand in a dynamical equilibrium. This mix of nucleons, Δ baryons, pions and, in significantly smaller amounts, other mesons is called *hadronic matter*.

Pions, since they are much lighter than the other hadrons, are primarily responsible for energy exchange inside hadronic matter. The energy density and temperature of hadronic matter produced in a collision of two atomic nuclei can be experimentally determined with the help of these pions. The temperature is found from the energy distribution of those pions which are emitted orthogonally to the beam direction. Their energy spectrum has the exponential behaviour expected of a Boltzmann distribution:

$$\frac{\mathrm{d}N}{\mathrm{d}E_{\mathrm{kin}}} \propto \mathrm{e}^{-E_{\mathrm{kin}}/kT} , \tag{19.2}$$

where E_{kin} is the kinetic energy of the pion. One finds experimentally that the temperature of the pionic radiation is never greater than $kT \approx 150\,\mathrm{MeV}$, no matter how high the energies of the colliding nuclei are. This may be understood as follows: hot nuclear matter expands and in doing so cools down. Below a temperature $kT \approx 150\,\mathrm{MeV}$, the hadronic interaction probability

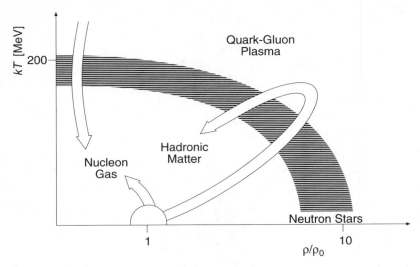

Fig. 19.8. Phase diagram for nuclear matter. Normal nuclei have $\varrho = \varrho_0$ $(= \varrho_N)$ and temperature $T = 0$. The arrows show the paths followed by nuclei in various heavy ion reactions. The short arrow symbolises the heating up of nuclei in peripheral collisions; the long arrow corresponds to relativistic heavy ion collisions, in which nuclear matter possibly crosses the quark-gluon plasma phase. The cooling of the universe at the time $T \approx 1\,\mu s$ is represented by the downwards pointing arrow.

of the pions, and thus energy exchange between them and other particles, decreases sharply. This process is referred to as the pions *freezing out*.[1]

Phase diagram for nuclear matter. The various phases of nuclear matter are summarised in Fig. 19.8. We want to clarify this phase diagram by comparing nuclear matter with usual matter (that composed of atoms or molecules). Cold nuclei have density ϱ_N and temperature $kT = 0$. A neutron star corresponds to a state with $kT = 0$, however, its density is about 3–10 times as big as that of nuclei.

If one supplies energy to a normal nucleus, it heats up and emits nucleons or small nuclei, mainly α clusters, just as a liquid droplet evaporates atoms or molecules. If, however, one confines the material, increasing the energy supplied leads to the excitation of internal degrees of freedom. In a molecular gas these are rotational and vibrational excitations. In nuclei nucleons can be excited into $\Delta(1232)$ resonances or to still higher nucleon states. We have called the mish-mash of nucleons and pions, which are then created by decays, *hadronic matter*.

---

[1] A similar process takes place in stars: the electromagnetic radiation in the interior of the sun is at many millions of K. On its way out it cools down via interactions with matter. What we observe is white light whose spectrum corresponds to the temperature of the solar surface. In contrast to hot nuclear matter, the sun is of course in equilibrium and is not expanding.

Quark-gluon plasma. The complete dissociation of atoms into electrons and atomic nuclei (a plasma) has its equivalent in the disintegration of nucleons and pions into quarks and gluons. Qualitatively the positions of the phase boundary in the temperature-density diagram (Fig. 19.8) may be understood as follows: at normal nuclear densities each nucleon occupies a volume of about 6 fm^3, whereas the actual volume of a nucleon itself is only about a tenth of this. If one then were to compress a cold nucleus (T=0) to ten times its usual density, the individual nucleons would overlap and cease to exist as individual particles. Quarks and gluons would then be able to move "freely" in the entire nuclear volume. If on the other hand one were to follow a path along the temperature axis, i.e., increase the temperature without thereby altering the nucleon density in the nucleus, then at a temperature of 200 MeV enough energy would be available to the individual nucleon-nucleon interactions to increase, via pion production, the hadronic density and the frequency of the collisions between them so much that it would be impossible to assign a quark or gluon to any particular hadron.

This state is referred to as a *quark-gluon plasma*. As we have already mentioned, this state, where the hadrons are dissolved, cannot be observed through the study of emitted hadrons. There are attempts to detect a quark-gluon plasma state via electromagnetic radiation. The coupling of photons to quarks is about two orders of magnitude smaller than that of strongly interacting matter is. Thus any electromagnetic radiation produced in any potential creation of a quark-gluon plasma, e.g., in relativistic heavy ion collisions, could be directly observed. It would not be cooled down in the expansion of the system.[2]

There is a great deal of interest in detecting a quark-gluon plasma because it would mean an experimental confirmation of our ideas of the structure of strongly interacting matter. If the assignment of quarks and gluons to individual hadrons were removed, the constituent quarks would lose their masses and turn into partonic quarks; one would be able to simulate the state of the universe at a very early stage in its history.

---

[2] The above analogy from astrophysics is also applicable here: the neutrinos which are created in fusion reactions in the solar interior are almost unhindered in their escape from the sun. Their energy spectrum thus corresponds to the temperature of where they were produced and not to that of the surface.

19.4 Particle Physics and Thermodynamics in the Early Universe

> In all societies men have constructed myths about the origins of the universe and of man. The aim of these myths is to define man's place in nature, and thus give him a sense of purpose and value.
>
> *John Maynard Smith* [Sm89]

The interplay between cosmology and particle physics during the last few decades has lead to surprising insights for both areas. In what follows we want to depict current ideas about the evolution of the universe and show what consequences this evolution has had for our modern picture of particle physics. We will here make use of the standard cosmological model, the big bang model, according to which the universe began as an infinitely hot and dense state. This fireball then expanded explosively and its temperature and density have continued to decrease till the present day. This expansion of an initially hot plasma of elementary particles was the origin of all nowadays known macroscopic and microscopic forms of matter: stars and galaxies; leptons, quarks, nucleons and nuclei. This model for the time development of the universe was motivated and then confirmed by two important experimental observations: the continuous expansion of the universe and the cosmic background radiation.

The expanding universe. The greatest part of the mass of the universe is located in galaxies. These spatially concentrated star systems are held together by the force of gravity and, depending upon their size, have masses of between 10^7 and 10^{13} solar masses. It is believed that there are about 10^{23} galaxies in the universe – a number comparable to the number of molecules in a mole.

With the help of large telescopes it is possible to measure the distance to and the velocities of galaxies which are a long way away from the earth. The velocity of a galaxy relative to the earth can be determined from the Doppler shift of atomic spectral lines, which are known from laboratory measurements. One so finds a shift of the observed lines into the red, i.e., the longer wave-length region. This corresponds to a motion of the galaxies away from us. This observation holds no matter what direction in the heavenly sphere the galaxy under observation is in. A determination of the distance to the galaxy is carried out by measuring its light intensity and estimating its luminosity; these quantities are related by the well-known $1/r^2$ law. Such distance estimates are particularly imprecise for very distant galaxies.

The measured velocities v of the observed galaxies are roughly linearly proportional to their separation d from the earth

$$v = H_0 \cdot d. \tag{19.3}$$

The proportionality constant, H_0, is, because of the uncertainties associated with the distance estimates, only known up to a factor of 2

$$H_0 = 50 \cdots 100 \,\text{km}\,\text{s}^{-1}/\text{MPc} \qquad (1\,\text{Pc} = 3.1 \cdot 10^{13}\,\text{km} = 3.3\,\text{light years}).$$

It is called the *Hubble constant* after the discoverer of this relationship. These observations taken together are interpreted as implying an isotropic expansion of the universe.

According to the big bang theory, the initial hot plasma filled the universe with extremely short wave-length electromagnetic radiation, which, though, increased its wave-length as the universe expanded and cooled. The observation, by Penzias and Wilson [Pe65], of this radiation in the microwave-length region, which we now call the *cosmic background radiation*, was therefore a very important confirmation of the big bang model. This microwave radiation corresponds to black body radiation at a temperature of 2.73 K and is measured in every direction of the universe as being extraordinarily isotropic (for reference see [PD98]).

A relation between the age and the size of the universe can be derived with the help of general relativity theory and the observed expansion of the universe. In the simplest model, the *Friedman model* of the expanding universe, one distinguishes between three cases which depend upon the average mass density of the universe: if the average density is greater than a critical density, then the mutual attraction of the galaxies will slow the expansion of the universe down and eventually produce a contraction. The universe will then collapse into a point (closed universe). If the average density is smaller than the critical density, gravitation cannot reverse the expansion. In such a case the universe will expand forever (open universe). If the average and critical densities are approximately the same the universe would asymptotically approach a limiting radius.

The density measured with optical methods is in fact smaller than the critical density. It is, however, suspected that one or more sorts of *dark matter* exist, which are not detectable with optical methods, and it cannot be excluded that the universe does after all possess the critical density. One conceivable sort of dark matter would be massive neutrinos. Experiments to measure the mass of the neutrino (Sect. 17.6) are of great importance for this suggestion. Even if neutrino masses were only a few eV/c^2 the large number of neutrinos in the universe would make a significant contribution to the mass, and hence the density, of the universe.

Since the universe is still in an early stage of its expansion the previous history of our universe would be similar in all three cases. The age of a universe with a sub-critical density is given by the inverse Hubble constant

$$t_0 = \frac{1}{H_0}\,, \tag{19.4}$$

and is between 10 and 20 thousand million years.

The first three minutes of the universe. In the initial phase of the universe all the (anti)particles and the gauge bosons were in thermodynamical equilibrium, i.e., there was so much thermal, and thus kinematical, energy available that all the (anti)particles could transform into each other at will. There was therefore no difference between quarks and leptons, which means that the strength of all the interactions was the same.

After about 10^{-35} s the temperature had decreased so much due to the expansion that a phase transition took place and the strong interaction decoupled from the electroweak interaction, i.e., the strongly interacting quarks barely interacted with the leptons any more. At this stage the ratio between the numbers of quarks and photons was fixed at about 10^{-9}.

After about 10^{-11} s, at a temperature $kT \approx 100$ GeV, a further phase transition took place in which the weak interaction decoupled from the electromagnetic interaction. We will discuss this process below.

When, after about 10^{-6} s, the continuous expansion of the universe had lowered its temperature down to $kT \approx 100$ MeV, which is the typical energy scale for hadronic excitations, the quarks formed bound states in the shape of baryons and mesons. The protons and neutrons so-produced were in thermal equilibrium due to weak processes.

After about 1 s and at a temperature $kT \approx 1$ MeV, the difference between the neutron and proton masses, the neutrinos had too little energy to maintain the state of equilibrium between the protons and neutrons. They decoupled from matter, i.e., they henceforth essentially no longer interacted at all and propagated freely through the universe. Meanwhile the ratio of protons to neutrons increased up to a value of 7.

After about three minutes of expansion the temperature had fallen to $kT \approx 100$ keV. From this moment the thermal equilibrium between nucleons and photons was broken, since the photon energies were no longer sufficient to break up the light nuclei, through photofission processes, into their constituents at the same pace as they were produced by nucleon fusion. In this phase the big bang nucleosynthesis of deuterium, helium and lithium nuclei took place.

Figure 19.9 schematically shows the early history of the universe from the electroweak phase transition once again. The curves represent the time (or temperature dependent) evolution of the energy density of radiation and matter. One can see the sharp drop in the energy density caused by the expansion of the universe. At temperatures of 10^{13} K the hadrons, and later the leptons, decouple from the radiation. At $T \approx 10^4$ K a matter dominated universe takes over from a previously radiation dominated universe. The

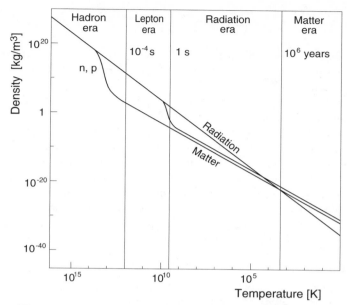

Fig. 19.9. The evolution of the energy density of the universe, as a function of temperature, after the electroweak phase transition ($T \approx 10^{15}\,\mathrm{K}$). In the early development of the universe radiation was in thermal equilibrium with matter and antimatter. Over a period of time matter decoupled from radiation and the matter and radiation energy densities developed different temperature dependences, so that the universe finally became matter dominated.

current temperature of the universe is 2.73 K, the temperature of the cosmic background radiation.

Below we want to delve further into some important events from this early history of the universe.

Matter-antimatter asymmetry. All observations show that the modern universe is made up solely of matter and there is no evidence for some parts of the universe being composed of antimatter. Since according to our ideas all (anti)particles at a very early stage of the universe were in thermal equilibrium, i.e., fermion-antifermion creation from gauge bosons was just as frequent as fermion-antifermion annihilation into gauge bosons, then if this symmetry had survived the development of the universe, there ought to be just as many fermions as antifermions or, more especially, as many quarks as antiquarks (which means as many baryons and antibaryons) in the universe. Furthermore there ought to be free photons which were produced in fermion-antifermion annihilation, but which due to the expansion and cooling of the universe could not go through the reverse reaction. One finds today that the ratio of baryons to photons is $3 \cdot 10^{-10}$. If all of these photons came from quark-antiquark annihilation, then a quark-antiquark asymmetry in the hot plasma of the early universe of

$$\Delta q = \frac{q - \overline{q}}{q + \overline{q}} = 3 \cdot 10^{-10} \tag{19.5}$$

would be sufficient to explain the current observed matter-antimatter asymmetry. The question is how did this small but decisive surplus of quarks arise in the early universe?

To generate a matter-antimatter asymmetry we have to fulfill three conditions: CP violation, baryon number violation and thermal non-equilibrium. In the framework of *grand unified theories*, GUT's, one can imagine that all of these conditions could be fulfilled. Consider the situation of the universe at time $t < 10^{-35}$ s. At this moment all (anti)fermions were equivalent, so they could be transformed into each other which could in certain reactions lead to a violation of baryon number. A hypothetical exchange particle, which mediates such a transition, is the X boson whose mass would be about 10^{14} GeV/c^2. These X bosons could be produced as real particles at sufficiently high energies and would decay into a quark and an electron, similarly the $\overline{X}$ boson decays into an antiquark and a positron. CP violation in the decay of the X boson would mean that the decay rates of the X and $\overline{X}$ bosons would not be exactly equal. In thermal equilibrium, i.e., at temperatures or energies above the mass of the X boson, the effect of CP violation on the baryon number would be eliminated since the creation and decay of the X and $\overline{X}$ bosons would be in equilibrium. This equilibrium would first be destroyed by the cooling of the universe and the asymmetry of the CP violating decay of the X boson would lead to a quark surplus, which eventually would be responsible for the matter-antimatter asymmetry we observe in the universe around us.

There are searches in progress for evidence of the existence of systems with CP violation and baryon number violation in the modern universe. As mentioned in Sect. 14.4 CP violation has been detected in K^0 decay. Experiments looking for proton decay have so far not yielded any evidence for baryon number violation.

Electroweak phase transitions. Let us now consider the universe at the age of just 10^{-11} s when it had a temperature of $kT \approx 100$ GeV. It is believed that one can reconstruct the development of the universe from what is now known of elementary particle physics back to this stage. Extrapolations further back into the past may be based on plausible assumptions but they are in no way proven.

It is believed that the the *electroweak phase transition* took place at this moment. Only after this phase transition did the now known properties of the elementary particles establish themselves. A loss of symmetry and an increase in order is characteristic of a phase transition of this type; just as in the phase transition from the paramagnetic to the ferromagnetic phase in iron when it drops below the Curie temperature. For temperatures equivalent to energies > 100 GeV, in other words before the phase transition, the photon, W and Z gauge bosons had similar properties and the distinction between the electromagnetic and weak forces was removed (symmetry!). In this state there was also no significant difference between electrons and neutrinos. Below the critical temperature this symmetry was, however, destroyed. This phenomenon, known in the standard model of elementary particle physics as *spontaneous symmetry breaking*, caused the W and Z bosons to acquire their large masses from so-called Higgs' fields and the elementary particles took on the properties that we are now familiar with (cf. Chap. 11.2).

Although today elementary particles may be accelerated up to energies $> 100 \, \mathrm{GeV}$ and the W and Z bosons have been experimentally produced and detected, it will not be possible to reproduce in the laboratory the high energy-densities of 10^8 times the nuclear density which reigned at the electroweak phase transition. We can therefore only try to reproduce and to demonstrate the traces left by the phase transition, i.e., the W, Z and Higgs bosons, so as to use them as witnesses of what went on in the initial stages of the universe.

Hadron formation. An additional phase transition took place when the universe was about $1 \, \mu s$ old. At this stage the universe had an equilibrium temperature $kT \approx 100 \, \mathrm{MeV}$. The hadrons constituted themselves in this phase from the previously free quarks and gluons (*quark-gluon plasma*). Mostly nucleons were formed in this way.

Since the masses of the u- and d-quarks are very similar, they first formed roughly the same numbers of protons and neutrons, which initially existed as free nucleons since the temperature was too high to permit the formation of nuclei. These protons and neutrons were in thermal equilibrium until the temperature of the universe had sunk so much that the reaction rates for neutron creation processes (e.g., $\bar{\nu}p \rightarrow e^{+}n$) were, as a consequence of the greater mass of the neutron, significantly less than that of the inverse processes of proton formation (e.g., $\bar{\nu}p \leftarrow e^{+}n$). Thenceforth the numerical ratio of neutrons to protons decreased.

There are currently attempts to simulate this transition from a quark-gluon plasma to a hadronic phase in heavy ion reactions. In these reactions one tries to first create a quark-gluon plasma through highly energetic collisions of ions, in which the matter density is briefly increased to a multiple of the usual nuclear density. In such a state the quarks should only feel the short range and not the long range part of the strong potential, since this last should be screened by their tightly packed neighbours. In such a case the quarks may be viewed as quasi-free and form a quark-gluon plasma. Such a quark-gluon plasma has, however, not yet been indubitably generated and a study of the transition to the hadronic phase is thus only possible in a rather limited fashion.

In the universe the transition from a quark-gluon plasma to the hadronic phase took place via the equilibrium temperature dropping at low matter densities. In the laboratory it is attempted to fleetingly create this transition by varying the matter density at high temperature (cf. Fig. 19.8 and Sect. 19.3).

Primordial synthesis of the elements. At $t = 200$ seconds in the cosmological calendar, the make up of baryonic matter was 88% protons and 12% neutrons. The creation of deuterium by the fusion of neutrons and protons was, until this stage, in equilibrium with the inverse reaction, the photodissociation of deuterium into a proton and a neutron, and the lifetime of the deuterons was extremely short. But now the temperature dropped below the level where the energy of the electromagnetic radiation sufficed to maintain the photodivision of the deuterons. Now long-lived deuterons were created by the reaction

$$n + p \rightarrow d + \gamma + 2.22\,\text{MeV}.$$

The lifetime of these deuterons was now limited by its fusion with protons and neutrons

$$p + d \quad \rightarrow \quad {}^3\text{He} + \gamma + 5.49\,\text{MeV}$$
$$n + d \quad \rightarrow \quad {}^3\text{H} + \gamma + 6.26\,\text{MeV}.$$

Finally the particularly stable ${}^4\text{He}$ nucleus was created in reactions like ${}^3\text{H}+\text{p}$, ${}^3\text{He} + \text{n}$, ${}^3\text{He} + \text{d}$ and $\text{d} + \text{d}$. The Li nuclei created by ${}^4\text{He} + {}^3\text{H} \rightarrow {}^7\text{Li} + \gamma + 2.47$ MeV were on the other hand immediately destroyed again by the highly exothermic reaction

$$^7\text{Li} + \text{p} \rightarrow 2\,{}^4\text{He} + 17.35\,\text{MeV}\,.$$

Essentially all of the neutrons ended the primordial nuclear synthesis phase inside ${}^4\text{He}$, which thus makes up about 24% of the mass of the universe.

Only traces of deuterium, ${}^3\text{He}$ and ${}^7\text{Li}$ are still present, so at that moment the greatest part of the baryonic mass must have been in the form of protons. Since there are no stable nuclei with masses $A = 5$ and $A = 8$ it was not possible at that stage of the universe's development to build up nuclei heavier than ${}^7\text{Li}$ through fusion processes. Such nuclei could only be produced much later in stellar interiors.

The big bang nuclear synthesis phase ended after about 30 minutes when the temperature had dropped so far that the Coulomb barrier prevented further fusion processes. The much later synthesis of heavy nuclei inside stars has not altered the composition of baryonic matter significantly. The ratio of hydrogen to helium which is observed in the present universe (cf. Fig. 2.2) is in excellent agreement with the theoretically calculated value. This is a strong argument in favour of the big bang model.

19.5 Stellar Evolution and Element Synthesis

The close weave linking nuclear physics and astrophysics stretches back to the thirties when Bethe, Weizsäcker and others tried to draw a quantitative balance between the energy emitted by the sun and the energy that could be released by the known nuclear reactions. It was, though, Eddington who in 1920 had recognised that nuclear fusion is the source of energy production in stars.

The basis for modern astrophysics was, however, laid by Fred Hoyle [Ho46] at the end of the forties. The research programme he proposed required a consistent treatment of astronomical observations, study of the plasma dynamics of stellar interiors and calculations of the sources of energy using the cross-sections for nuclear reactions measured in laboratories. Stellar evolution and the creation of the elements had to be treated together. The observed abundance of the elements around us had to be explicable from element synthesis in the early stages of the universe and from nuclear reactions in stars and this would thus be a decisive test of the consistency of stellar evolution models. The results of this programme were presented by E. Burbidge, G. Burbidge, Fowler and Hoyle [Bu57].

Stars are produced by the contraction of interstellar gas and dust. This matter is almost solely composed of primordial hydrogen and helium. The contraction heats up the centre of the star. When the temperature and pressure are sufficiently large to render nuclear fusion possible, radiation is produced whose pressure prevents a further contraction of the star. The virial theorem for the gravitational force law implies a fall-off in the temperature of stars from their centres to their exteriors. This means that at any separation from the centre of a star the average kinetic energy of an atom is half the size of its potential energy. The energy produced in nuclear reactions is primarily transported by radiation to the surface. The matter in the star is not greatly mixed up in the process. During the life of the star its chemical composition changes in the regions where the nuclear reactions take place, in other words most of all in the heart of the star.

Fusion reactions. A star in equilibrium produces as much energy through nuclear reactions as it radiates. The equilibrium state is thus highly dependent upon the rate of the fusion reactions. Energy may be released by fusing light nuclei together. It is especially effective to fuse hydrogen isotopes together to form ^{4}He, since the difference between its binding energy per nucleon, 7.07 MeV, and that of its neighbours is especially large (cf. Fig. 2.4). We will treat this reaction in more detail below. Fusion processes demand a sufficiently high temperature, or energy, for the reaction partners to spring over the hurdle of the Coulomb barrier. It is not necessary that the energy of the nuclei involved is actually above the barrier, rather what really matters, in analogy to α-decay, is the probability, e^{-2G}, that the Coulomb barrier may be tunneled through. The Gamow factor, G, depends upon the relative velocities and the charge numbers of the reaction partners. It is given by (see chapter 3.2)

$$G \approx \frac{\pi \alpha Z_1 Z_2}{v/c} \, . \tag{19.6}$$

Fusion reactions in stars too normally take place below the Coulomb barrier and through the tunnel effect.

The reaction rate per unit volume is according to (4.3) and (4.4) given by

$$\dot{N} = n_1 n_2 \langle \sigma v \rangle \tag{19.7}$$

where n_1 and n_2 are the particle densities of the two fusion partners. We have written the average value $\langle \sigma v \rangle$ since the velocity distribution in a hot stellar plasma is given by a Maxwell-Boltzmann distribution

$$n(v) \propto e^{-mv^2/2kT} = e^{-E/kT} \tag{19.8}$$

and the cross-section σ of the fusion reaction depends strongly, through the Gamow factor, upon the relative velocity of the reaction partners. This average value must be calculated by integration over v. Figure 19.10 schematically shows the convolution of the Gamow factor with a Maxwell distribution. The overlap of the distributions fixes the reaction rate and the energy range for which fusion reactions are possible. This depends upon the plasma temperature and the charges of the fusion partners. The higher the charge numbers, the higher the temperatures at which fusion reactions become possible.

In this way the lightest nuclide in the solar interior, hydrogen, is burnt up, i.e., fused together. When this is used up, the temperature has to increase drastically for helium and, later, other heavier elements to be able to fuse together. The length of the various burn-phases depends upon the mass of the star in question. For heavier stars the pressure and thus the density of the plasma at the centre is higher and so the reaction rate is higher compared to lighter stars. Thus heavier stars are shorter lived than heavy ones.

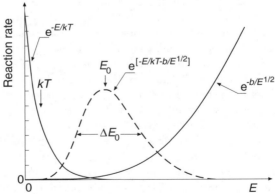

Fig. 19.10. Schematic representation of the convolution of a Maxwell distribution $\exp\{-E/kT\}$ with a Gamow factor $\exp\{-b/E^{1/2}\}$ as used to calculate the rate of fusion reactions. The product of the curves is proportional to the fusion probability (*dashed curve*). Fusion essentially takes place in a very narrow energy interval with width ΔE_0. The integral over this curve is proportional to the total reaction rate.

Burning hydrogen. In the formation phase of stars with masses greater than about one tenth of a solar mass, the temperatures inside the stars reach values of $T > 10^7$ K, and thus the first nuclear fusion processes are possible. In the early part of their lives stars gain their energy by burning hydrogen into helium in the *proton-proton cycle:*

$$
\begin{aligned}
\mathrm{p} + \mathrm{p} &\rightarrow \mathrm{d} + \mathrm{e}^+ + \nu_\mathrm{e} + 0.42\ \mathrm{MeV} \\
\mathrm{p} + \mathrm{d} &\rightarrow {}^3\mathrm{He} + \gamma + 5.49\ \mathrm{MeV} \\
{}^3\mathrm{He} + {}^3\mathrm{He} &\rightarrow \mathrm{p} + \mathrm{p} + \alpha + 12.86\ \mathrm{MeV} \\
\mathrm{e}^+ + \mathrm{e}^- &\rightarrow 2\gamma + 1.02\ \mathrm{MeV}.
\end{aligned}
$$

All in all, in the net reaction $4\mathrm{p} \rightarrow \alpha + 2\mathrm{e}^+ + 2\nu_\mathrm{e}$, 26.72 MeV of energy is released. Of this 0.52 MeV is on average taken by neutrinos and thus lost to the star. The first reaction is the slowest in the cycle since it requires not only the fusion of two protons but also the simultaneous transformation of a proton into a neutron via a weak interaction process. This reaction thus determines the lifetime of the star in the first stage of its career. There are various possible branches to the proton-proton cycle, but they are of little importance for energy production in stars.

As long as the supplies of hydrogen are adequate the star remains stable. For our sun this period will last about 10^{10} years, of which about half are already gone. Larger stars with higher central densities and temperatures burn faster. If in such stars ^{12}C is already present, then the *carbon cycle* can take place:

$$
{}^{12}_{6}\mathrm{C} \xrightarrow{\mathrm{p}} {}^{13}_{7}\mathrm{N} \xrightarrow{\beta^+} {}^{13}_{6}\mathrm{C} \xrightarrow{\mathrm{p}} {}^{14}_{7}\mathrm{N} \xrightarrow{\mathrm{p}} {}^{15}_{6}\mathrm{O} \xrightarrow{\beta^+} {}^{15}_{7}\mathrm{N} \xrightarrow{\mathrm{p}} {}^{12}_{6}\mathrm{C} + \alpha . \tag{19.9}
$$

The amount of carbon which was transformed at the beginning of the cycle is again available for further use at the end and thus it acts as a catalyst. The net reaction is as in the proton-proton cycle, $4\mathrm{p} \rightarrow \alpha + 2\mathrm{e}^+ + 2\nu_\mathrm{e}$, and the amount of energy released is also 26.72 MeV. The carbon cycle can take place much faster than the proton-proton cycle. But this new cycle only starts at higher temperatures due to the greater Coulomb barrier.

Burning helium. Once the hydrogen supplies have dried up, the core of the star, which is now composed of helium, cannot withstand the pressure and collapses. For stars much smaller than the sun the gravitational pressure is not great enough to ignite further fusion reactions. Without the radiative pressure, the star collapses under its own gravity to a planet sized sphere. Fermi pressure is the first thing to stop the collapse and the star becomes a white dwarf.

Heavier stars heat up until they reach a temperature of about 10^8 K and a density of 10^8 kg/m^3. Helium burning then starts up. There is still some hydrogen in the outermost regions of the star, which is heated up by the helium burning in the hot central region until in this layer hydrogen burning commences. The outer mantle swells up through the radiation pressure. Since the surface area increases the surface temperature drops, even though the energy production is increasing in this stage. The colour of the star turns red and it becomes a red giant.

A synthesis of nuclei heavier than ^{4}He appears to be impossible because there are no stable nuclei with $A = 5$ and $A = 8$. ^{8}Be has a lifetime of only 10^{-16} s, and ^{5}He und ^{5}Li are still less stable. But in 1952 E. Salpeter showed how heavy nuclei could be produced by helium fusion [Sa52].

At high temperatures around 10^8 K, which are present in stellar interiors, the unstable ^{8}Be nucleus can be formed from helium-helium fusion and equilibrium for the reaction ^{4}He $+^4$ He $\leftrightarrow$ ^{8}Be is created. This reaction is only possible in sufficient amounts at such high temperatures, since as well as the Coulomb barrier an energy level difference of 90 keV must be overcome (Fig. 19.11). At a density of 10^8 kg/m^3 in the interior of the star an equilibrium concentration of one ^{8}Be nucleus for 10^9 ^{4}He nuclei is produced. This miniscule proportion would be enough to produce sizeable amounts of carbon via ^{4}He $+$ ^{8}Be $\rightarrow$ ^{12}C* if there were a 0^+ state in ^{12}C a little above the production threshold over which a resonant reaction can take place. Shortly after this suggestion was made such a state at an excitation energy of 7.654 MeV was indeed found [Co57]. This state decays with a probability of $4 \cdot 10^{-4}$ into the ^{12}C ground state (Fig. 19.11). Although this state is 287 keV above the ^{8}Be $+ \alpha$ threshold, it can indeed be populated by reaction partners from the high energy tail of the Maxwell velocity distribution. The net reaction of helium fusion into carbon is thus

$$3\,^4\text{He} \rightarrow\ ^{12}\text{C} + 2\gamma + 7.37 \text{ MeV}.$$

This so-called $3\,\alpha$-process plays a key role in building up the heavier elements of the universe. Approximately 1 % of all the nuclei in the universe are heavier than helium and they were practically all created in the $3\,\alpha$-process.

Burning into iron. When the helium supplies have been used up and the star is primarily made up of ^{12}C, then stars with masses of the order of the solar mass turn into white dwarves.

More massive stars go through further phases of development. According to the temperature α-particles can fuse with ^{12}C, ^{16}O, ^{20}Ne etc., or carbon, oxygen, neon and silicon can simply fuse with each other.

As an example let us mention the reactions

$$
\begin{aligned}
^{12}\text{C} + {}^{12}\text{C} \quad &\rightarrow \quad {}^{20}_{10}\text{Ne} + \alpha + 4.62 \text{ MeV} \\
&\rightarrow \quad {}^{23}_{11}\text{Na} + \text{p} + 2.24 \text{ MeV} \\
&\rightarrow \quad {}^{23}_{12}\text{Mg} + \text{n} - 2.61 \text{ MeV} \\
&\rightarrow \quad {}^{16}_{8}\text{O} + 2\alpha - 0.11 \text{ MeV}.
\end{aligned}
$$

Other reactions follow the same pattern and populate all the elements between carbon and iron.

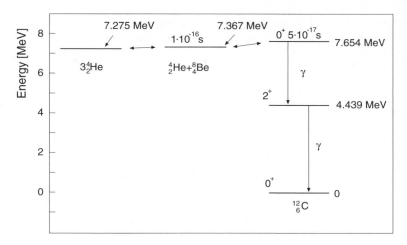

Fig. 19.11. Energy levels of the system: $3\,\alpha$, $\alpha + {}^8$Be and ^{12}C. Just above the ground states of the $3\,\alpha$ system and of the $\alpha + {}^8$Be system there is a 0^+ state in the ^{12}C nucleus, which can be created through resonant fusion of ^{4}He nuclei. This excited state decays with a $0.04\,\%$ probability into the ^{12}C ground state.

The heavier the fusing nuclei are, the greater is the Coulombic repulsion and so the temperature must then be higher for fusion to take place. Since the temperature is greatest at the centre and falls-off towards the surface, an onion-like stellar structure is formed. At the centre of the star iron is synthesised, towards the edges ever lighter elements are made. In the outermost layers the remnants of hydrogen and helium are burnt off.

The burning of the heavier nuclei takes place at ever shorter time scales, since the centre of the star needs to be ever hotter, but simultaneously the energy gained per nucleon-fusion decreases as the mass number increases (Fig. 2.4). The final phase, the fusion of silicon to form iron, lasts for only a matter of days [Be85]. The process of nuclear fusion in stars concludes with the formation of iron since iron has the largest binding energy per nucleon.

When the centre of the star is made of iron, there is no further source of energy available. There is neither radiative pressure nor thermal motion to withstand gravity. The star collapses. The outer material of the star collapses as if in free fall to the centre. Through this implosion the nuclear matter at the centre reaches a tremendous density and temperature which leads to an enormous explosion. The star emits at a stroke more energy than it has previously created in its entire life. This is called a supernova. The greater part of the stellar matter is then flung out into interstellar space and can later be used as building material for new stars. If the mass of the remaining stellar core is smaller than the mass of the sun, the star ends its life as a white dwarf. If it is between one and two solar masses a neutron star is born. The matter from still heavier remnants ends up as a black hole.

Synthesis of heavier nuclei. Nuclei heavier than iron are synthesised by neutron accumulation. We distinguish between two processes.

The slow process (s-process). In the burning phase of the star neutrons are produced in nuclear reactions such as, e.g.,

$$^{22}_{10}\text{Ne} + \alpha \rightarrow {}^{25}_{12}\text{Mg} + \text{n} - 0.48 \text{ MeV} \tag{19.10}$$

or

$$^{13}_{6}\text{C} + \alpha \rightarrow {}^{16}_{8}\text{O} + \text{n} - 0.91 \text{ MeV}. \tag{19.11}$$

Through repeated neutron captures, neutron rich isotopes are produced. If the isotopes are unstable under β-decay, they decay into their most stable isobar (Fig. 3.2, 3.3). Thus the synthesis of heavier and heavier elements can proceed along a stability valley (Fig. 3.1). A limit is, however, reached at lead. Nuclei above lead are α-unstable. Isotopes built up by the slow process then decay again into α-particles and lead.

The rapid process (r-process). This process takes place during a supernova explosion when neutron fluxes of 10^{32} m^{-2}s^{-1} can be reached and the successive accumulation of many neutrons is much quicker than β- or α-decay processes. Elements heavier than lead can be produced in this process. The upper limit for the creation of transuranic elements is determined by spontaneous fission.

All the elements (apart from hydrogen and helium) which make up the earth and ourselves came originally from the interior of stars and were (probably several times in fact) released through supernova explosions. Even the absolute amounts as well as the distribution of the elements which are heavier than helium may be calculated from the age of the universe and from cross-sections measured in laboratories. The results are in excellent agreement with the measured values of the abundance of the elements (Fig. 2.2). This is definitely one of the great triumphs of the joint efforts of astro and nuclear physicists.

Problems

1. **Sun**
   The solar mass is $M_\odot \approx 2 \cdot 10^{30}$ kg ($3.3 \cdot 10^5$ times the mass of the earth). The chemical composition of the solar surface is 71% hydrogen, 27% helium and 2% heavier elements (expressed as parts by mass). The luminosity of the sun is $4 \cdot 10^{26}$ W.
   a) How much hydrogen is converted into helium every second?
   b) How much mass does the sun lose in the same period?
   c) What fraction of the original hydrogen content has been converted into helium since the creation of the sun ($5 \cdot 10^9$ yrs)?
   d) How large was the loss of mass in the same period?
   e) Model calculations indicate that the sun will burn hydrogen at a similar rate for a further $5 \cdot 10^9$ years. A shortage of hydrogen will then force it into a red giant state. Motivate this time scale.

2. **Solar neutrinos**
   The most important method of energy production in the sun is the fusion of protons into ^{4}He nuclei. This predominantly takes place through the reactions $p + p \rightarrow d + e^+ + \nu_e$, $p + d \rightarrow {}^3He + \gamma$ and $^3He + {}^3He \rightarrow {}^4He + p + p$. A total of 28.3 MeV is released for every ^{4}He nucleus produced. 90 % of this energy exits as electromagnetic radiation and the rest is mostly converted into the kinetic energy of neutrinos (typically 0.4 MeV) [Ba89].
   a) What is the flux of solar neutrinos at the earth (distance from the sun: $a = 1.5 \cdot 10^8$ km)?
   b) In a tunnel in the Abruzzi the GALLEX experiment measures neutrinos through the reaction $^{71}_{31}Ga + \nu_e \rightarrow {}^{71}_{32}Ge + e^-$. The cross-section of this reaction is about $2.5 \cdot 10^{-45}$ cm^2. One looks for radioactive ^{71}Ge atoms (lifetime, $\tau = 16$ days) which are produced in a tank containing 30 t of dissolved gallium (40 % ^{71}Ga, 60 % ^{69}Ga) chloride [An92]. About 50 % of the neutrinos have an energy above the reaction threshold. One extracts all the germanium atoms from the tank. Estimate how many ^{71}Ge atoms are produced each day and after three weeks? How many if one waits "forever"?

3. **Supernova**
   A neutron star with mass, $M = 1.5 M_\odot$ ($\approx 3.0 \cdot 10^{30}$ kg), and radius $R \approx 10$ km is the remnant of a supernova. The stellar material originates from the iron core ($R \gg 10$ km) of the supernova.
   a) How much energy was released during the lifetime of the original star by converting hydrogen into iron? (The binding energy of ^{56}Fe is $B = 8.79$ MeV/nucleon.) *NB:* Since after the implosion only a part of the original iron core remains in the neutron star, the calculation should be performed only for this mass.
   b) How much energy was released during the implosion of the iron core into a neutron star?
   c) In what form was the energy radiated off?

20. Many-Body Systems in the Strong Interaction

> How many bodies are required before we have a problem?
> G. E. Brown points out that this can be answered by a
> look at history. In eighteenth-century Newtonian mecha-
> nics, the three-body problem was insoluble. With the birth
> of relativity around 1910 and quantum electrodynamics in
> 1930, the two- and one-body problems became insoluble.
> And within modern quantum field theory, the problem of
> zero bodies (vacuum) is insoluble. So, if we are out after
> exact solutions, no bodies at all is already too many!
>
> *R. D. Mattuck* [Ma76]

In the second part of this book we have described how many-body systems
may be built out of quarks. The strong interaction is responsible for the
binding of these systems, which should be contrasted with the binding of
atoms, molecules and solids which are held together by the electromagnetic
interaction.

The systems which are built out of quarks – hadrons and nuclei – are
complex quantum-mechanical systems. This complexity manifests itself in
the systems' many, apparently mutually incompatible facets. Some aspects
of these systems may be understood in a single particle picture, while some
indicate the existence of large sub-structures and others are explained as
collective effects of the entire system and finally some are chaotic and only
amenable to a statistical description. Each of these concepts, however, only
describes a single aspect of these systems.

Quasi-particles. At sufficiently low excitation energies, many-body systems, even if they possess a complicated internal structure, may often be
described as systems of so-called *quasi-particles*: instead of treating the elementary building blocks, together with their vast variety of mutual interactions, one works with "effective particles" (e. g., electrons and holes in semi-
conductors). A large part of the interactions of the fundamental constituents
with each other is thus incorporated into the internal structure of the quasi-
particles which then, in consequence, only weakly interact with each other.

Collective states. Another group of elementary low-energy excitations are
the so-called collective states, where many building blocks of a system interfere coherently. Examples of this are lattice vibrations in a crystal (phonons)
and waves on the surface of an atomic nucleus.

Chaotic phenomena. For greater excitation energies all many-particle systems become more and more complex, until they can no longer be descri-

bed quantitatively in terms of elementary excitations. Statistical phenomena, which have a universal character, and are thus independent of the details of the interaction, are observed.

Hadrons. Little is so far known about the structure of hadrons. Their elementary constituents are gluons and quarks. However, in order to actually observe these experimentally, measurements at "infinitely" large momentum transfers would be necessary. Therefore even in deep inelastic scattering one only ever observes effective quarks, i.e., many-particle systems. The success of QCD lies in the fact that it is able to quantitatively explain the dependence of the structure functions on the resolution. However, the absolute shape of the structure functions, i.e., hadronic structure, cannot yet be predicted even at large momentum transfers.

The structure of the nucleons depends, however, on the behaviour of quarks at relatively small momenta, since the energies of the excited states are only a few hundred MeV. At such low momentum transfers the coupling constant α_s is so large that the standard QCD perturbative expansion is no longer applicable and we have to deal with a genuine many-particle system.

It has been seen that the spectrocopic properties of hadrons can be described simply in terms of constituent quarks and that one does not need to take the gluons into account. Constituent quarks are complex objects and not elementary particles: we have to understand them as *quasi-particles*. Their properties (e. g., their masses, sizes and magnetic moments) are distinctly different from those of the elementary quarks. It seems that a certain order in hadronic spectroscopy can be obtained by introducing these quasi-particles. The group-theoretical classification of excited states is in fact very successful, but the dynamics are not well understood. It is also not evident whether complex hadronic excitations can be described in the constituent quark model.

Excited states of hadrons made out of light quarks are known only up to about 3 GeV. The resonances get broader and are more closely packed together as their energy increases. At energies $\gtrsim 3$ GeV, no further resonance structures can be recognized. This could perhaps be a region where chaotic phenomena might be expected. However, they cannot be observed because of the large width of the resonances.

Collective phenomena have also not yet been observed in hadrons. This may be due to the fact that the number of effective constituents is too small to produce coherent phenomena.

Forces of the strong interaction. *Elementary* particles (quarks and leptons) interact through *elementary* forces which are mediated by the exchange of gluons, photons and the W and Z bosons. The forces between systems with internal structure (atoms, nucleons, constituent quarks) are of a more complicated nature and are themselves many-particle phenomena (e.g., the Van der Waals force or covalent binding forces).

To a first approximation the forces of the strong interaction between nucleons or between constituent quarks may be parametrised by effective forces.

These are short-ranged and may be, depending upon spin and isospin, either attractive or repulsive. For constituent quarks the short distance interaction seems to be adequately described by one-gluon exchange with an effective coupling constant α_s while at large distances many-gluon exchange is parameterised by a confinement potential. Two-gluon exchange (Van der Waals force) and two-quark exchange (covalent bond) presumably play a minor role in the interaction between two nucleons.

The short-range repulsion is, on the one hand, a consequence of the symmetry of the quark wave function of the nucleon, and, on the other hand, of chromomagnetic repulsion. The dominant part of the attractive nuclear force is mediated by the exchange of $q\bar{q}$ pairs. It is not surprising that these pairs can be identified with the light mesons.

Within the nucleus, this force is also strongly modified by many-body effects (e. g., the Pauli principle). Hence in nuclear physics calculations, phenomenological forces, whose forms and parameters have to be fitted to experimental results, are frequently employed.

Nuclei. The idea that nuclei are composed of nucleons is somewhat naive. It is more realistic to conceive of the constituents of the nucleus as quasi-nucleons. The properties of these quasi-particles are similar to those of the nucleons if they are close to the Fermi surface. Some low energy nuclear phenomena (spin, magnetic moments, excitation energies) can be described by the properties of individual, weakly bound nucleons in the outermost shells or by holes in an otherwise closed shell.

Strongly bound nucleons cannot be assigned to individual states of the shell model. This can be seen, for example, in the very broad states observed in quasi-elastic scattering. In contradistinction, a strongly bound Λ particle inside the nucleus can, it seems, be adequately described as a quasi-particle even in deeply bound states.

Even larger structures in the nucleus may behave like quasi-particles. Pairs of neutrons or protons can couple in the nucleus to form $J^P = 0^+$ pairs, i. e., quasi-particles with boson properties. This pairing is suspected to lead to superfluid phenomena in nuclei, analogous to Cooper pairs in superconductors and atomic pairs in superfluid ^{3}He. As we have seen, the moments of inertia of rotational states can be qualitatively described in a two-fluid model composed of a normal and of a superfluid phase.

Some nuclear properties can be understood as collective excitations. Such effects can most clearly be observed in heavy nuclei. For example, giant dipole resonances can be interpreted as density oscillations. A nucleus, since it is a finite system, may also undergo shape oscillations. In analogy to solid state physics, quadrupole excitations are described in terms of phonons. The rotational bands of deformed nuclei have an especially collective nature.

At higher energies the collective and quasi-particle character of the excitations is lost. This is the start of the domain of configuration admixtures, where states are built from superpositions of collective and/or particle-hole wave functions. At even higher excitation energies the nuclear level density increases exponentially with the excitation energy and a quantitative description of the individual levels becomes impossible. The great complexity of the levels makes a new description using statistical methods possible.

Digestive. In our approach to complex systems we have tried to let ourselves be guided by our understanding of more elementary systems. This helped us to gain a deeper insight into the architecture of more complex systems, and yet we had to introduce new *effective* building blocks, which mutually interact via effective forces, to obtain a *quantitative* treatment of complex phenomena.

Thus in hadron spectroscopy, we used constituent quarks, and not the quarks from the underlying theory of QCD; the interactions between nucleons are best described in terms of meson exchange, not by the exchange of gluons and quarks; in the nucleus effective forces are usually employed instead of the forces known from the nucleon-nucleon interaction and the richness of collective states in nuclei are, even though we have sketched the connection to the shell model, quantitatively better described in terms of collective variables and not in terms of single-particle excitations. This all means that the best description always seems to come from the framework of an "effective theory" chosen according to our experimental resolution. This is by no means a peculiarity of the complex systems of the strong interactions, but is a general property of many-body systems.

Our modern struggles to improve our understanding are fought on two fronts: physicists are testing whether the modern standard model of elementary particle physics is indeed fundamental or itself "just" an effective theory, and are simultaneously trying to improve our understanding of the regularities of the complex systems of the strong interaction.

And it shall be, when thou hast made an end of reading
this book, that thou shalt bind a stone to it, and cast it
into the midst of Euphrates:

Jeremiah 51. 63

A. Appendix

In the main body of this book we have described particle and nuclear physics and the underlying interactions concisely and in context. We have here and there elucidated the basic principles and methods of the experiments that have led us to this knowledge. We now want to briefly describe the individual tools of experimental physics – the particle accelerators and detectors – whose invention and development have often been a sine qua non for the discoveries discussed here. More detailed discussions may be found in the literature [Kl92a, Le94, Wi93].

A.1 Accelerators

Particle accelerators provide us with different types of particle beams whose energies (at the time of writing) can be anything up to a TeV (10^6 MeV). These beams serve on the one hand as "sources" of energy which if used to bombard nuclei can generate a variety of excited states or indeed new particles. On the other hand they can act as "probes" with which we may investigate the structure of the target particle.

The most important quantity, whether we want to generate new particles or excite a system into a higher state, is the centre of mass energy $\sqrt{s}$ of the reaction under investigation. In the reaction of a beam particle a with total energy E_a with a target particle b which is at rest this is

$$\sqrt{s} = \sqrt{2E_a m_b c^2 + (m_a^2 + m_b^2)c^4} \,. \tag{A.1}$$

In high energy experiments where the particle masses may be neglected in comparison to the beam energy this simplifies to

$$\sqrt{s} = \sqrt{2E_a m_b c^2} \,. \tag{A.2}$$

The centre of mass energy for a stationary target only, we see, grows with the square root of the beam particle's energy.

If a beam particle with momentum p is used to investigate the structure of a stationary target, then the best possible resolution is characterised by its reduced de Broglie wavelength $\lambda = \hbar/p$. This is related to the energy E through (4.1).

All accelerators essentially consist of the following: a particle source, a structure to actually do the accelerating and an evacuated beam pipe. It should also be possible to focus and deflect the particle beam. The accelerating principle is always the same: charged particles are accelerated if they are exposed to an electric field. A particle with charge Ze which traverses a potential difference U receives an amount of energy, $E = ZeU$. In the following we wish to briefly present the three most important types of accelerators.

Electrostatic accelerators. In these accelerators the relation $E = ZeU$ is directly exploited. The main components of an electrostatic accelerator are a high voltage generator, a terminal and an evacuated beam pipe. In the most common sort, the *Van de Graaff accelerator*, the terminal is usually a metallic sphere which acts as a capacitor with capacitance C. The terminal is charged by a rotating, insulated band and this creates a high electric field. From an earthed potential positive charges are brought onto the band and then stripped off onto the terminal. The entire set up is placed inside an earthed tank which is filled with an insulating gas (e.g., SF_6) to prevent premature discharge. The voltage $U = Q/C$ which may be built up in this way can be as much as $15\,\mathrm{MV}$. Positive ions, produced in an ion source, at the terminal potential now traverse inside the beam pipe the entire potential difference between the terminal and the tank. Protons can in this way reach kinetic energies up to $15\,\mathrm{MeV}$.

Energies twice as high may be attained in tandem Van de Graaff accelerators (Fig. A.1). Here the accelerating potential is used twice over. Negative ions are first produced at earth potential and then accelerated along a beam pipe towards the terminal. A thin foil, or similar, placed there strips some of the electrons off the ions and leaves them positively charged. The accelerating voltage now enters the game again and protons may in this way attain kinetic energies of up to $30\,\mathrm{MeV}$. Heavy ions may lose several electrons at once and consequently reach even higher kinetic energies.

Van de Graaff accelerators can provide reliable, continuous particle beams with currents of up to $100\,\mu\mathrm{A}$. They are very important workhorses for nuclear physics. Protons and both light and heavy ions may be accelerated in them up to energies at which nuclear reactions and nuclear spectroscopy may be systematically investigated.

Linear accelerators. GeV-type energies may only be attained by repeatedly accelerating the particle. Linear accelerators, which are based upon this principle, are made up of many accelerating tubes laid out in a straight line and the particles progress along their central axis. Every pair of neighbouring tubes have oppositely arranged potentials such that the particles between them are accelerated, while the interior of the tubes is essentially field free (Wideröe type). A high frequency generator changes the potentials with a period such that the particles between the tubes always feel an accelerating force. After passing through n tubes the particles will have kinetic energy

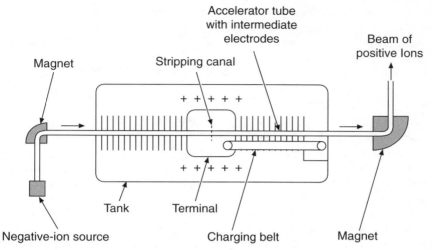

Fig. A.1. Sketch of a tandem Van de Graaff accelerator. Negative ions are accelerated from the left towards the terminal where some of their electrons are stripped off and they become positively charged. This causes them to now be accelerated away from the terminal and the potential difference between the terminal and the tank is traversed for a second time.

$E = nZeU$. Such accelerators cannot produce continuous particle beams; they accelerate packets of particles which are in phase with the generator frequency.

Since the generator frequency is fixed, the lengths of the various stages need to be adjusted to fit the speed of the particles as it passes through (Fig. A.2). If we have an electron beam this last subtlety is only relevant for the first few acceleration steps, since the small electron mass means that their velocity is very soon nearly equal to the speed of light. On the other hand

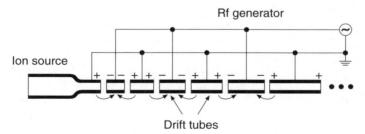

Fig. A.2. Sketch of the fundamentals of a (Wideröe type) linear accelerator. The potentials of the tubes shown are for one particular moment in time. The particles are accelerated from the source to the first drift tube. The lengths L_i of the tubes and the generator frequency ω must be adjusted to each other so that we have $L_i = v_i\,\pi/\omega$ where v_i is the particle velocity at the ith tube. This depends both upon the generator voltage and the type of particle being accelerated.

the tube lengths generally need to be continually altered along the entire length of proton linear accelerators. The final energy of a linear accelerator is determined by the number of tubes and the maximal potential difference between them.

At present the largest linear accelerator in the world, where many important experiments on deep inelastic scattering off nucleons have been carried out, is the roughly 3 km long electron linear accelerator at the Stanford Linear Accelerator Center (SLAC). Here electrons pass through around 100 000 accelerating stages to reach energies of about 50 GeV.

Synchrotrons. While particles pass through each stage of a linear accelerator just once, synchrotrons, which have a circular form, may be used to accelerate particles to high energies by passing them many times through the same accelerating structures.

The particles are kept on their circular orbits by magnetic fields. The accelerating stages are mostly only placed at a few positions upon the circuit. The principle of the synchrotron is to synchronously change the generator frequency ω of the accelerating stages together with the magnetic field B in such a way that the particles, whose orbital frequencies and momenta p are increasing as a result of the acceleration, always feel an accelerating force and are simultaneously kept on their assigned orbits inside the vacuum pipe. This means that the following constraints must be simultaneously fulfilled:

$$\omega = n \cdot \frac{c}{R} \cdot \frac{pc}{E} \qquad n = \text{positive integer} \qquad (A.3)$$

$$B = \frac{p}{ZeR}, \qquad (A.4)$$

where R is the radius of curvature of the synchrotron ring. Technical limitations upon the B and ω available mean that one has to inject preaccelerated particles into synchrotron rings whereupon they can be brought up to their preassigned final energy. Linear accelerators or smaller synchrotrons are used in the preacceleration stage. Synchrotrons also only produce packets of particles and do not deliver continuous beams.

High particle intensities require well focused beams close to the ideal orbit. Focusing is also of great importance in the transport of the beam from the preaccelerator to the main stage and from there to the experiment (injection and extraction). Magnetic lenses, made from quadrupole magnets, are used to focus the beam in high energy accelerators. The field of a quadrupole magnet focuses charged particles in one plane on its central axis and defocuses them on the other plane perpendicular to it. An overall focusing in both planes may be achieved by putting a second quadrupole magnet, whose poles are rotated relative to those of the first one through 90°, after the first magnet. This principle of *strong focusing* is similar to the optical combination of thin diverging and converging lenses which always effectively focuses. Figure A.3 depicts the essentials of a synchrotron and the focusing effects of such quadrupole doublets.

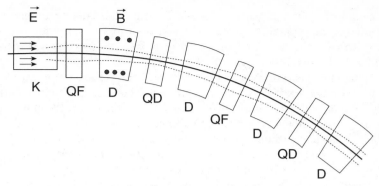

Fig. A.3. Section (to scale) of a synchrotron from above. The essential accelerating and magnetic structures are shown together with the beam pipe (*continuous line*). High frequency accelerator tubes (K) are usually only placed at a few positions around the synchrotron. The fields of the dipole magnets (D), which keep the particles on their circular paths, are perpendicular to the page. Pairs of quadrupole magnets form doublets which focus the beam. This is indicated by the dotted lines which (exaggeratedly) show the shape of the beam envelope. The quadrupoles marked QF have a focusing effect in the plane of the page and the QD quadrupoles a defocusing effect.

Particles accelerated in synchrotrons lose some of their energy to *synchrotron radiation*. This refers to the emission of photons by any charged particle which is forced onto a circular path and is thus radially accelerated. The energy lost to synchrotron radiation must be compensated by the accelerating stages. This loss is for highly relativistic particles

$$-\Delta E = \frac{4\pi\alpha\hbar c}{3R}\beta^3\gamma^4 \qquad \text{where} \quad \beta = \frac{v}{c} \approx 1 \quad \text{and} \quad \gamma = \frac{E}{mc^2}, \qquad (A.5)$$

per orbit – it increases in other words with the fourth power of the particle energy E. The mass dependence means that this rate of energy loss is about 10^{13} times larger for electrons than for protons of the same energy. The maximal energy in modern electron synchrotrons is thus about $100\,\text{GeV}$. Synchrotron radiation does not play an important role for proton beams. The limit on their final energy is set by the available field strengths of the dipole magnets which keep the protons in the orbit. Proton energies up to a TeV may be achieved with superconducting magnets.

There are two types of experiment which use particles accelerated in synchrotrons. The beam may, after it has reached its final energy, be deflected out of the ring and led off towards a stationary target. Alternatively the beam may be stored in the synchrotron until it is either loosed upon a thin, internal target or collided with another beam.

Storage rings. The centre of mass energy of a reaction involving a stationary target only grows with the square root of the beam energy (A.2). Much higher centre of mass energies may be obtained for the same beam energies if we employ colliding particle beams. The centre of mass energy for a head on collision of two particle beams with energy E is $\sqrt{s} = 2E$ – i.e., it increases linearly with the beam energy.

The particle density in particle beams, and hence the reaction rate for the collision of two beams, is very tiny; thus they need to be repeatedly collided in any experiment with reasonable event rates. High collision rates may, e.g., be obtained by continuously operating two linear accelerators and colliding the particle beams they produce. Another possibility is to store particle beams, which were accelerated in a synchrotron, at their final energy and at the accelerating stages just top up the energy they lose to synchrotron radiation. These stored particle beams may be then used for collision experiments.

Consider as an example the HERA ring at the Deutsche Elektronen-Synchrotron (German Electron Synchrotron, DESY) in Hamburg. This is made up of two separate storage rings of the same diameter which run parallel to each other at about 1 m separation. Electrons are accelerated up to about 30 GeV and protons to about 920 GeV before storage. The beam tubes come together at two points, where the detectors are positioned, and the oppositely circling beams are allowed to collide there.

Construction is rather simpler if one wants to collide particles with their antiparticles (e.g., electrons and positrons or protons and antiprotons). In such cases only one storage ring is needed and these equal mass but oppositely charged particles can simultaneously run around the ring in opposite directions and may be brought to collision at various interaction points. Examples of these are the LEP ring (Large Electron Positron Ring) at CERN where 86 GeV electrons and positrons collide and the Sp$\bar{\text{p}}$S (Super Proton Antiproton Synchrotron) where 310 GeV protons and antiprotons are brought violently together. Both of these machines are to be found at the European Nuclear Research Centre CERN just outside Geneva.

An example of a research complex of accelerators is shown in Fig. A.4; that of DESY. A total of seven preaccelerators service the DORIS and HERA storage rings where experiments with electrons, positrons and protons take place. Two preaccelerator stages are needed for the electron-positron ring DORIS where the beams each have a maximal energy of 5.6 GeV. Three such stages are required for the electron-proton ring HERA (30 GeV electrons and 820 GeV protons). DORIS also serves as an source of intensive synchrotron radiation and is used as a research instrument in surface physics, chemistry, biology and medicine.

A.2 Detectors

The construction and development of detectors for particle and nuclear physics has, as with accelerator physics, developed into an almost independent branch of science. The demands upon the quality and complexity of these

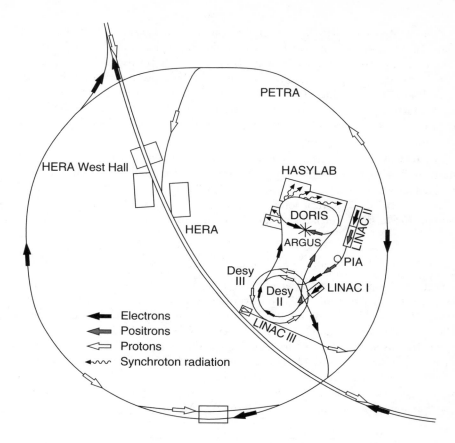

Fig. A.4. The accelerator complex at the German Electron Synchrotron, DESY, in Hamburg. The DORIS and HERA storage rings are serviced by a chain of pre-accelerators. Electrons are accelerated up to 450 MeV in the LINAC I or LINAC II linear accelerators before being injected into the DESY II synchrotron, where they may reach up to 9 GeV. Thence they either pass into DORIS or the PETRA synchrotron. PETRA acts as a final preaccelerator for HERA and electron energies of up to 14 GeV may be attained there. Before HERA was commissioned PETRA worked as an electron-positron storage ring with a beam energy of up to 23.5 GeV. Positrons are produced with the help of electrons accelerated in LINAC II and are then accumulated in the PIA storage ring before their injection into DESY II where they are further accelerated and then led off to DORIS. Protons are accelerated in LINAC III up to 50 MeV and then preaccelerated in the proton synchrotron DESY III up to 7.5 GeV before being injected into PETRA. There they attain 40 GeV before being injected into HERA. The HERA ring, which is only partially shown here, has a circumference of 6336 m, while the circumference of PETRA is 2300 m and that of DESY II(III) is around 300 m. (*Courtesy of DESY*)

detectors increase with the ever higher particle energies and currents involved. This has necessarily led to a strong specialisation among the detectors. There are now detectors to measure times, particle positions, momenta and energies and to identify the particles involved. The principles underlying the detectors are mostly based upon the electromagnetic interactions of particles with matter, e.g., ionisation processes. We will therefore first briefly delineate these processes before showing how they are applied in the individual detectors.

Interaction of particles with matter. If charged particles pass through matter they lose energy through collisions with the medium. A large part of this corresponds to interactions with the atomic electron clouds which lead to the atoms being excited or ionised. The energy lost to ionisation is described by the *Bethe-Bloch formula* [Be30, Bl33]. Approximately we have [PD94]

$$-\frac{\mathrm{d}E}{\mathrm{d}x} = \frac{4\pi}{m_e c^2} \frac{n z^2}{\beta^2} \left(\frac{e^2}{4\pi\varepsilon_0}\right)^2 \left[\ln \frac{2m_e c^2 \beta^2}{I \cdot (1 - \beta^2)} - \beta^2\right] \tag{A.6}$$

where $\beta = v/c$, ze and v are the charge and speed of the particle, n is the electron density and I is the average excitation potential of the atoms (typically 16 eV$\cdot Z^{0.9}$ for nuclear charge numbers $Z > 1$). The energy loss thus depends upon the charge and speed of the particle (Fig. A.5) but not upon its mass. It decreases for small velocities as $1/v^2$, reaches a minimum around $p/m_0 c \approx 4$ and then increases only logarithmically for relativistic velocities. The energy loss to ionisation per length $\mathrm{d}x$ traversed normalised to the density ϱ of the matter at the ionisation minimum, and also for higher particle energies, is roughly $1/\varrho \cdot \mathrm{d}E/\mathrm{d}x \approx 2\,\mathrm{MeV}/(\mathrm{g\,cm}^{-2})$.

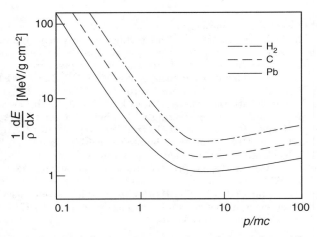

Fig. A.5. Rough sketch of the average energy loss of charged particles to ionisation processes in hydrogen, carbon and lead. The energy loss divided by the density of the material is plotted against $p/mc = \beta\gamma$ for the particle in a log-log plot. The specific energy loss is greater for lighter elements than for heavy ones.

Electrons and positrons lose energy not just to ionisation but also to a further important process: *bremsstrahlung*. Electrons braking in the field of a nucleus radiate energy in the form of photons. This process strongly depends upon the material and the energy: it increases roughly linearly with energy and quadratically with the charge number Z of the medium. Above a critical energy E_c, which may be coarsely parameterised by $E_c \approx 600\,\mathrm{MeV}/Z$, bremsstrahlung energy loss is more important for electrons than is ionisation. For such high energy electrons an important material parameter is the *radiation length* X_0. This describes the distance over which the electron energy decreases due to bremsstrahlung by a factor of e. High energy electrons are best absorbed in materials with high charge numbers Z; e.g., lead, where the radiation length is just 0.56 cm.

While charged particles traversing matter lose energy slowly to electromagnetic interactions before finally being absorbed, the interaction of a photon with matter takes place at a point. The intensity I of a photon beam therefore decreases exponentially with the thickness ℓ of the matter traversed:

$$I = I_0 \cdot \mathrm{e}^{-\mu\ell} \,. \tag{A.7}$$

The absorption coefficient μ depends upon the photon energy and the type of matter.

The interaction of photons with matter essentially takes place via one of three processes: the *photoelectric effect*, the *Compton effect* and *pair production*. These processes depend strongly upon the medium and the energy involved. The photoelectric effect dominates at low energies in the keV range, the Compton effect for energies from several 100 keV to a few MeV while in high energy experiments only pair production is of any importance. Here the photon is converted inside the nuclear field to an electron-positron pair. This is the dominant process above several MeV. In this energy range the photon can also be described by the radiation length X_0: the conversion length λ of a high energy photon is $\lambda = 9/7 \cdot X_0$. The energy dependence of these three processes in lead is illustrated in Fig. A.6.

We wish to briefly mention two further processes which are useful in particle identification: the radiation of Cherenkov light and nuclear reactions. *Cherenkov radiation* is photon emission from charged particles that cross through a medium with a velocity greater than the speed of light in that medium. These photons are radiated in a cone with angle

$$\theta = \arccos \frac{1}{\beta n} \tag{A.8}$$

around the path of the charged particle (n is the refractive index of the medium). The energy loss to Cherenkov radiation is small compared to that through ionisation.

Nuclear reactions are important for detecting neutral hadrons such as neutrons that do not participate in any of the above processes. Possible reactions are nuclear fission and neutron capture (eV – keV range), elastic and inelastic scattering (MeV range) and hadron production (high energies).

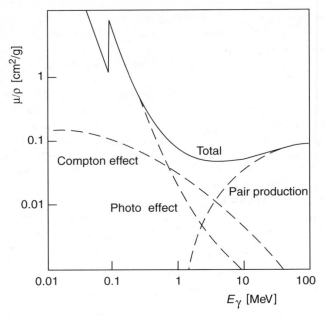

Fig. A.6. The photon absorption coefficient μ in lead divided by the density plotted against the photon energy. The dashed lines are the contributions of the individual processes; the photoelectric effect, the Compton effect and pair production. Above a few MeV pair production plays the dominant role.

Measuring positions. The ability to measure the positions and momenta of particles is important in order to reconstruct the kinematics of reactions. The most common detectors of the paths of particles exploit the energy lost by charged particles to ionisation.

Bubble chambers, spark chambers, and *streamer chambers* show us where particles pass through by making their tracks visible so that they may be photographed. These pictures have a high illustrative value and possess a certain aesthetic appeal. Many new particles were discovered in bubble chambers in particular in the 1950's and 1960's. These detectors are nowadays only used for special applications.

Proportional counters consist of flat, gas-filled forms in which many thin, parallel wires ($r \approx 10\,\mu$m) are arranged. The wires are maintained at a positive potential of a few kV and are typically arranged at separations of about 2 mm. Charged particles passing through the gas ionise the gas atoms in their paths and the so-released electrons drift off to the anode wires (Fig. A.7). The electric field strengths around the thin wires are very high and so the primary electrons are accelerated and reach kinetic energies such that they themselves start to ionise the gas atoms. A charge avalanche is let loose which leads to a measurable voltage pulse on the wire. The arrival time and amplitude of

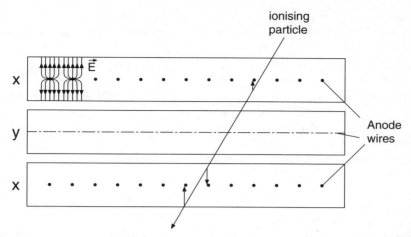

Fig. A.7. Group of three proportional chambers. The anode wires of the layers marked x point into the page, while those of the y layer run at right angles to these (*dashed line*). The cathodes are the edges of the chambers. A positive voltage applied to the anode wires generates a field like the one sketched in the upper left hand corner. A particle crossing through the chamber ionises the gas in its path and the electrons drift along the field lines to the anode wire. In the example shown a signal would be obtained from one wire in the upper x plane and from two in the lower x layer.

the pulse are registered electronically. The known position of the wire tells us where the particle passed by. The spatial resolution in the direction perpendicular to the wires is of the order of half the wire separation. An improved resolution and a reconstruction of the path in all three spatial coordinates is in practice obtained by using several layers of proportional counters with the wires pointing in different directions.

Drift chambers function similarly to proportional chambers. The wires are, however, at a few centimetres separation. The position of the particle's path x is now obtained from the time of the voltage pulse t_{wire} on the wire relative to the time t_0 that the particle crossed through the detector. This latter time has to be measured in another detector. Ideally we should have the linear relation

$$x = x_{\text{wire}} + v_{\text{drift}} \cdot (t_{\text{wire}} - t_0) , \tag{A.9}$$

if the electric field due to additional electrodes, and hence the drift velocity v_{drift} of the released electrons in the gas, are very homogenous. Drift chambers' spatial resolution can be as good as $50\,\mu$m. Several layers are again required for a three dimensional reconstruction. Wire chambers are very useful for reconstructing paths over large areas. They may be made to cover several square metres.

Silicon strip detectors are made out of silicon crystals with very thin electrodes attached to them at separations of about, e.g., $20\,\mu$m. A charged particle crossing the wafer produces electron-hole pairs, in silicon this only requires 3.6 eV per pair. An external voltage collects the charge at the electrodes where it is registered. Spatial resolutions less than $10\,\mu$m may be reached in this way.

Measuring momenta. The momenta of charged particles may be determined with the help of strong magnetic fields. The Lorentz force causes these particles to follow circular orbits which may then be, e.g., measured in bubble chamber photographs or reconstructed from several planes of wire chambers. A „rule of thumb" for the momentum component $p_\perp$ perpendicular to the magnetic field may be obtained from the measured radius of curvature of the particle path R and the known, homogenous magnetic field B:

$$p_\perp \approx 0.3 \cdot B \cdot R \; \left[\frac{\text{GeV}/c}{\text{T m}}\right] . \tag{A.10}$$

Magnetic spectrometers are used to indirectly determine the radius of curvature from the angle which the particle is deflected through in the magnetic field; one measures the particle's path before and after the magnets. This method of measuring the momenta actually has smaller errors than a direct determination of the radius of curvature would have. The relative accuracy of these measurements typically decreases with increasing momenta as $\delta(p)/p \propto p$. This is because the particle path becomes straighter at high momenta.

Measuring energies. A measurement of the energy of a particle usually requires the particle to be completely absorbed by some medium. The absorbed energy is transformed into ionisation, atomic excitations or perhaps Cherenkov light. This signal which may, with the help of suitable devices, be transformed into a measurable one is proportional to the original energy of the particle. The energy resolution depends upon the statistical fluctuations of the transformation process.

Semiconductor detectors are of great importance in nuclear physics. Electron-hole pairs created by charged particles are separated by an external voltage and then detected as voltage pulses. In germanium only 2.8 eV is required to produce an electron-hole pair. In silicon 3.6 eV is needed. Semiconductor detectors are typically a few millimetres thick and can absorb light nuclei with energies up to a few tens of MeV. Photon energies are determined through the photoelectric effect – one measures the signal of the absorbed photoelectron. The large number N of electron-hole pairs that are produced means that the energy resolution of such semiconductor counters is excellent, $\delta E/E \propto \sqrt{N}/N$. For 1 MeV particles it is between 10^{-3} and 10^{-4}.

Electromagnetic calorimeters may be used to measure the energies of electrons, positrons and photons above about 100 MeV. One exploits the cascade of secondary particles that these particles produce via repeated bremsstrahlung and pair production processes inside the material of the calorimeter. The production of such a measurable ionisation or visible signal is illustrated in Fig. A.8. The complete absorption of such a shower in a calorimeter takes place, depending upon the energy involved, over a distance of about 15–25 times the radiation length X_0. We will consider the example of homogenous calorimeters made of NaI(Tl) crystals or lead glass.

NaI doped with small amounts of thallium is an inorganic scintillator in which charged particles produce visible wavelength photons. These photons may then be converted into an electric pulse with the help of photomultipliers. Calorimeters are made from large crystals of NaI(Tl) with photomultipliers attached to their backs (see Fig. 13.4). The relative energy resolution typically has values of the order of $\delta E/E \approx 1-2\,\%\,/\sqrt[4]{E\,[\text{GeV}]}$. NaI(Tl) is also of

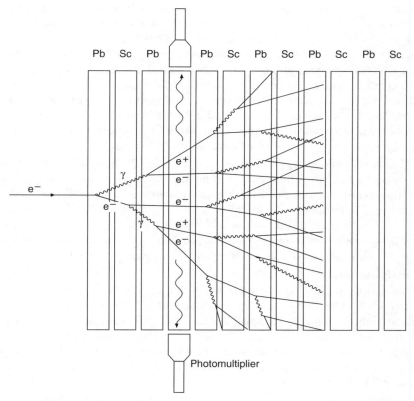

Fig. A.8. Sketch of particle cascade formation inside a calorimeter. An electromagnetic cascade inside a sampling calorimeter made out of layers of lead and scintillator is depicted. The lead acts as an absorber material where the bremsstrahlung and pair production processes primarily take place. The opening angles are, for purposes of clarity, exaggerated in the diagram. The particle tracks are for the same reason not continued on into the rearmost layers of the detector. Electrons and positrons in the scintillator produce visible scintillation light, which through total reflection inside the scintillator is led off to the sides (*large wavy lines*) where it is detected by photomultipliers. The total amount of scintillator light measured is proportional to the energy of the incoming electron.

great importance for nuclear-gamma spectroscopy, and hence for energies $\lesssim$ 1 MeV, since it has a large photon absorption coefficient, particularly for the photoelectric effect.

Cascade particles in lead glass produce Cherenkov light which may also be registered with the help of photomultipliers. Lead glass calorimeters may be built up from a few thousand lead glass blocks, which can cover several square metres. The transverse dimension of these blocks is adjusted to the transverse extension of electromagnetic showers, typically a few centimetres. Energy resolution is typically around $\delta E/E \approx 3-5\,\%\,/\sqrt{E\,[\text{GeV}]}$.

Hadronic calorimeters may be used to measure hadronic energies. These produce a shower of secondary particles (mostly further hadrons) in inelastic reactions. Such hadronic showers have, compared to electromagnetic showers, a larger spatial extension and display much larger fluctuations in both the number and type of secondary particles involved. *Sampling calorimeters* made up of alternating layers of a pure absorber material (e.g., iron, uranium) and a detector material (e.g., an organic scintillator) are used to measure hadron energies. Only a small fraction of the original particle's energy is deposited in the detector material. The energy resolution of hadronic calorimeters is, both for this reason and because of the large fluctuations in the number of secondary particles, only about $\delta E/E \approx 30-80\,\%\,/\sqrt{E\,[\text{GeV}]}$.

Momentum and energy measurements are interchangeable for highly relativistic particles (5.6). The accuracy of momentum measurements in magnetic spectrometers decreases linearly with particle momentum, while the precision of energy measurements in calorimeters increases as $1/\sqrt{E}$. Depending upon the particle type and the particular detector configuration it can make sense for particles with momenta above 50–100 GeV/c to measure momenta indirectly through a more accurate energy measurement in a calorimeter.

Identifying particles. The mass and the charge of a particle generally suffice to identify it. The sign of a particle's charge may be easily read off from the particle's deflection in a magnetic field, but a direct measurement of the particle's mass is mostly impossible. There is therefore no general particle identification recipe; rather lots of different methods, which often use other particle properties, are available. In the following we will briefly list those methods which are used in particle physics for particles with momenta above about 100 MeV/c.

- Short lived particles may be identified from their decay products with the help of the method of invariant masses (cf. Sec. 15.1).
- The presence of neutrinos is usually only detected by measuring a deficit of energy or momentum in a reaction.
- Electrons and photons are recognised through their characteristic electromagnetic showers in calorimeters. We may distinguish between them by putting an ionisation detector (e.g., a scintillator or a wire chamber) in front of the calorimeter – of the two only an electron will leave an ionisation trail.

– Muons are identified by their exceptional penetrative powers. They primarily lose energy to ionisation and may be detected with the help of ionisation chambers placed behind lead plates, which will absorb all other charged particles.

– Charged hadrons, such as pions, kaons and protons, are the most difficult particles to distinguish. For them not only a momentum measurement is required but also a further independent measurement is needed – which one is best suited depends upon the particle's momentum.

• The time of flight between two ionisation detectors may be measured for momenta below $1\,\mathrm{GeV}/c$, since the velocity depends for a fixed momentum upon the mass. A further possibility is to measure the loss of energy to ionisation – this depends upon the particle velocity. In this range it varies as $1/v^2$.

• This latter approach may be extended to $1.5\,$–$50\,\mathrm{GeV}/c$ momenta (where the energy loss only increases logarithmically as $\beta = v/c$) if the measurements are performed repeatedly.

• Various sorts of *Cherenkov counters* may be used in the range up to about $100\,\mathrm{GeV}/c$. Threshold Cherenkov counters require a material with a refractive index n so arranged that only specific particles with a particular momentum can produce Cherenkov light (cf. A.8). In ring imaging Cherenkov counters (RICH) the opening angle of the Cherenkov photons is measured for all the particles and their speed may be calculated from this. If their momentum is known then this determines their identity.

• *Transition radiation detectors* may be used for $\gamma = E/mc^2 \gtrsim 100$. Transition radiation is produced when charged particles cross from one material to another which has a different dielectric constant. The intensity of the radiation depends upon γ. Thus an intensity measurement can enable us to distinguish between different hadrons with the same momenta. This is in fact the only way to identify such particles if the energy of the hadron is above $100\,\mathrm{GeV}$. Transition radiation may also be employed to distinguish between electrons and pions. The tiny mass of the electrons means that this is already possible for energies around $1\,\mathrm{GeV}$.

– Neutron detection is a special case. (n, α) and (n, p) nuclear reactions are used to identify neutrons — from those with thermal energies to those with momenta up to around $20\,\mathrm{MeV}/c$. The charged reaction products have fixed kinetic energies and these may be measured in scintillation counters or gas ionisation counters. For momenta between $20\,\mathrm{MeV}/c$ and $1\,\mathrm{GeV}/c$ one looks for protons from elastic neutron-proton scattering. The proton target is generally part of the material of the detector itself (plastic scintillator, counter gas). At higher momenta only hadron calorimeter measurements are available to us. The identification is then, however, as a rule not unambiguous.

A detector system. We wish to present as an example of a system of detectors the ZEUS detector at the HERA storage ring. This detector measures the reaction products in high energy electron-proton collisions with centre of mass energies up to about 314 GeV (Fig. A.9). It is so arranged that apart from the beam pipe region the reaction zone is hermetically covered. Many different detectors, chosen to optimise the measurement of energy and momentum and the identification of the reaction products, make up the whole. The most important components are the wire chambers, which are arranged directly around the reaction point, and, just outside these, a uranium-scintillator calorimeter where the energies of electrons and hadrons are measured to a high precision.

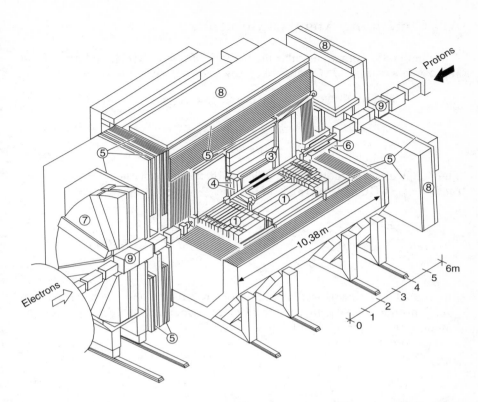

Fig. A.9. The ZEUS detector at the HERA storage ring in DESY. The electrons and protons are focused with the help of magnetic lenses (*9*) before they are made to collide at the interaction point in the centre of the detector. The tracks of charged reaction products are registered in the vertex chamber (*3*) which surrounds the reaction point and also in the central track chamber (*4*). These drift chambers are surrounded by a superconducting coil which produces a magnetic field of up to 1.8 T. The influence of this magnetic field on the electron beam which passes through it must be compensated by additional magnets (*6*). The next layer is a uranium-scintillator calorimeter (*1*) where the energies of electrons, photons and also of hadrons may be measured to a great accuracy. The iron yoke of the detector (*2*), into which the magnetic flux of the central solenoid returns, also acts as an absorber for the backwards calorimeter, where the energy of those high energy particle showers that are not fully absorbed in the central uranium calorimeter may be measured. Large area wire chambers (*5*), positioned behind the iron yoke, surround the whole detector and are used to betray the passage of any muons. These chambers may be used to measure the muons' momenta since they are inside either the magnetic field of the iron yoke or an additional 1.7 T toroidal field (*7*). Finally a thick reinforced concrete wall (*8*) screens off the experimental hall as far as is possible from the radiation produced in the reactions. (*Courtesy of DESY*)

A.3 Combining Angular Momenta

The combination of two angular momenta $|j_1 m_1\rangle$ and $|j_2 m_2\rangle$ to form a total angular momentum $|JM\rangle$ must obey the following selection rules:

$$|j_1 - j_2| \leq J \leq j_1 + j_2 , \tag{A.11}$$

$$M = m_1 + m_2 , \tag{A.12}$$

$$J \geq |M| . \tag{A.13}$$

The coupled states may be expanded with the help of the *Clebsch-Gordan coefficients* (CGC) $(j_1 j_2 m_1 m_2 | JM)$ in the $|j_1 j_2 JM\rangle$ basis:

$$|j_1 m_1\rangle \otimes |j_2 m_2\rangle = \sum_{\substack{J=|j_1-j_2| \\ M=m_1+m_2}}^{J=j_1+j_2} (j_1 j_2 m_1 m_2 | JM) \cdot |j_1 j_2 JM\rangle. \tag{A.14}$$

The probability that the combination of two angular momenta $|j_1 m_1\rangle$ and $|j_2 m_2\rangle$ produces a system with total angular momentum $|JM\rangle$ is thus the square of the corresponding CGC's.

The corollary

$$|j_1 j_2 JM\rangle = \sum_{\substack{m_1=-j_1 \\ m_2=M-m_1}}^{m_1=+j_1} (j_1 j_2 m_1 m_2 | JM) \cdot |j_1 m_1\rangle \otimes |j_2 m_2\rangle , \tag{A.15}$$

also holds. For a system $|JM\rangle$, which has been produced from a combination of two angular momenta j_1 and j_2, the square of the CGC's gives the probability that the individual angular momenta may be found in the states $|j_1 m_1\rangle$ and $|j_2 m_2\rangle$.

Equations (A.14) and (A.15) may also be applied to isospin. Consider, for example, the Δ^+ baryon ($I = 3/2$, $I_3 = +1/2$) which can decay into $p + \pi^0$ or $n + \pi^+$. The branching ratio can be found to be

$$\frac{B(\Delta^+ \to p + \pi^0)}{B(\Delta^+ \to n + \pi^+)} = \frac{|(\frac{1}{2} \ 1 \ +\frac{1}{2} \ 0 \ | \ \frac{3}{2} \ +\frac{1}{2})|^2}{|(\frac{1}{2} \ 1 \ -\frac{1}{2} \ +1 \ | \ \frac{3}{2} \ +\frac{1}{2})|^2} = \frac{\left(\sqrt{\frac{2}{3}}\right)^2}{\left(\sqrt{\frac{1}{3}}\right)^2} = 2. \tag{A.16}$$

The CGC's are listed for combinations of low angular momenta. The values for $j_1 = 1/2$ and $j_2 = 1$ may be found with the help of the general phase relation

$$(j_2 j_1 m_2 m_1 | JM) = (-1)^{j_1+j_2-J} \cdot (j_1 j_2 m_1 m_2 | JM) . \tag{A.17}$$

$j_1 = 1/2$			$j_2 = 1/2$	
m_1	m_2	J	M	CGC
1/2	1/2	1	1	$+1$
1/2	−1/2	1	0	$+\sqrt{1/2}$
1/2	−1/2	0	0	$+\sqrt{1/2}$
−1/2	1/2	1	0	$+\sqrt{1/2}$
−1/2	1/2	0	0	$-\sqrt{1/2}$
−1/2	−1/2	1	−1	$+1$

$j_1 = 1$			$j_2 = 1/2$	
m_1	m_2	J	M	CGC
1	1/2	3/2	3/2	$+1$
1	−1/2	3/2	1/2	$+\sqrt{1/3}$
1	−1/2	1/2	1/2	$+\sqrt{2/3}$
0	1/2	3/2	1/2	$+\sqrt{2/3}$
0	1/2	1/2	1/2	$-\sqrt{1/3}$
0	−1/2	3/2	−1/2	$+\sqrt{2/3}$
0	−1/2	1/2	−1/2	$+\sqrt{1/3}$
−1	1/2	3/2	−1/2	$+\sqrt{1/3}$
−1	1/2	1/2	−1/2	$-\sqrt{2/3}$
−1	−1/2	3/2	−3/2	$+1$

$j_1 = 1$			$j_2 = 1$	
m_1	m_2	J	M	CGC
1	1	2	2	$+1$
1	0	2	1	$+\sqrt{1/2}$
1	0	1	1	$+\sqrt{1/2}$
1	−1	2	0	$+\sqrt{1/6}$
1	−1	1	0	$+\sqrt{1/2}$
1	−1	0	0	$+\sqrt{1/3}$
0	1	2	1	$+\sqrt{1/2}$
0	1	1	1	$-\sqrt{1/2}$
0	0	2	0	$+\sqrt{2/3}$
0	0	1	0	0
0	0	0	0	$-\sqrt{1/3}$
0	−1	2	−1	$+\sqrt{1/2}$
0	−1	1	−1	$+\sqrt{1/2}$
−1	1	2	0	$+\sqrt{1/6}$
−1	1	1	0	$-\sqrt{1/2}$
−1	1	0	0	$-\sqrt{1/3}$
−1	0	2	−1	$+\sqrt{1/2}$
−1	0	1	−1	$-\sqrt{1/2}$
−1	−1	2	−2	$+1$

A.4 Physical Constants

Table A.1. Physical constants [Co87, La95, PD98]. The numbers in brackets signify the uncertainty in the last decimal places. The sizes of c, μ_0 (and hence ε_0) are defined by the units "metre" and "ampere" [Pe83]. These constants are therefore error free.

Constants	Symbol	Value
Speed of light	c	$2.997\,924\,58 \cdot 10^8\,\mathrm{m\,s^{-1}}$
Planck's constant	h	$6.626\,075\,5\,(40) \cdot 10^{-34}\,\mathrm{J\,s}$
	$\hbar = h/2\pi$	$1.054\,572\,66\,(63) \cdot 10^{-34}\,\mathrm{J\,s}$
		$= 6.582\,122\,0\,(20) \cdot 10^{-22}\,\mathrm{MeV\,s}$
	$\hbar c$	$197.327\,053\,(59)\,\mathrm{MeV\,fm}$
	$(\hbar c)^2$	$0.389\,379\,66\,(23)\,\mathrm{GeV^2\,mbarn}$
Atomic mass unit	$u = M_{12\mathrm{C}}/12$	$931.494\,32\,(28)\,\mathrm{MeV}/c^2$
Mass of the proton	M_p	$938.272\,31\,(28)\,\mathrm{MeV}/c^2$
Mass of the neutron	M_n	$939.565\,63\,(28)\,\mathrm{MeV}/c^2$
Mass of the electron	m_e	$0.510\,999\,06\,(15)\,\mathrm{MeV}/c^2$
Elementary charge	e	$1.602\,177\,33\,(49) \cdot 10^{-19}\,\mathrm{A\,s}$
Dielectric constant	$\varepsilon_0 = 1/\mu_0 c^2$	$8.854\,187\,817 \cdot 10^{-12}\,\mathrm{A\,s/V\,m}$
Permeability of vacuum	μ_0	$4\pi \cdot 10^{-7}\,\mathrm{V\,s/A\,m}$
Fine structure constant	$\alpha = e^2/4\pi\varepsilon_0\hbar c$	$1/137.035\,989\,5\,(61)$
Class. electron radius	$r_\mathrm{e} = \alpha\hbar c/m_\mathrm{e}c^2$	$2.817\,940\,92\,(38) \cdot 10^{-15}\,\mathrm{m}$
Compton wavelength	$\lambda_\mathrm{e} = r_\mathrm{e}/\alpha$	$3.861\,593\,23\,(35) \cdot 10^{-13}\,\mathrm{m}$
Bohr radius	$a_0 = r_\mathrm{e}/\alpha^2$	$5.291\,772\,49\,(24) \cdot 10^{-11}\,\mathrm{m}$
Bohr magneton	$\mu_\mathrm{B} = e\hbar/2m_\mathrm{e}$	$5.788\,382\,63\,(52) \cdot 10^{-11}\,\mathrm{MeV\,T^{-1}}$
Nuclear magneton	$\mu_\mathrm{N} = e\hbar/2m_\mathrm{p}$	$3.152\,451\,66\,(28) \cdot 10^{-14}\,\mathrm{MeV\,T^{-1}}$
Magnetic moment	μ_e	$1.001\,159\,652\,193\,(10)\,\mu_\mathrm{B}$
	μ_p	$2.792\,847\,386\,(63)\,\mu_\mathrm{N}$
	μ_n	$-1.913\,042\,75\,(45)\,\mu_\mathrm{N}$
Avogadro's number	N_A	$6.022\,136\,7\,(36) \cdot 10^{23}\,\mathrm{mol^{-1}}$
Boltzmann's constant	k	$1.380\,658\,(12) \cdot 10^{-23}\,\mathrm{J\,K^{-1}}$
		$= 8.617\,385\,(73) \cdot 10^{-5}\,\mathrm{eV\,K^{-1}}$
Gravitational constant	G	$6.672\,59\,(85) \cdot 10^{-11}\,\mathrm{N\,m^2\,kg^{-2}}$
	$G/\hbar c$	$6.707\,11\,(86) \cdot 10^{-39}\,(\mathrm{GeV}/c^2)^{-2}$
Fermi constant	$G_\mathrm{F}/(\hbar c)^3$	$1.166\,39\,(1) \cdot 10^{-5}\,\mathrm{GeV^{-2}}$
Weinberg angle	$\sin^2\theta_\mathrm{W}$	$0.231\,24\,(24)$
Mass of the $\mathrm{W}^\pm$	M_W	$80.41\,(10)\,\mathrm{GeV}/c^2$
Mass of the Z^0	M_Z	$91.187\,(7)\,\mathrm{GeV}/c^2$
Strong coupling const.	$\alpha_\mathrm{s}(M_\mathrm{Z}^2 c^2)$	$0.119\,(2)$

Solutions to the Problems

Chapter 2

1. Proton repulsion in ^{3}He:

$$
\begin{aligned}
V_C &= \frac{-\hbar c \alpha}{R} = (M_{^3He} - M_{^3H}) \cdot c^2 - (M_n - M_p) \cdot c^2 \\
&= E_\beta^{max} - (M_n - M_p - m_e) \cdot c^2 .
\end{aligned}
$$

This yields $R = 1.88$ fm. The β-decay recoil and the difference between the atomic binding energies may be neglected.

Chapter 3

1. a) At Saturn we have $t/\tau = 4\,\mathrm{yrs}/127\,\mathrm{yrs}$ and we require

$$
N_0 \frac{1}{\tau} e^{-t/\tau} \cdot 5.49\,\mathrm{MeV} \cdot 0.055 = 395\,\mathrm{W}
$$

   power to be available. This implies $N_0 = 3.4 \cdot 10^{25}$ nuclei, which means 13.4 kg ^{238}Pu.
   b) At Neptune (after 12 years) 371 W would be available.
   c) The power available from radiation decreases as $1/r^2$. Hence at Saturn 395 W power would require an area of $2.5 \cdot 10^3$ m^2 and 371 W at Neptune could be produced by an area of $2.3 \cdot 10^4$ m^2. This would presumably lead to construction and weight problems.

2. a) Applying the formula $N = N_0 e^{-\lambda t}$ to both uranium isotopes leads to

$$
\frac{99.28}{0.72} = \frac{e^{-\lambda_{238} t}}{e^{-\lambda_{235} t}} \quad \text{which yields:} \quad t = 5.9 \cdot 10^9 \text{ years.}
$$

   Uranium isotopes, like all heavy $(A \gtrsim 56)$ elements, are produced in supernova explosions. The material which is so ejected is used to build up new stars. The isotopic analysis of meteorites leads to the age of the solar system being $4.55 \cdot 10^9$ years.
   b) After $2.5 \cdot 10^9$ years, $(1 - e^{-\lambda t})$ of the nuclei will have decayed. This is 32 %.

c) Equation (2.8) yields that a total of 51 MeV is released in the ^{238}U$\rightarrow$ ^{206}Pb decay chain. In spontaneous fission 190 MeV is set free.

3. a)
$$A_2(t) = N_{0,1} \cdot \lambda_1 \cdot \frac{\lambda_2}{\lambda_2 - \lambda_1} \left(e^{-\lambda_1 t} - e^{-\lambda_2 t} \right)$$
for large times t, because of $\lambda_1 \ll \lambda_2$:
$$A_2(t) = N_{0,1} \cdot \lambda_1.$$

b) The concentration of ^{238}U in concrete can thus be found to be

room volume V: 400 m^3 $\Longrightarrow \varrho_U = \dfrac{V \cdot A}{V_B \cdot \lambda_{238}} = 1.5 \cdot 10^{21} \dfrac{\text{atoms}}{\text{m}^3}$.
eff. concrete volume V_B: 5.4 m^3

4. Nuclear masses for fixed A depend quadratically upon Z. From the definitions in (3.6) the minimum of the parabola is at $Z_0 = \beta/2\gamma$. The constant a_a in β and γ is part of the asymmetry term in the mass formula (2.8) and, according to (17.12), does not depend upon the electromagnetic coupling constant α. The "constant" a_c, which describes the Coulomb repulsion and enters the definition of γ, is on the other hand proportional to α and may be written as: $a_c = \kappa\alpha$. Inserting this into $Z_0 = \beta/2\gamma$ yields

$$Z_0 = \frac{\beta}{2\left(a_a/A + \kappa\alpha/A^{1/3}\right)} \qquad \Longrightarrow \qquad \frac{1}{\alpha} = \frac{2\kappa A Z_0}{A^{1/3}\left(A\beta - 2a_a Z_0\right)}.$$

Assuming that the minimum of the mass formula is exactly at the given Z one finds $1/\alpha$ values of 128, 238 and 522 for the $^{186}_{74}$W, $^{186}_{82}$Pb and $^{186}_{88}$Ra nuclides. Stable $^{186}_{94}$Pu cannot be obtained just by "twiddling" α.

5. The energy E released in ^A_ZX $\rightarrow$ $^{A-4}_{Z-2}$Y $+ \alpha$ is
$$E = B(\alpha) - \delta B \qquad \text{where} \quad \delta B = B(\text{X}) - B(\text{Y}).$$

Note that we have here neglected the difference in the atomic binding energies. If we further ignore the pairing energy, which only slightly changes, we obtain
$$\begin{aligned} E &= B(\alpha) - \frac{\partial B}{\partial Z}\delta Z - \frac{\partial B}{\partial A}\delta A = B(\alpha) - 2\frac{\partial B}{\partial Z} - 4\frac{\partial B}{\partial A} \\ &= B(\alpha) - 4a_v + \frac{8}{3}a_s\frac{1}{3A^{1/3}} + 4a_c\frac{Z}{A^{1/3}}\left(1 - \frac{Z}{3A}\right) - a_a\left(1 - \frac{2Z}{A}\right)^2. \end{aligned}$$

Putting in the parameters yields $E > 0$ if $A \gtrsim 150$. Natural α-activity is only significant for $A \gtrsim 200$, since the lifetime is extremely long for smaller mass numbers.

6. The mother nucleus and the α particle are both 0^+ systems which implies that the spin J and parity P of a daughter nucleus with orbital angular momentum L and spatial wave function parity $(-1)^L$ must combine to 0^+. This means that $J^P = 0^+, 1^-, 2^+, 3^-, \cdots$ are allowed.

Chapter 4

1. a) In analogy to (4.5) the reaction rate must obey $\dot{N} = \sigma \dot{N}_d n_t$, where $\dot{N}_d$ signifies the deuteron particle current and n_t is the particle areal density of the tritium target. The neutron rate found in any solid angle element $d\Omega$ must then obey

$$d\dot{N} = \frac{d\sigma}{d\Omega} d\Omega \, \dot{N}_d \, n_t = \frac{d\sigma}{d\Omega} \frac{F}{R^2} \frac{I_d}{e} \frac{\mu_t}{m_t} N_A ,$$

where e is the elementary electric charge, m_t is the molar mass of tritium and N_A is Avogadro's number.

Inserting the numbers yields $d\dot{N} = 1444$ neutrons/s.

b) Rotating the target away from the orthogonal increases the effective particle area density "seen" by the beam by a factor of $1/\cos\theta$. A rotation through $10°$ thus increases the reaction rate by 1.5%.

2. The number N of beam particles decreases according to (4.5) with the distance x covered as $e^{-x/\lambda}$ where $\lambda = 1/\sigma n$ is the absorbtion length.

a) Thermal neutrons in cadmium: We have

$$n_{Cd} = \varrho_{Cd} \frac{N_A}{A_{Cd}}$$

where the atomic mass of cadmium is given by $A_{Cd} = 112.40 \, \mathrm{g \, mol^{-1}}$. We thus obtain

$$\lambda_{n,Cd} = 9 \, \mu m .$$

b) For highly energetic photons in lead one may find in an analogous manner ($A_{Pb} = 207.19 \, \mathrm{g \, mol^{-1}}$)

$$\lambda_{\gamma,Pb} = 2.0 \, \mathrm{cm} .$$

c) Antineutrinos predominantly react with the electrons in the earth. Their density is

$$n_{e,earth} = \varrho_{earth} \left(\frac{Z}{A} \right)_{earth} N_A .$$

We therefore obtain

$$\lambda_{\bar{\nu}/earth} = 6.7 \cdot 10^{16} \, \mathrm{m} ,$$

which is about $5 \cdot 10^9$ times the diameter of the planet.

Note: the number of beam particles only decreases exponentially with distance if *one* reaction leads to the beam particles being absorbed; a criterion which is fulfilled in the above examples. The situation is different if $k \gg 1$ reactions are needed (e.g., α particles in air). In such cases the range is almost constant $L = k/\sigma n$.

Chapter 5

1. a) From $Q^2 = -(p - p')^2$ and (5.13) one finds

$$Q^2 = 2M(E - E'),$$

with M the mass of the heavy nucleus. This implies that Q^2 is largest at the smallest value of E', i.e., $\theta = 180°$. The maximal momentum transfer is then from (5.15)

$$Q^2_{max} = \frac{4E^2 M}{Mc^2 + 2E},$$

b) From (5.15) we find for $\theta = 180°$ that the energy transfer $\nu = E - E'$ is

$$\nu = E\left(1 - \frac{1}{1 + 2\frac{E}{Mc^2}}\right) = \frac{2E^2}{Mc^2 + 2E}.$$

The energy of the backwardly scattered nucleus is then

$$E'_{nucleus} = Mc^2 + \nu = Mc^2 + \frac{2E^2}{Mc^2 + 2E}$$

and its momentum is

$$|\boldsymbol{P}'| = \sqrt{Q^2_{max} + \frac{\nu^2}{c^2}} = \sqrt{\frac{4ME^2}{Mc^2 + 2E} + \frac{4E^4}{c^2(Mc^2 + 2E)^2}}.$$

c) The nuclear Compton effect may be calculated with the help of $\Delta\lambda = \frac{h}{Mc}(1 - \cos\theta)$. The same result as for electron scattering is obtained since we have neglected the electron rest mass in a) and b) above.

2. Those α particles which directly impinge upon the ^{56}Fe nucleus are absorbed. Elastically scattered α particles correspond to a "shadow scattering" which may be described as Fraunhofer diffraction upon a disc. The diameter D of the disc is found to be

$$D = 2(\sqrt[3]{4} + \sqrt[3]{56}) \cdot 0.94 \text{ fm} \approx 10 \text{ fm}.$$

In the literature D is mostly parameterised by the formula $D = 2\sqrt[3]{A} \cdot 1.3$ fm, which gives the same result. The wavelength of the α particles is $\lambda = h/p$, where p is to be understood as that in the centre of mass system of the reaction. Using $pc = 840$ MeV one finds $\lambda = 1.5$ fm.
The first minimum is at $\theta = 1.22 \lambda/D \approx 0.18 \approx 10.2°$. The intensity distribution of the diffraction is given by the Bessel function j_0. The further minima correspond to the nodes of this Bessel function.
The scattering angle ought, however, to be given in the laboratory frame and is given by $\theta_{lab} \approx 9.6°$.

3. The smallest separation of the α particles from the nucleus is $s(\theta) = a + \frac{a}{\sin\theta/2}$ for the scattering angle θ. The parameter a is obtained from $180°$ scattering, since the kinetic energy is then equal to the potential energy:

$$E_{\mathrm{kin}} = \left| \frac{zZe^2\hbar c}{4\pi\varepsilon_0\hbar c2a} \right| .$$

For 6 MeV α scattering off gold, we have $a = 19$ fm and $s = 38$ fm. For deviations from Rutherford scattering to occur, the α-particles must manage to get close to the nuclear forces, which can first happen at a separation $R = R_\alpha + R_{\mathrm{Au}} \approx 9$ fm. A more detailed discussion is given in Sec. 18.4. Since $s \gg R$ no nuclear reactions are possible between 6 MeV α particles and gold and no deviation from the Rutherford cross-section should therefore be expected. This would only be possible for much lighter nuclei.

4. The kinetic energy of the electrons may be found as follows:

$$\frac{\hbar}{\sqrt{2M_\alpha E_\alpha^{\mathrm{kin}}}} \approx \lambda_\alpha \overset{!}{=} \lambda_e \approx \frac{\hbar c}{E_e^{\mathrm{kin}}} \implies E_e^{\mathrm{kin}} \approx \sqrt{2M_\alpha c^2 E_\alpha^{\mathrm{kin}}} = 211\,\mathrm{MeV} .$$

The momentum transfer is maximal for scattering through $180°$. Neglecting the recoil we have

$$|\boldsymbol{q}|_{\mathrm{max}} = 2|\boldsymbol{p}_e| = \frac{2\hbar}{\lambda_e} \approx 2\sqrt{2M_\alpha E_\alpha^{\mathrm{kin}}} = 423\,\mathrm{MeV}/c ,$$

and the variable α in Table 5.1 may be found with the help of (5.56) to be

$$\alpha_{\mathrm{max}} = \frac{|\boldsymbol{q}|_{\mathrm{max}} R}{\hbar} = \frac{423\,\mathrm{MeV} \cdot 1.21 \cdot \sqrt[3]{197}\,\mathrm{fm}}{197\,\mathrm{MeV\,fm}} = 15.1 .$$

The behaviour of the function $3\,\alpha^{-3}\,(\sin\alpha - \alpha\cos\alpha)$ from Table 5.1 is such that it has 4 zero points in the range $0 < \alpha \leq 15.1$.

5. Electrons oscillate most in the field of the X-rays since $M_{\mathrm{nuclear}} \gg m_e$. As in the H atom, the radial wave function of the electrons also falls off exponentially in He. Hence, just as for electromagnetic electron scattering off nucleons, a dipole form factor is observed.

6. If a 511 keV photon is Compton scattered through $30°$ off an electron at rest, the electron receives momentum, $p_e = 0.26\,\mathrm{MeV}/c$. From the virial theorem an electron bound in a helium atom must have kinetic energy $E_{\mathrm{kin}} = -E_{\mathrm{pot}}/2 = -E_{\mathrm{tot}} = 24\,\mathrm{eV}$, which implies that the momentum of the Compton electron is smeared out with $\Delta p \approx \pm 5 \cdot 10^{-3}\,\mathrm{MeV}/c$ which corresponds to an angular smearing of $\Delta\theta_e \approx \Delta p/p = \pm 20\,\mathrm{mrad} \approx \pm 1°$.

Chapter 6

1. The form factor of the electron must be measured up to $|q| \approx \hbar/r_0 = 200 \text{ GeV}/c$. One thus needs $\sqrt{s} = 200 \text{ GeV}$, i.e., 100 GeV. colliding beams. For a target at rest, $2m_e c^2 E = s$ implies that $4 \cdot 10^7 \text{ GeV}$ (!) would be needed.

2. Since the pion has spin zero, the magnetic form factor vanishes and we have (6.10):

$$\frac{d\sigma(e\pi \to e\pi)}{d\Omega} = \left(\frac{d\sigma}{d\Omega}\right)_{\text{Mott}} G^2_{E,\pi}(Q^2)$$

$$G^2_{E,\pi}(Q^2) \approx \left(1 - \frac{Q^2 \langle r^2 \rangle_\pi}{6 \hbar^2}\right)^2 = 1 - 3.7 \frac{Q^2}{\text{GeV}^2/c^2}$$

Chapter 7

1. a) The photon energy in the electron rest frame is obtained through a Lorentz transformation with dilatation factor $\gamma = 26.67 \text{ GeV}/m_e c^2$. This yields $E_i = 2\gamma E_\gamma = 251.6 \text{ keV}$ for $E_\gamma = 2\pi\hbar c/\lambda = 2.41 \text{ eV}$.

   b) Photon scattering off a stationary electron is governed by the Compton scattering formula:

$$E_f(\theta) = \left(\frac{1 - \cos\theta}{mc^2} + \frac{1}{E_i}\right)^{-1},$$

   where $E_f(\theta)$ is the energy of the photon after the scattering and θ is the scattering angle. Scattering through 90° (180°) leads to $E_f = 168.8$ (126.8) keV.

   After the reverse transformation into the laboratory system, we have the energy E'_γ:

$$E'_\gamma(\theta) = \gamma E_f(\theta) (1 - \cos\theta) = \gamma \left(\frac{1}{mc^2} + \frac{1}{E_i(1 - \cos\theta)}\right)^{-1}.$$

   For the two cases of this example, E'_γ takes on the values 8.80 (13.24) GeV. The scattering angle in the laboratory frame θ_{lab} is also 180°, i.e., the outgoing photon flies exactly in the direction of the electron beam. Generally we have

$$\theta_{\text{lab}} = \pi - \frac{1}{\gamma \tan\frac{\theta}{2}}.$$

   c) For $\theta = 90°$ this yields $\theta_{\text{lab}} = \pi - 1/\gamma = \pi - 19.16 \mu\text{rad}$. The spatial resolution of the calorimeter must therefore be better than 1.22 mm.

2. Comparing the coefficients in (6.5) and (7.7) yields

$$\frac{2W_1}{W_2} = 2\tau \qquad \text{where} \qquad \tau = \frac{Q^2}{4m^2c^2},$$

and m is the mass of the target. Replacing W_1 by F_1/Mc^2 and W_2 by F_2/ν means that we can write

$$\frac{\nu}{Mc^2} \cdot \frac{F_1}{F_2} = \frac{Q^2}{4m^2c^2}.$$

Since we consider elastic scattering off a particle with mass m we have $Q^2 = 2m\nu$ and thus

$$m = \frac{Q^2}{2\nu} = x \cdot M \qquad \text{since} \qquad x = \frac{Q^2}{2M\nu}.$$

Inserting this mass into the above equation yields (7.13).

3. a) The centre of mass energy of the electron-proton collision calculated from

$$s = (p_{\mathrm{p}}c + p_{\mathrm{e}}c)^2 = m_{\mathrm{p}}^2c^4 + m_{\mathrm{e}}^2c^4 + 2(E_{\mathrm{p}}E_{\mathrm{e}} - \boldsymbol{p}_{\mathrm{p}} \cdot \boldsymbol{p}_{\mathrm{e}}c^2) \approx 4E_{\mathrm{p}}E_{\mathrm{e}}$$

is $\sqrt{s} = 314\text{GeV}$, if we neglect the electron and proton masses. For a stationary proton target ($E_{\mathrm{p}} = m_{\mathrm{p}}c^2$; $\boldsymbol{p}_{\mathrm{p}} = 0$) the squared centre of mass energy of the electron-proton collision is found to be $s \approx 2E_{\mathrm{e}}m_{\mathrm{p}}c^2$. The electron beam energy would have to be

$$E_{\mathrm{e}} = \frac{s}{2m_{\mathrm{p}}c^2} = 52.5 \text{ TeV}$$

to attain a centre of mass energy $\sqrt{s} = 314\,\text{GeV}$.

b) Consider the underlying electron-quark scattering reaction $\mathrm{e}(E_{\mathrm{e}}) + \mathrm{q}(xE_{\mathrm{p}}) \to \mathrm{e}(E_{\mathrm{e}}') + \mathrm{q}(E_{\mathrm{q}}')$, where the bracketed quantities are the particle energies. Energy and momentum conservation yield the following three relations:

(1) $E_{\mathrm{e}} + xE_{\mathrm{p}} = E_{\mathrm{e}}' + E_{\mathrm{q}}'$ overall energy
(2) $E_{\mathrm{e}}' \sin\theta/c = E_{\mathrm{q}}' \sin\gamma/c$ transverse momentum
(3) $(xE_{\mathrm{p}} - E_{\mathrm{e}})/c = (E_{\mathrm{q}}' \cos\gamma - E_{\mathrm{e}}' \cos\theta)/c$ longitudinal momentum

Q^2 may be expressed in terms of the electron parameters E_{e}, E_{e}' and θ as (6.2)

$$Q^2 = 2E_{\mathrm{e}}E_{\mathrm{e}}'(1 - \cos\theta)/c^2.$$

We now want to replace E_{e}' with the help of (1)–(3) by E_{e}, θ and γ. After some work we obtain

$$E'_e = \frac{2E_e \sin\gamma}{\sin\theta + \sin\gamma - \sin(\theta - \gamma)}$$

and thus

$$Q^2 = \frac{4E_e^2 \sin\gamma(1 - \cos\gamma)}{[\sin\theta + \sin\gamma - \sin(\theta - \gamma)]c^2}.$$

Experimentally the scattering angle γ of the scattered quark may be expressed in terms of the energy-weighted average angle of the hadronisation products

$$\cos\gamma = \frac{\sum_i E_i \cos\gamma_i}{\sum_i E_i}.$$

c) The greatest possible value of Q^2 is $Q^2_{\max} = s/c^2$. This occurs for electrons scattering through $\theta = 180°$ (backwards scattering) when the energy is completely transferred from the proton to the electron, $E'_e = E_p$. At HERA $Q^2_{\max} = 98\,420\ \text{GeV}^2/c^2$, while for experiments with a static target and beam energy $E = 300\ \text{GeV}$ we have $Q^2_{\max} = 2E_e m_p \approx 600\ \text{GeV}^2/c^2$. The spatial resolution is $\Delta x \simeq \hbar/Q$ which for the cases at hand is $0.63 \cdot 10^{-3}\ \text{fm}$ and $7.9 \cdot 10^{-3}\ \text{fm}$ respectively, i.e., a thousandth or a hundredth of the proton radius. In practice the fact that the cross-section falls off very rapidly at large Q^2 means that measurements are only possible up to about $Q^2_{\max}/2$.

d) The minimal value of Q^2 is obtained at the minimal scattering angle ($7°$) and for the minimal energy of the scattered electron ($5\ \text{GeV}$). From (6.2) we obtain $Q^2_{\min} \approx 2.2\ \text{GeV}^2/c^2$. The maximal value of Q^2 is obtained at the largest scattering angle ($178°$) and maximal scattering energy ($820\ \text{GeV}$). This yields $Q^2_{\max} \approx 98\,000\ \text{GeV}^2/c^2$. The corresponding values of x are obtained from $x = Q^2/2Pq$, where we have to substitute the four-momentum transfer q by the four-momenta of the incoming and scattered electron. This gives us $x_{\min} \approx 2.7 \cdot 10^{-5}$ and $x_{\max} \approx 1$.

e) The transition matrix element and hence the cross-section of a reaction depend essentially upon the coupling constants and the propagator (4.23, 10.3). We have

$$\sigma_{\text{em}} \propto \frac{e^2}{Q^4}, \qquad\qquad \sigma_{\text{weak}} \propto \frac{g^2}{(Q^2 + M_W^2 c^2)^2}.$$

Equating these expressions and using $e = g\sin\theta_W$ (11.14f) implies that the strengths of the electromagnetic and weak interactions will be of the same order of magnitude for $Q^2 \approx M_W^2 c^2 \approx 10^4\ \text{GeV}^2/c^2$.

4. a) The decay is isotropic in the pion's centre of mass frame (marked by a circumflex) and we have $\hat{\boldsymbol{p}}_\mu = -\hat{\boldsymbol{p}}_\nu$. Four-momentum conservation $p_\pi^2 = (p_\mu + p_\nu)^2$ implies

$$|\hat{\boldsymbol{p}}_\mu| \;=\; \frac{m_\pi^2 - m_\mu^2}{2m_\pi}\,c \qquad \approx \quad 30 \quad \mathrm{MeV}/c \qquad \text{and thus}$$

$$\hat{E}_\mu \;=\; \sqrt{\hat{p}_\mu^2 c^2 + m_\mu^2 c^4} \;\approx\; 110 \quad \mathrm{MeV}\,.$$

Using $\beta \approx 1$ and $\gamma = E_\pi/m_\pi c^2$, the Lorentz transformations of $\hat{E}_\mu$ into the laboratory frame for muons emitted in the direction of the pion's flight ("forwards") and for those emitted in the opposite direction ("backwards") are

$$E_\mu = \gamma\left(\hat{E}_\mu \pm \beta|\hat{\boldsymbol{p}}_\mu|c\right) \Longrightarrow \left\{ \begin{array}{l} E_{\mu,\mathrm{max}} \approx E_\pi\,, \\ E_{\mu,\mathrm{min}} \approx E_\pi (m_\mu/m_\pi)^2\,. \end{array} \right.$$

The muon energies are therefore: $200\,\mathrm{GeV} \lesssim E_\mu \lesssim 350\,\mathrm{GeV}$.

b) In the pion centre of mass frame the muons are $100\,\%$ longitudinally polarised because of the parity violating nature of the decay (Sec. 10.5). This polarisation must now be transformed into the laboratory frame. Consider initially just the "forwards" decays: the pion and muon momenta are parallel to the direction of the transformation. Such a Lorentz transformation will leave the spin unaffected and we see that these muons will also be $100\,\%$ longitudinally polarised, i.e., $P_{\mathrm{long}} = 1.0$. Similarly for decays in the "backwards" direction we have $P_{\mathrm{long}} = -1.0$. The extremes of the muon energies thus lead to extreme values of the polarisation. If we select at intermediate muon energies we automatically vary the longitudinal polarisation of the muon beam. For example $260\,\mathrm{GeV}$ muon beams have $P_{\mathrm{long}} = 0$. The general case is given in [Le68]. P_{long} depends upon the muon energy as

$$P_{\mathrm{long}} = \frac{u - \left[(m_\mu^2/m_\pi^2)(1-u)\right]}{u + \left[(m_\mu^2/m_\pi^2)(1-u)\right]} \qquad \text{where} \qquad u = \frac{E_\mu - E_{\mu,\mathrm{min}}}{E_{\mu,\mathrm{max}} - E_{\mu,\mathrm{min}}}\,.$$

5. The squared four momentum of the scattered parton is $(q + \xi P)^2 = m^2 c^2$, where m is the mass of the parton. Expanding and multiplying with x^2/Q^2 yields

$$\frac{x^2 M^2 c^2}{Q^2}\xi^2 + x\xi - x^2\left(1 + \frac{m^2 c^2}{Q^2}\right) = 0$$

Solving the quadratic equation for ξ and employing the approximate formula given in the question yields the result we were asked to obtain. For $m = xM$ we have $x = \xi$. In a rapidly moving frame of reference we also have $x = \xi$, since the masses m and M can then be neglected.

Chapter 8

1. a) From $x = Q^2/2M\nu$ we obtain $x \gtrsim 0.003$.

b) The average number of resolved partons is given by the integral over the parton distributions from x_{min} to 1. The normalisation constant, A, has to be chosen such that the number of valence quarks is exactly 3. One finds:

	Sea quarks	Gluons
$x > 0.3$	0.005	0.12
$x > 0.03$	0.4	4.9

Chapter 9

1. a) The relation between the event rate $\dot{N}$, the cross-section σ and the luminosity $\mathcal{L}$ is from (4.13) $\dot{N} = \sigma \cdot \mathcal{L}$. Therefore using (9.5)

$$\dot{N}_{\mu^+\mu^-} = \frac{4\pi\alpha^2\hbar^2c^2}{3 \cdot 4E^2} \cdot \mathcal{L} = 0.14/\text{s}.$$

At this centre of mass energy, $\sqrt{s} = 8\,\text{GeV}$, it is possible to produce pairs of u, d, s and c-quarks. The ratio R defined in (9.10) can therefore be calculated using (9.11) and we so obtain $R = 10/3$. This implies

$$\dot{N}_{\text{hadrons}} = \frac{10}{3} \cdot \dot{N}_{\mu^+\mu^-} = 0.46/\text{s}.$$

b) At $\sqrt{s} = 500\,\text{GeV}$ pair creation of all 6 quark flavours is possible. The ratio is thus $R = 5$. To reach a statistical accuracy of 10 % one would need to detect 100 events with hadronic final states. From $N_{\text{hadrons}} = 5 \cdot \sigma_{\mu^+\mu^-} \cdot \mathcal{L} \cdot t$ we obtain $\mathcal{L} = 8 \cdot 10^{33}\,\text{cm}^{-2}\,\text{s}^{-1}$. Since the cross-section falls off sharply with increasing centre of mass energies, future e^+e^--accelerators will need to have luminosities of an order 100 times larger than present day storage rings.

2. a) From the supplied parameters we obtain $\delta E = 1.9\,\text{MeV}$ and thus $\delta W = \sqrt{2}\,\delta E = 2.7\,\text{MeV}$. Assuming that the natural decay width of the Υ is smaller than δW, the measured decay width, i.e., the energy dependence of the cross-section, merely reflects the uncertainty in the beam energy (and the detector resolution). This is the case here.

b) Using $\lambda = \hbar/|\boldsymbol{p}| \approx (\hbar c)/E$ we may re-express (9.8) as

$$\sigma_f(W) = \frac{3\pi\hbar^2c^2\Gamma_{e^+e^-}\Gamma_f}{4E^2\,[(W - M_\Upsilon c^2)^2 + \Gamma^2/4]}.$$

In the neighbourhood of the (sharp) resonance we have $4E^2 \approx M_\Upsilon^2 c^4$. From this we obtain

$$\int \sigma_f(W)\,\mathrm{d}W = \frac{6\pi\hbar^2c^2\Gamma_{e^+e^-}\Gamma_f}{M_\Upsilon^2 c^4\,\Gamma}.$$

The measured quantity was $\int \sigma_f(W)\,dW$ for $\Gamma_f = \Gamma_{had}$. Using $\Gamma_{had} = \Gamma - 3\Gamma_{\ell^+\ell^-} = 0.925\Gamma$ we find $\Gamma = 0.051\,\mathrm{MeV}$ for the total natural decay width of the Υ. The true height of the resonance ought therefore to be $\sigma(W = M_\Upsilon) \approx 4100\,\mathrm{nb}$ (with $\Gamma_f = \Gamma$). The experimentally observed peak was, as a result of the uncertainty in the beam energy, less than this by a factor of over 100 (see Part a).

Chapter 10

1. $p + \overline{p} \to \ldots$ strong interaction.
   $p + K^- \to \ldots$ strong interaction.
   $p + \pi^- \to \ldots$ baryon number not conserved, so reaction impossible.
   $\overline{\nu}_\mu + p \to \ldots$ weak interaction, since neutrino participates.
   $\nu_e + p \to \ldots$ L_e not conserved, so reaction impossible
   $\Sigma^0 \to \ldots$ electromagnetic interaction, since photon radiated off.

2. a) • $\mathcal{C}|\gamma\rangle = -1|\gamma\rangle$. The photon is its own antiparticle. Its C-parity is -1 since it couples to electric charges which change their sign under the $\mathcal{C}$-parity transformation.
   • $\mathcal{C}|\pi^0\rangle = +1|\pi^0\rangle$, since $\pi_0 \to 2\gamma$ and C-parity is concerved in the electromagnetic interaction.
   • $\mathcal{C}|\pi^+\rangle = |\pi^-\rangle$, not a $\mathcal{C}$-eigenstate.
   • $\mathcal{C}|\pi^-\rangle = |\pi^+\rangle$, not a $\mathcal{C}$-eigenstate.
   • $\mathcal{C}(|\pi^+\rangle - |\pi^-\rangle) = (|\pi^-\rangle - |\pi^+\rangle) = -1(|\pi^-\rangle - |\pi^+\rangle)$, $\mathcal{C}$-eigenstate.
   • $\mathcal{C}|\nu_e\rangle = |\overline{\nu}_e\rangle$, not a $\mathcal{C}$-eigenstate.
   • $\mathcal{C}|\Sigma^0\rangle = |\overline{\Sigma}^0\rangle$, not a $\mathcal{C}$-eigenstate.

   b) • $\mathcal{P}\boldsymbol{r} = -\boldsymbol{r}$
   • $\mathcal{P}\boldsymbol{p} = -\boldsymbol{p}$
   • $\mathcal{P}\boldsymbol{L} = \boldsymbol{L}$ since $\boldsymbol{L} = \boldsymbol{r} \times \boldsymbol{p}$
   • $\mathcal{P}\boldsymbol{\sigma} = \boldsymbol{\sigma}$, since $\boldsymbol{\sigma}$ is also angular momentum;
   • $\mathcal{P}\boldsymbol{E} = -\boldsymbol{E}$, positive and negative charges are (spatially) flipped by $\mathcal{P}$ the field vector thus changes its direction;
   • $\mathcal{P}\boldsymbol{B} = \boldsymbol{B}$, magnetic fields are created by moving charges, the sign of the direction of motion and of the position vector are both flipped (cf. Biot-Savart law: $B \propto q\boldsymbol{r} \times \boldsymbol{v}/|\boldsymbol{r}|^3$).
   • $\mathcal{P}(\boldsymbol{\sigma} \cdot \boldsymbol{E}) = -\boldsymbol{\sigma} \cdot \boldsymbol{E}$
   • $\mathcal{P}(\boldsymbol{\sigma} \cdot \boldsymbol{B}) = \boldsymbol{\sigma} \cdot \boldsymbol{B}$
   • $\mathcal{P}(\boldsymbol{\sigma} \cdot \boldsymbol{p}) = -\boldsymbol{\sigma} \cdot \boldsymbol{p}$
   • $\mathcal{P}(\boldsymbol{\sigma} \cdot (\boldsymbol{p}_1 \times \boldsymbol{p}_2)) = \boldsymbol{\sigma} \cdot (\boldsymbol{p}_1 \times \boldsymbol{p}_2)$

3. a) Since pions have spin 0, the spin of the f_2-meson must be transferred into orbital angular momentum for the pions, i.e., $\ell = 2$. Since $P = (-1)^\ell$, the parity of the f_2-meson is $P = (-1)^2 * P_\pi^2 = +1$. Since the parity and $\mathcal{C}$-parity transformations of the f_2-decay both lead to the

same state (spatial exchange of π^+/π^- and exchange of the π-charge states) we have $C = P = +1$ for the f_2-meson.

b) A decay is only possible if P and C are conserved by it. Since $C|\pi^0\rangle|\pi^0\rangle = +1|\pi^0\rangle|\pi^0\rangle$ and the angular momentum argument of a) remains valid ($\ell = 2 \to P = +1$), the decay $f_2 \to \pi^0\pi^0$ is allowed. For the decay into two photons we have: $C|\gamma\rangle|\gamma\rangle = +1$. The total spin of the two photons must be $2\hbar$ and the z-component $S_z = \pm 2$. Therefore one of the two photons must be left handed and the other right handed. (Sketch the decay in the centre of mass system and draw in the momenta and spins of the photons!) Only a linear combination of $S_z = +2$ and $S_z = -2$ can fulfill the requirement of parity conservation, e.g., the state $(|S_z = +2\rangle + |S_z = -2\rangle)$. Applying the parity operator to this state yields the eigenvalue $+1$. This means that the decay into two photons is also possible.

4. a) The pion decays in the centre of mass frame into a charged lepton with momentum $\boldsymbol{p}$ and a neutrino with momentum $-\boldsymbol{p}$. Energy conservation supplies $m_\pi c^2 = \sqrt{m_\ell^2 c^4 + |\boldsymbol{p}|^2 c^2} + |\boldsymbol{p}|c$. For the charged lepton we have $E_\ell^2 = m_\ell^2 c^4 + |\boldsymbol{p}|^2 c^2$. Taking $v/c = |\boldsymbol{p}|c/E_\ell$ one obtains from the above relations

$$1 - \frac{v}{c} = \frac{2m_\ell^2}{m_\pi^2 + m_\ell^2} = \begin{cases} 0.73 & \text{for } \mu^+ \\ 0.27 \cdot 10^{-4} & \text{for } e^+ . \end{cases}$$

b) The ratio of the squared matrix elements is

$$\frac{|\mathcal{M}_{\pi e}|^2}{|\mathcal{M}_{\pi\mu}|^2} = \frac{1 - v_e/c}{1 - v_\mu/c} = \frac{m_e^2}{m_\mu^2} \frac{m_\pi^2 + m_\mu^2}{m_\pi^2 + m_e^2} = 0.37 \cdot 10^{-4} .$$

c) We need to calculate $\varrho(E_0) = dn/dE_0 = dn/d|\boldsymbol{p}| \cdot d|\boldsymbol{p}|/dE_0 \propto |\boldsymbol{p}|^2 d|\boldsymbol{p}|/dE_0$. From the energy conservation equation (see Part a) we find $d|\boldsymbol{p}|/dE_0 = 1 + v/c = 2m_\pi^2/(m_\pi^2 + m_\ell^2)$ and $|\boldsymbol{p}_\ell| = c(m_\pi^2 - m_\ell^2)/(2m_\pi)$. Putting it together we get

$$\frac{\varrho_e(E_0)}{\varrho_\mu(E_0)} = \frac{(m_\pi^2 - m_e^2)^2}{(m_\pi^2 - m_\mu^2)^2} \frac{(m_\pi^2 + m_e^2)^2}{(m_\pi^2 + m_\mu^2)^2} = 3.49 .$$

Therefore the phase space factor for the decay into the positron is larger.

d) The ratio of the partial decay widths now only depends upon the masses of the particles involved and turns out to be

$$\frac{\Gamma(\pi^+ \to e^+\nu)}{\Gamma(\pi^+ \to \mu^+\nu)} = \frac{m_e^2}{m_\mu^2} \frac{(m_\pi^2 - m_e^2)^2}{(m_\pi^2 - m_\mu^2)^2} = 1.28 \cdot 10^{-4} .$$

This value is in good agreement with the experimental result.

Chapter 11

1. The total width Γ_{tot} of Z^0 may be written as

$$\Gamma_{\text{tot}} = \Gamma_{\text{had}} + 3\Gamma_\ell + N_\nu \Gamma_\nu$$

and $\Gamma_\nu/\Gamma_\ell = 1.99$ (see text). From (11.9) it follows that

$$\sigma_{\text{had}}^{\text{max}} = \frac{12\pi(\hbar c)^2}{M_Z^2} \frac{\Gamma_e \Gamma_{\text{had}}}{\Gamma_{\text{tot}}}.$$

Solving for Γ_{tot} and inserting it into the above formula yields from the experimental results $N_\nu = 2.96$. Varying the experimental results inside the errors only changes the calculated value of N_ν by about ± 0.1.

Chapter 13

1. The reduced mass of positronium is $m_e/2$. From (13.4) we thus find the ground state ($n = 1$) radius to be

$$a_0 = \frac{2\hbar}{\alpha m_e c} = 1.1 \cdot 10^{-10}\,\text{m}.$$

The range of the weak force may be estimated from Heisenberg's uncertainty relation:

$$R \approx \frac{\hbar}{M_W c} = 2.5 \cdot 10^{-3}\,\text{fm}.$$

At this separation the weak and electromagnetic couplings are of the same order of magnitude. The masses of the two particles, whose bound state would have the Bohr radius R, would then be

$$M \approx \frac{2\hbar}{\alpha R c} \approx 2 \cdot 10^4\,\text{GeV}/c^2.$$

This is equivalent to the mass of $4 \cdot 10^7$ electrons or $2 \cdot 10^4$ protons. This vividly shows just how weak the weak force is.

2. From (18.1) the transition probability obeys $1/\tau \propto E_\gamma^3 |\langle r_{fi}\rangle|^2$. If m is the reduced mass of the atomic system, we have $|\langle r_{fi}\rangle| \propto 1/m$ and $E_\gamma \propto m$. $1/\tau = m/m_e \cdot 1/\tau_H$ implies $\tau = \tau_H/940$ for protonium.

3. The transition frequency in positronium $f_{e^+e^-}$ is given by

$$\frac{f_{e^+e^-}}{f_H} = \frac{7}{4} \frac{g_e}{g_p} \frac{m_p}{m_e} \frac{|\psi(0)|^2_{e^+e^-}}{|\psi(0)|^2_H}$$

Using (13.4) one finds $|\psi(0)|^2 \propto m_{\text{red}}^3 = [(m_1 \cdot m_2)/(m_1 + m_2)]^3$. One so obtains $f_{e^+e^-} = 204.5\,\text{GHz}$. One can analogously find $f_{\mu^+e^-} = 4.579\,\text{GHz}$.

The deviations from the measured values (0.5 % and 2.6 % respectively) are due to higher order QED corrections to the level splitting. These are suppressed by a factor of the order $\alpha \approx 0.007$.

4. a) The average decay length is $s = v\tau_{\text{lab}} = c\beta\gamma\tau$ where $\gamma = E_{\text{B}}/m_{\text{B}}c^2 = 0.5\, m_\Upsilon/m_{\text{B}}$ and $\beta\gamma = \sqrt{\gamma^2 - 1}$. One thus obtains $s = 0.028\,\text{mm}$.

   b) From $0.2\,\text{mm} = c\beta\gamma\tau = \tau \cdot |\boldsymbol{p}_{\text{B}}|/m_{\text{B}}$ we obtain $|\boldsymbol{p}_{\text{B}}| = 2.3\,\text{GeV}/c$.

   c) From the assumption, $m_{\text{B}} = 5.29\,\text{GeV}/c^2 = m_\Upsilon/2$, the B-mesons do not have any momentum in the centre of mass frame. In the laboratory frame, $|\boldsymbol{p}_{\text{B}}| = 2.3\,\text{GeV}/c$ and thus $|\boldsymbol{p}_\Upsilon| = 2|\boldsymbol{p}_{\text{B}}|$. We obtain from this $E_\Upsilon = \sqrt{m_\Upsilon^2 c^4 + p_\Upsilon^2 c^4} = 11.6\,\text{GeV}$.

   d) From four-momentum conservation $p_\Upsilon = p_{\text{e}^+} + p_{\text{e}^-}$ we obtain (setting $m_{\text{e}} = 0$) $E_\Upsilon = E_{\text{e}^+} + E_{\text{e}^-}$ and $\boldsymbol{p}_\Upsilon c = E_{\text{e}^+} + E_{\text{e}^-}$ from this we get $E_{\text{e}^+} = 8.12\,\text{GeV}$ and $E_{\text{e}^-} = 3.44\,\text{GeV}$ (or vice-versa).

Chapter 14

1. Angular momentum conservation requires $\ell = 1$, since pions are spin 0. In the ($\ell = 1$) state, the wave function is antisymmetric, but two identical bosons must have a totally symmetric wave function.

2. The branch in the denominator is Cabibbo-suppressed and from (10.19) we thus expect: $R \approx 20$.

3. a) From the decay law $N(t) = N_0\, e^{-t/\tau}$ we obtain the fraction of the decaying particles to be $F = (N_0 - N)/N_0 = 1 - e^{-t/\tau}$. In the laboratory frame we have $t_{\text{lab}} = d/(\beta c)$ and $\tau_{\text{lab}} = \gamma\tau^*$, where τ^* is the usual lifetime in the rest frame of the particle. We thus obtain

$$F = 1 - \exp\left(-\frac{d}{\beta c\gamma\tau^*}\right) = 1 - \exp\left(-\frac{d}{\sqrt{1 - \frac{m^2 c^4}{E^2}}\, c\, \frac{E}{mc^2}\, \tau^*}\right),$$

   and from this we find $F_\pi = 0.9\,\%$ und $F_{\text{K}} = 6.7\,\%$.

   b) From four momentum conservation we obtain, e.g., for pion decay $p_\mu^2 = (p_\pi - p_\nu)^2$ and upon solving for the neutrino energy get

$$E_\nu = \frac{m_\pi^2 c^4 - m_\mu^2 c^4}{2(E_\pi - |\boldsymbol{p}_\pi|c\,\cos\theta)}.$$

   At $\cos\theta = 1$ we have maximal E_ν, while for $\cos\theta = -1$ it is minimal. We can so obtain $E_\nu^{\max} \approx 87.5\,\text{GeV}$ and $E_\nu^{\min} \approx 0\,\text{GeV}$ (more precisely: 11 keV) in pion decay. In the case of kaon decay, we obtain $E_\nu^{\max} \approx 191\,\text{GeV}$ and $E_\nu^{\min} \approx 0\,\text{GeV}$ (more precisely: 291 keV).

Chapter 15

1. b) All of the neutral mesons made out of u- and d-quarks (and similarly the $s\bar{s}$ (ϕ) meson) are very short lived; $c\tau < 100$ nm. The dilatation factor γ that they would need to have in order to traverse a distance of several centimetres in the laboratory frame is simply not available at these beam energies. Since mesons with heavy quarks (c, b) cannot be produced, as not enough energy is available, the only possible mesonic decay candidate is the K_S^0. Similarly the only baryons that come into question are the Λ^0 and the $\overline{\Lambda}^0$. The primary decay modes of these particles are $K_S^0 \rightarrow \pi^+\pi^-$, $\Lambda^0 \rightarrow p\pi^-$ and $\overline{\Lambda}^0 \rightarrow \bar{p}\pi^+$.

   c) We have for the mass M_X of the decayed particle from (15.1)

$$M_X^2 = m_+^2 + m_-^2 + 2\sqrt{p_+^2/c^2 + m_+^2}\sqrt{p_-^2/c^2 + m_-^2}$$
$$- \tfrac{2}{c^2}|\boldsymbol{p}_+||\boldsymbol{p}_-|\cos\sphericalangle\,(\boldsymbol{p}_+,\boldsymbol{p}_-)$$

where the masses and momenta of the decay products are denoted by $m_\pm$ and $\boldsymbol{p}_\pm$ respectively. Consider the first pair of decay products: the hypothesis that we have a $K_S^0 \rightarrow \pi^+\pi^-$ ($m_\pm = m_{\pi\pm}$) decay leads to $M_X = 0.32\,\mathrm{GeV}/c^2$ which is inconsistent with the true K^0 mass ($0.498\,\mathrm{GeV}/c^2$). The hypothesis $\Lambda^0 \rightarrow p\pi^-$ ($m_+ = m_p$, $m_- = m_{\pi^-}$) leads to $M_X = 1.11\,\mathrm{GeV}/c^2$ which is in very good agreement with the mass of the Λ^0. The $\overline{\Lambda}^0$ possibility can, as with the K^0 hypothesis, be confidently excluded. Considering the second pair of decay particles we similarly find: K^0 hypothesis, $M_X = 0.49\,\mathrm{GeV}/c^2$; Λ^0 hypothesis, $M_X = 2.0\,\mathrm{GeV}/c^2$; the $\overline{\Lambda}^0$ hypothesis also leads to a contradiction. In this case we are dealing with the decay of a K^0.

   d) Conservation of strangeness in the strong interaction means that as well as the Λ^0, which is made up of a uds quark combination, a further hadron with an $\bar{s}$-quark must be produced. The observed K_S^0 decay means that this was a K^0 ($\bar{s}d$).[1] Charge and baryon number conservation now combine to imply that the most likely total reaction was

$$p + p \rightarrow K^0 + \Lambda^0 + p + \pi^+.$$

We cannot, however, exclude additional, unobserved neutral particles or very short lived intermediate states (such as a Δ^{++}).

2. Let us consider the positively charged Σ particles $|\Sigma^+\rangle = |u^\uparrow u^\uparrow s^\downarrow\rangle$ and $|\Sigma^{+*}\rangle = |u^\uparrow u^\uparrow s^\uparrow\rangle$. Since the spins of the two u-quarks are parallel, we have

$$\sum_{\substack{i,j=1\\i<j}}^{3} \frac{\boldsymbol{\sigma}_i \cdot \boldsymbol{\sigma}_j}{m_i m_j} = \frac{\boldsymbol{\sigma}_u \cdot \boldsymbol{\sigma}_u}{m_u^2} + 2\frac{\boldsymbol{\sigma}_u \cdot \boldsymbol{\sigma}_s}{m_u m_s}.$$

---

[1] Both the K^0 and the $\overline{K}^0$ can decay as K_S^0 (cf. Sec. 14.4).

We first inspect

$$2\,\boldsymbol{\sigma}_{\mathrm u}\cdot\boldsymbol{\sigma}_{\mathrm s} = \sum_{\substack{i,j=1\\i<j}}^{3}\boldsymbol{\sigma}_i\cdot\boldsymbol{\sigma}_j - \boldsymbol{\sigma}_{\mathrm u}\cdot\boldsymbol{\sigma}_{\mathrm u}\,.$$

We already know the first term on the r.h.s. from (15.10). It is -3 for $S=1/2$ baryons and $+3$ for $S=3/2$ baryons. The second term is $+1$. This yields

$$\Delta M_{\mathrm{ss}} = \begin{cases} \dfrac{4}{9}\dfrac{\hbar^3}{c^3}\pi\alpha_{\mathrm s}\,|\psi(0)|^2\left(\dfrac{1}{m_{\mathrm{u,d}}^2} - \dfrac{4}{m_{\mathrm{u,d}}m_{\mathrm s}}\right) & \text{for the } \Sigma \text{ states}, \\[4mm] \dfrac{4}{9}\dfrac{\hbar^3}{c^3}\pi\alpha_{\mathrm s}\,|\psi(0)|^2\left(\dfrac{1}{m_{\mathrm{u,d}}^2} + \dfrac{2}{m_{\mathrm{u,d}}m_{\mathrm s}}\right) & \text{for the } \Sigma^* \text{ states}. \end{cases}$$

The average mass difference between the Σ and Σ^* baryons is about 200 MeV/c^2. With the mass formula (15.12) we have

$$\begin{aligned} M_{\Sigma^*} - M_\Sigma &= \Delta M_{\mathrm{ss}}(\Sigma^*) - \Delta M_{\mathrm{ss}}(\Sigma) = \tfrac{4}{9}\tfrac{\hbar^3}{c^3}\pi\alpha_{\mathrm s}\,|\psi(0)|^2\,\tfrac{6}{m_{\mathrm{u,d}}m_{\mathrm s}} \\ &\approx 200\,\mathrm{MeV}/c^2\,, \end{aligned}$$

where we assume that $\psi(0)$ is the same for both states. We thus obtain ($m_{\mathrm{u,d}} = 363\,\mathrm{MeV}/c^2$, $m_{\mathrm s} = 538\,\mathrm{MeV}/c^2$)

$$\alpha_{\mathrm s}\,|\psi(0)|^2 = 0.61\,\mathrm{fm}^{-3}\,.$$

Inserting a hydrogen atom-type wave function, $|\psi(0)|^2 = 3/4\pi r^3$, and $\alpha_{\mathrm s} \approx 1$, yields a rough approximation for the average separation r of the quarks in such baryons: $r \approx 0.8\,\mathrm{fm}$.

3. The Λ is an isospin singlet ($I=0$). To a first approximation the decay is just the quark transition $\mathrm s \to \mathrm u$, which changes the isospin by $1/2$. Thus the pion-nucleon system must be a $I=1/2$ state. Charge conservation implies that the third component is $I_3^{\mathrm N} + I_3^\pi = -1/2$. The matrix elements of the decay of the Λ^0 are proportional to the squares of the Clebsch-Gordan coefficients:

$$\frac{\sigma(\Lambda^0 \to \pi^- + \mathrm p)}{\sigma(\Lambda^0 \to \pi^0 + \mathrm n)} = \frac{(1\,\tfrac{1}{2}\,-1\,+\tfrac{1}{2}\,|\,\tfrac{1}{2}\,-\tfrac{1}{2})^2}{(1\,\tfrac{1}{2}\,0\,-\tfrac{1}{2}\,|\,\tfrac{1}{2}\,-\tfrac{1}{2})^2} = \frac{(-\sqrt{2/3})^2}{(\sqrt{1/3})^2} = 2.$$

4. The probability that a muon be captured from a 1s state into a ^{12}C nucleus is

$$\frac{1}{\tau_{\mu\mathrm C}} = \frac{2\pi}{\hbar}\left|\langle ^{12}\mathrm B\,\mathrm e^{\mathrm i p_\nu r}\,|\sum_i g_{\mathrm A}\sigma_i I_-\,|^{12}\mathrm C\psi_\mu(r)\rangle_{(r=0)}\right|^2 \int \frac{p_\nu^2\,\mathrm dp_\nu\,\mathrm d\Omega}{(2\pi\hbar)^3\,\mathrm dE_\nu}\,.$$

Since carbon has $J^P = 0^+$ and boron $J^P = 1^+$, this is a purely axial vector transition. We further have $dp_\nu/dE_\nu = 1/c$, $\int d\Omega = 4\pi$ and $|\psi_\mu(r=0)|^2 = 3/(4\pi r_\mu^3)$. The radius of the ^{12}C muonic atomic is found to be

$$r_\mu = \frac{a_{\text{Bohr}} m_e}{Z m_\mu} = 42.3\,\text{fm},$$

and the energy is

$$E_\nu = m_\mu c^2 - 13.3\,\text{MeV} \approx 90\,\text{MeV}.$$

This yields the absorption probability

$$\frac{1}{\tau_{\mu C}} = \frac{2\pi}{\hbar c} \frac{4\pi c E_\nu^2}{(2\pi)^3(\hbar c)^3} \left| \langle {}^{12}\text{B}| \sum_i g_A \sigma_i I_- |{}^{12}\text{C}\rangle \right|^2 |\psi(0)|^2.$$

These are all known quantities except for the matrix element. This may be extracted from the known lifetime of the $^{12}\text{B} \to {}^{12}\text{C} + e^- + \bar\nu_e$ decay:

$$\frac{1}{\tau_{^{12}\text{B}}} = \frac{1}{2\pi^3 \hbar^7 c^6} \left| \langle {}^{12}\text{C}| \sum_i g_A \sigma_i I_+ |{}^{12}\text{B}\rangle \right|^2 E_{\text{max}}^5.$$

We thus finally obtain

$$\frac{1}{\tau_{\mu C}} \approx 1.5 \cdot 10^4\,\text{s}^{-1}.$$

The total decay probability of the muon decay in ^{12}C is the sum of the probabilities of the free muon decaying and of its being captured by the nucleus:

$$\frac{1}{\tau} = \frac{1}{\tau_\mu} + \frac{1}{\tau_{\mu C}}.$$

5. These branching ratios depend primarily upon two things: a) the phase space and b) the fact that the strangeness changes in the first case (Cabibbo suppression) but not in the latter. A rough estimate may be obtained by assuming that the matrix elements are, apart from Cabibbo suppression, identical. From (10.19) and (15.49) one finds

$$\frac{W(\Sigma^- \to n)}{W(\Sigma^- \to \Lambda^0)} \approx \frac{\sin^2\theta_C}{\cos^2\theta_C} \cdot \left(\frac{E_1}{E_2}\right)^5 = \frac{1}{20}\left(\frac{257\,\text{MeV}}{81\,\text{MeV}}\right)^5 \approx 16.$$

This agreement is not bad at all, considering the coarseness of our approximation. In the decay $\Sigma^+ \to n + e^+ + \nu_e$ we would need two quarks to change their flavours; (suu) → (ddu).

6. a) Baryon number conservation means that baryons can neither be annihilated nor created but rather only transformed into each other. Hence only the relative parities of the baryons have any physical meaning.

   b) The deuteron is a ground state p-n system, i.e., $\ell = 0$. Its parity is therefore $\eta_{\mathrm{d}} = \eta_{\mathrm{p}}\eta_{\mathrm{n}}(-1)^0 = +1$. Since quarks have zero orbital angular momentum in nucleons, the quark intrinsic parities must be positive.

   c) The downwards cascade of pions into the ground state may be seen from the characteristic X rays.

   d) Since the deuteron has spin 1, the d-π system is in a state with total angular momentum $J = 1$. The two final state neutrons are identical fermions and so must have an antisymmetric spin-orbit wave function. Only $^3P_1$ of the four possible states with $J = 1$, $^3S_1$, $^1P_1$, $^3P_1$ and $^3D_1$ fulfills this requirement.

   e) From $\ell_{nn} = 1$, we see that the pion parity must be $\eta_\pi = \eta_{\mathrm{n}}^2(-1)^1/\eta_{\mathrm{d}} = -1$.

   f) The number of quarks of each individual flavour $(N_{\mathrm{q}} - N_{\bar{\mathrm{q}}})$ is separately conserved in parity conserving interactions. The quark parities can therefore be separately chosen. One could thus choose, e.g., $\eta_{\mathrm{u}} = -1, \eta_{\mathrm{d}} = +1$, giving the proton a positive and the neutron a negative parity. The deuteron would then have a negative parity and the charged pions a positive one. The π^0 as a $u\bar{u}/d\bar{d}$ mixed state would though keep its negative parity. Particles like (π^+, π^0, π^-) or (p, n) although inside the same isospin multiplets would then have distinct parities – a rather unhelpful convention. For $\eta_{\mathrm{n}} = \eta_{\mathrm{p}} = -1$, on the other hand, isospin symmetry would be fulfilled. The parities of nucleons and odd nuclei would then be the opposite of the standard convention, while those of mesons and even nuclei would be unchanged. The Λ and Λ_{c} parities are just those of the s- and c-quarks and may be chosen to be positive.

Chapter 16

1. The ranges, $\lambda \approx \hbar c/mc^2$, are: 1.4 fm (1π), 0.7 fm (2π), 0.3 fm (ϱ, ω). Two pion exchange with vacuum quantum numbers, $J^P = 0^+, I = 0$, generates a scalar potential which is responsible for nuclear binding. Because of its negative parity, the pion is emitted with an angular momentum, $\ell = 1$. The spin dependence of this component of the nuclear force is determined by this. Similar properties hold for the ϱ and ω. The isospin dependence is determined by the isospin of the exchange particle; $I = 1$ for the π and ϱ and $I = 0$ for the ω. Since isospin is conserved in the strong interaction, the isospin of interacting particles is coupled, just as is the case with angular momentum.

2. Taking (16.1), (16.2) and (16.6) into account we obtain

$$\sigma = 4\pi \left(\frac{\sin kb}{k} \right)^2 .$$

At low energies, where the $\ell = 0$ partial wave dominates, we obtain in the $k \to 0$ limit, the total cross-section, $\sigma = 4\pi b^2$.

Chapter 17

1. At constant entropy S the pressure obeys

$$p = - \left(\frac{\partial U}{\partial V} \right)_S ,$$

where V is the volume and U is the internal energy of the system. In the Fermi gas model we have from (17.9):

$$U = \frac{3}{5} A E_{\mathrm{F}} \qquad \text{and hence} \qquad p = -\frac{3}{5} A \frac{\partial E_{\mathrm{F}}}{\partial V} .$$

From (17.3) we find for $N = Z = A/2$:

$$A = 2 \frac{V p_{\mathrm{F}}^3}{3\pi^2 \hbar^3} = 2 \frac{V (2 M E_{\mathrm{F}})^{3/2}}{3\pi^2 \hbar^3} \qquad \Longrightarrow \qquad \frac{\partial E_{\mathrm{F}}}{\partial V} = -\frac{2 E_{\mathrm{F}}}{3V} .$$

The Fermi pressure is then

$$p = \frac{2A}{5V} E_{\mathrm{F}} = \frac{2}{5} \varrho_{\mathrm{N}} E_{\mathrm{F}} ,$$

where ϱ_{N} is the nucleon density. This implies for $\varrho_{\mathrm{N}} = 0.17$ nucleons/fm^3 and $E_{\mathrm{F}} \approx 33$ MeV

$$p = 2.2 \, \mathrm{MeV/fm}^3 = 3.6 \cdot 10^{27} \, \mathrm{bar}.$$

2. a) We only consider the odd nucleons. The even ones are all paired off in the ground state. The first excited state is produced either by (I) the excitation of the unpaired nucleon into the next subshell or (II) by the pairing of this nucleon with another which is excited from a lower lying subshell.

	$^{7}_{3}$Li	$^{23}_{11}$Na	$^{33}_{16}$S	$^{41}_{21}$Sc	$^{83}_{36}$Kr	$^{93}_{41}$Nb
Ground state	$1\mathrm{p}^3_{3/2}$	$1\mathrm{d}^3_{5/2}$	$1\mathrm{d}^1_{3/2}$	$1\mathrm{f}^1_{7/2}$	$1\mathrm{g}^{-3}_{9/2}$	$1\mathrm{g}^1_{9/2}$
Excited (I)	$1\mathrm{p}^1_{1/2}$	$2\mathrm{s}^1_{1/2}$	$(1\mathrm{f}^1_{7/2})$	$(2\mathrm{p}^1_{3/2})$	$(1\mathrm{g}^1_{7/2})$	$(1\mathrm{g}^1_{7/2})$
Excited (II)	$(1\mathrm{s}^{-1}_{1/2})$	$1\mathrm{p}^{-1}_{1/2}$	$2\mathrm{s}^{-1}_{1/2}$	$1\mathrm{d}^{-1}_{3/2}$	$2\mathrm{p}^{-1}_{1/2}$	$2\mathrm{p}^{-1}_{1/2}$
J_0^P experiment	$3/2^-$	$3/2^+$	$3/2^+$	$7/2^-$	$9/2^+$	$9/2^+$
J_0^P model	$3/2^-$	$5/2^+$	$3/2^+$	$7/2^-$	$9/2^+$	$9/2^+$
J_1^P experiment	$1/2^-$	$5/2^+$	$1/2^+$	$3/2^+$	$7/2^+$	$1/2^-$
J_1^P case (I)	$1/2^-$	$1/2^+$	$(7/2^-)$	$(3/2^-)$	$(7/2^+)$	$(7/2^+)$
J_1^P case (II)	$(1/2^+)$	$1/2^-$	$1/2^+$	$3/2^+$	$1/2^-$	$1/2^-$

Those states whose excitation would be beyond a "magic" boundary are shown here in brackets. This requires a lot of energy and so is only to be expected for higher excitations. As one sees, the predictive powers of the shell model are good for those nuclei where the unfilled subshell is only occupied by a single nucleon.

b) The $(\text{p-}1\text{p}_{3/2}^1; \text{n-}1\text{p}_{3/2}^1)$ in $_3^6\text{Li}$ implies $J^P = 0^+, 1^+, 2^+, 3^+$. $_{19}^{40}\text{K}$ has from $(\text{p-}1\text{d}_{3/2}^{-1}; \text{n-}1\text{f}_{7/2}^1)$ a possible coupling to $2^-, 3^-, 4^-, 5^-$.

3. a) An ^{17}O nucleus may be viewed as being an ^{16}O nucleus with an additional neutron in the $1\text{f}_{5/2}$ shell. The energy of this level is thus $B(^{16}\text{O}) - B(^{17}\text{O})$. The $1\text{p}_{1/2}$ shell is correspondingly at $B(^{15}\text{O}) - B(^{16}\text{O})$. The gap between the shells is thus

$$E(1\text{f}_{5/2}) - E(1\text{p}_{1/2}) = 2B(^{16}\text{O}) - B(^{15}\text{O}) - B(^{17}\text{O}) = 11.5\,\text{MeV}.$$

b) One would expect the lowest excitation level with the "right" quantum numbers to be produced by exciting a nucleon from the topmost, occupied shell into the one above. For ^{16}O this would be the $J^P = 3^-$ state, which is at 6.13 MeV, and could be interpreted as $(1\text{p}_{1/2}^{-1}, 1\text{d}_{5/2})$. The excitation energy is, however, significantly smaller than the theoretical result of 11.5 MeV. It seems that collective effects (state mixing) are making themselves felt. This is confirmed by the octupole radiation transition probability, which is an order of magnitude above what one would expect for a single particle excitation.

c) The $1/2^+$ quantum numbers make it natural to interpret the first excited state of ^{17}O as $2\text{s}_{1/2}$. The excitation energy is then the gap between the shells.

d) Assuming (more than a little naively) that the nuclei are homogenous spheres with identical radii, one finds from (2.11) that the difference in the binding energies implies the radius is $(3/5) \cdot 16\hbar\alpha c/3.54\,\text{MeV} = 3.90\,\text{fm}$, which is much larger than the value of 3.1 fm, which follows from (5.56). In the shell model one may interpret each of these nuclei as an ^{16}O nucleus with an additional nucleon. The valence nucleon in the $\text{d}_{5/2}$ shell thus has a larger radius than one would expect from the above simple formula which does not take shell effects into account.

e) The larger Coulomb repulsion means that the potential well felt by the protons in ^{17}F is shallower than that of the neutrons in ^{17}O. As a result the wave function of the excited, "additional" proton in ^{17}F is more spread out than that of the equivalent "additional" neutron in ^{17}O and the nuclear force felt by the neutron is stronger than that acting upon the proton. This difference is negligible for the ground state since the nucleon is more strongly bound.

4. At the upper edge of the closed shells which correspond to the magic numbers 50 and 82 we find the closely adjacent $2\text{p}_{1/2}$, $1\text{g}_{9/2}$ and the $2\text{d}_{3/2}$, $1\text{h}_{11/2}$, $3\text{s}_{1/2}$ levels respectively. It is thus natural that for nuclei with

nucleon numbers just below 50 or 82 the transition between the ground state and the first excited state is a single particle transition ($g_{9/2} \leftrightarrow p_{1/2}$ andr $h_{11/2} \leftrightarrow d_{3/2}, s_{1/2}$ respectively). Such processes are 5th order (M4 or E5) and hence extremely unlikely [Go51].

5. a) The spin of the state is given by the combination of the unpaired nucleons which are in the (p-$1f_{7/2}$, n-$1f_{7/2}$) state.

   b) The nuclear magnetic moment is just the sum of the magnetic moments of the neutron in the $f_{7/2}$ shell $-1.91\,\mu_N$ and of the proton in the $f_{7/2}$ shell $+5.58\,\mu_N$. From (17.36) we would expect a g factor of 1.1.

6. a) In the de-excitation $i \to f$ of an Sm nucleus *at rest* the atom receives a recoil energy of $p_{Sm}^2/2M$ where $|p_{Sm}| = |p_\gamma| \approx (E_i - E_f)/c$. In the case at hand this is 3.3 eV. The same amount of energy is lost when the photon is absorbed by another Sm nucleus.

   b) If we set the matrix element in (18.1) to one, this implies a lifetime of $\tau = 0.008$ ps, which is equivalent to $\Gamma = 80$ meV. In actual measurements one finds $\tau = 0.03$ ps, i.e., $\Gamma = 20$ meV [Le78], which is of a similar size. Since the width of the state is much smaller than the energy shift of $2 \cdot 3.3$ eV, no absorption can take place. Thermal motion will change $|p_{Sm}|$ by roughly $\pm\sqrt{M \cdot kT}$. At room temperature this corresponds to smearing the energy by ± 0.35 eV, which is also insufficient.

   c) If the Sm atom emits a neutrino before the deexcitation, then $|p_{Sm}|$ is changed by $\pm|p_\nu| = \pm E_\nu/c$. If the emission directions of the neutrino and of the photon are opposite to each other, then the energy of the radiated photon is 3.12 eV larger than the excitation energy $E_i - E_f$. This corresponds to the classical Doppler effect. In this case resonant fluorescence is possible for the γ radiation. The momentum direction of the neutrino can be determined in this fashion (for details, see [Bo72]).

7. The three lowest proton shells in the ^{14}O nucleus, the $1s_{1/2}$, $1p_{3/2}$ and $1p_{1/2}$, are fully occupied as are the two lowest neutron shells. The $1p_{1/2}$ shell is, however, empty (sketched on p. 269). Thus one of the two valence nucleons (one of the protons in their $1p_{1/2}$ shell) can transform into a neutron at the equivalent level and with the same wave function (super allowed β-decay). We thus have $\int \psi_n^* \psi_p = 1$. This is a $0^+ \to 0^+$ transition, i.e., a pure Fermi decay. Each of the two protons contributes a term to the matrix element equal to the vector part of (15.39). The total is therefore $|\mathcal{M}_{fi}|^2 = 2g_V^2/V^2$. Equation (15.47) now becomes

$$\frac{\ln 2}{t_{1/2}} = \frac{1}{\tau} = \frac{m_e^5 c^4}{2\pi^3 \hbar^7} \cdot 2g_V^2 \cdot f(E_0)\,.$$

Using the vectorial coupling (15.56) one finds the half-life is 70.373 s – which is remarkably close to the experimental value. Note: the quantum numbers and definite shell structure here means that this is one of the few cases where a nuclear β-decay can be calculated exactly. In practice this decay is used to determine the strength of the vectorial coupling.

Chapter 18

1. a) In the collective model of giant resonances we consider Z protons and N neutrons whose mutual vibrations are described by a harmonic oscillator. The Hamiltonian may be written as

$$\mathcal{H} = \frac{p^2}{2m} + \frac{m\omega^2}{2}x^2 \qquad \text{where} \qquad \hbar\omega = 80\,A^{-1/3}\,\text{MeV},$$

and $m = A/2M_N$ is the reduced mass. The solution of the Schrödinger equation yields the lowest lying oscillator states [Sc95]

$$\psi_0 = \frac{1}{\sqrt[4]{\pi}\sqrt{x_0}} \cdot e^{-(x/x_0)^2/2} \qquad \text{where} \quad x_0 = \sqrt{\hbar/m\omega},$$

$$\psi_1 = \frac{1}{\sqrt[4]{\pi}\sqrt{x_0}} \cdot \sqrt{2}\left(\frac{x}{x_0}\right) e^{-(x/x_0)^2/2}.$$

The average deviation is

$$x_{01} := \langle\psi_0|x|\psi_1\rangle = \frac{\sqrt{2}}{\sqrt{\pi}}x_0 \int \left(\frac{x}{x_0}\right)^2 e^{-(x/x_0)^2}\,\mathrm{d}\frac{x}{x_0} = \frac{\sqrt{2}}{\sqrt{\pi}}x_0\,.$$

For ^{40}Ca we have $x_0 = 0.3$ fm and $x_{01} = 0.24$ fm.

b) The matrix element is Zx_{01}. Its square is therefore 23 fm^2.

c) The single particle excitations have about half the energy of the giant resonance, i.e., $\hbar\omega \approx 40A^{-1/3}$ MeV. The reduced mass in this case is approximately the nucleon mass, since the nucleon moves in the mean field of the heavy nucleus. This increases x_0, and thus x_{01}, by a factor of $\sqrt{40}$. The 24 nucleons in the outermost shell each contribute to the square of the matrix element with an effective charge $e/2$. The square of the matrix element is so seen to be 27.6 fm^2. The agreement with the result of b), i.e., the model where the protons and neutrons oscillate collectively, is very good.

2. See Sec. 17.4 (17.38ff) and (18.49f).
   From

$$\delta = \frac{a-b}{\langle R \rangle}, \qquad \langle R \rangle = ab^{2\,1/3}.$$

and the nucleon density, which in the nucleus is roughly $\varrho_N \approx 0.17$ nucleons/fm^3, it follows that $a \approx 8$ fm, $b \approx 6$ fm.

3. The Fermi velocity is $v_F = p_F/\sqrt{M_N^2 + p_F^2/c^2} = 0.26\,c$. The angular velocity is

$$\omega = \frac{|\mathbf{L}|}{\Theta} \approx \frac{60\hbar}{AM_N(a^2 + b^2)^2/5} = 0.95 \cdot 10^{21}\,\mathrm{s}^{-1}\,,$$

where $a = 2b = \sqrt[3]{4}R$, and we have employed the value of R from (5.56). The speed is $v = a \cdot \omega$ and is about $0.03\,c$ or around $12\,\%$ of the Fermi velocity. The high rotational velocity causes a Coriolis force which is responsible for breaking up the nucleon pairs.

Chapter 19

1. a) In the reaction $4p \to \alpha + e^+ + 2\nu_e$, $26.72\,\mathrm{MeV}$ of energy is released. The neutrinos carry off $0.52\,\mathrm{MeV}$, and so $26.2\,\mathrm{MeV}$ remains to heat up the sun. The number of hydrogen atoms which are converted into helium every second is:

$$\dot{N}_p = 4 \cdot \frac{4 \cdot 10^{26}\,\mathrm{W}}{26.2 \times 1.6 \cdot 10^{-13}\,\mathrm{Ws}} \approx 0.4 \cdot 10^{39}\,\mathrm{atoms/s}\,.$$

   b) $0.4 \cdot 10^{10}\,\mathrm{kg/s}$
   c) $\approx 7\%$
   d) ≈ 130 terrestrial masses
   e) Nuclear reactions take place in the interior of the sun, primarily at radii $r < R_\odot/4$. By burning off $7\,\%$ of the hydrogen the helium concentration in the interior of the sun is increased by about $50\,\%$. Doubling this concentration means that hydrogen burning is no longer efficient: helium burning starts up and the sun swells into a red giant.

2. a) Two neutrinos are produced for every ^{4}He nucleus created

$$\Phi_\nu = \frac{\dot{N}_\nu}{4\pi a^2} = \frac{2 \cdot \dot{E}_\odot}{B_{\mathrm{He}} \cdot 4\pi a^2} = 5.9 \cdot 10^{10}\,\mathrm{cm}^{-2}\mathrm{s}^{-1}\,.$$

   b) The number of ^{71}Ga nuclei is found to be

$$N_{^{71}\mathrm{Ga}} = \frac{\text{total mass of gallium}}{\text{average mass per atom}} \cdot \text{proportion of } ^{71}\mathrm{Ga}$$

$$= \frac{3 \cdot 10^4\,\mathrm{kg}}{(0.40 \cdot 71 + 0.60 \cdot 69) \cdot 931.5 \cdot 1.6 \cdot 10^{-13}\mathrm{J}/c^2} \cdot 0.40$$

$$= 1.0 \cdot 10^{29}\,,$$

and from this we can find the reaction rate:

$$\dot{N}_{\mathrm{Reaktion}} = N_{^{71}\mathrm{Ga}} \cdot \sigma_{\nu\mathrm{Ge}} \cdot \Phi_\nu \cdot \varepsilon$$

$$= 1.0 \cdot 10^{29} \cdot 2.5 \cdot 10^{-45}\mathrm{cm}^2 \cdot 5.9 \cdot 10^{10}\mathrm{cm}^{-2}\mathrm{s}^{-1} \cdot 0.5$$

$$= 0.7/\mathrm{day}\,.$$

Since $N(t) = \dot{N}_{\text{reaction}} \tau (1 - e^{-t/\tau})$ we expect 8 Ge atoms after 3 weeks and after a very long time we would expect to have 11 atoms.

Note: The cross-section depends strongly upon the energy. The value quoted is an average weighted according to the energy spectrum of solar neutrinos.

3. a) The number of neutrons in the neutron star is $N_n = 1.8 \cdot 10^{57}$. The energy released by fusing N_n protons into ^{56}Fe is $2.6 \cdot 10^{45}$ J.

   b) We neglect the gravitational energy of the iron core in the original star, (since $R \gg 10\,\text{km}$). Thus the energy released during the implosion is the gravitational energy of the neutron star minus the energy needed to transform the iron into free neutrons (this last is the energy which was originally released during the fusion of hydrogen into iron):

$$E_{\text{Implosion}} \approx \frac{3GM^2}{5R} - 2.6 \cdot 10^{45}\,\text{J} = 3.3 \cdot 10^{46}\,\text{J}.$$

The energy released via the implosion during the supernova explosion is more than ten times larger than the fusion energy. Although only about $20 \ldots 50\,\%$ of the matter of the original star ends up in the neutron star, the fusion energy released during the entire lifetime of the star is slightly less than the energy released in the supernova explosion.

   c) Most of the energy is radiated off as neutrino emission:

$$e^+ + e^- \to \bar{\nu}_e + \nu_e, \bar{\nu}_\mu + \nu_\mu, \bar{\nu}_\tau + \nu_\tau.$$

The positrons in this process are generated in the reaction:

$$p + \bar{\nu}_e \to e^+ + n.$$

Neutrinos can, however, also be directly produced in:

$$p + e^- \to n + \nu_e.$$

The last two processes are responsible for the transformation of the protons in ^{56}Fe.

References

[Ab95a]   S. Abachi et al.: Phys. Rev. Lett. **74** (1995) 2632
[Ab95b]   F. Abe et al.: Phys. Rev. Lett. **74** (1995) 2626
          First evidence for the top quark: F. Abe et al.: Phys. Rev. Lett. **73** (1994)
          225
[Ab97]   H. Abele et al.: Phys. Lett. **B407** (1997) 212
[Al60]   M. Alston et al.: Phys. Rev. Lett. **5** (1960) 520
[Al77]   G. Altarelli, G. Parisi: Nucl. Phys. **B126** (1977) 298
[Al87a]   C. Albajar et al.: Phys. Lett. **B186** (1987) 247
[Al87b]   H. Albrecht et al.: Phys. Lett. **B192** (1987) 245
[Al90]   H. Albrecht et al.: Phys. Lett. **B234** (1990) 409
[Al91]   H. Albrecht et al.: Phys. Lett. **B255** (1991) 297
[Al92a]   H. Albrecht et al.: Z. Phys **C55** (1992) 357
[Al92b]   J. Alitti et al.: Phys. Lett. **B276** (1992) 354
[Am84]   S. R. Amendolia et al.: Phys. Lett. **B146** (1984) 116
[Am86]   S. R. Amendolia et al.: Phys. Lett. **B178** (1986) 435
[Am87]   U. Amaldi et al.: Phys. Rev. **D36** (1987) 1385
[Am92a]   P. Amaudruz et al.: Phys. Lett. **B295** (1992) 159
[Am92b]   P. Amaudruz et al.: Nucl. Phys. **B371** (1992) 3
[AM93]   *The 1993 Atomic Mass Evaluation*
          G. Andi, A. H. Wapstra: Nucl. Phys. **A565** (1993) 1
[Am95]   P. Amaudruz et al.: Z. Phys. **C51** (1991) 387;
          korrigierte Daten: P. Amaudruz et al.: Nucl. Phys. **B441** (1995) 3
[An87]   R. Ansari et al.: Phys. Lett. **B186** (1987) 440
[An92]   P. Anselmann et al.: Phys. Lett. **B285** (1992) 376;
          Phys. Lett. **B314** (1993) 445; Phys. Lett. **B327** (1994) 377
[Ar54]   A. Arima, H. Horie: Progr. Theor. Phys. **11** (1954) 509;
          A. Arima, H. Horie: Progr. Theor. Phys. **12** (1954) 623
[Ar83]   G. Arnison et al.: Phys. Lett. **B122** (1983) 103;
          Phys. Lett. **B126** (1983) 398; Phys. Lett. **B166** (1986) 484
[Ar88]   M. Arneodo et al.: Phys. Lett. **B211** (1988) 493
[Ar93]   M. Arneodo et al.: Phys. Lett. **B309** (1993) 222
[Ar94]   M. Arneodo: Phys. Rep. **240** (1994) 301
[At82]   W. B. Atwood: *Lectures on Lepton Nucleon Scattering and Quantum
          Chromodynamics,* Progress in Physics Vol. 4 (Birkhäuser, Boston, Basel,
          Stuttgart 1982)
[Au74a]   J. J. Aubert et al.: Phys. Rev. Lett. **33** (1974) 1404
[Au74b]   J.-E. Augustin et al.: Phys. Rev. Lett. **33** (1974) 1406
[Ba57]   J. Bardeen, L. N. Cooper, J. R. Schrieffer: Phys. Rev. **108** (1957) 1157
[Ba68]   W. Bartel et al.: Phys. Lett. **B28** (1968) 148
[Ba78]   W. Bacino et al.: Phys. Rev. Lett. **41** (1978) 13
[Ba80]   J. D. Barrow, J. Silk: Scientific American **242** (April 1980) 98

[Ba83a]   M. Banner et al.: Phys. Lett. **B122** (1983) 476
[Ba83b]   P. Bagnaia et al.: Phys. Lett. **B129** (1983) 130
[Ba85]   J. W. Bartel et al.: Phys. Lett. **B161** (1985) 188
[Ba88]   B. C. Barish, R. Stroynowski: Phys. Rep. **157** (1988) 1
[Ba89]   J. N. Bahcall: *Neutrino Astrophysics* (Cambridge University Press, Cambridge 1989)
[Be30]   H. A. Bethe: Ann. Physik **5** (1930) 325
[Be36]   H. A. Bethe: Rev. Mod. Phys. **8** (1936) 139
[Be67]   J. B. Bellicard et al.: Phys. Rev. Lett. **19** (1967) 527
[Be69]   H. Behrens, J. Janecke: *Numerical Tables for Beta Decay and Electron Capture,* Landolt-Bernstein, new Series, vol. I/4, (Springer, Berlin 1969)
[Be75]   B. L. Berman, S. C. Fultz: Rev. Mod. Phys. **47** (1975) 713
[Be77]   V. Bechtold et al.: Phys. Lett. **B72** (1977) 169
[Be85]   H. A. Bethe, G. Brown: Scientific American **252** (Mai 1985) 40
[Be87]   H.-J. Behrend et al.: Phys. Lett. **B191** (1987) 209
[Be90a]   R. Becker-Szendy et al.: Phys. Rev. **D42** (1990) 2974
[Be90b]   A. C. Benvenuti et al.: Phys. Lett. **B237** (1990) 592
[Bi92]   I. G. Bird: Dissertation, Vrije Universiteit te Amsterdam (1992)
[Bl33]   F. Bloch: Ann. Physik **16** (1933) 285
[Bl69]   E. D. Bloom et al.: Phys. Rev. Lett. **23** (1969) 930
[Bo53]   A. Bohr, B. R. Mottelson: Mat. Fys. Medd. Dan. Vid. Selsk. **27** Nr. 16 (1953)
[Bo69]   A. Bohr, B. R. Mottelson: *Nuclear Structure* (Benjamin, New York 1969)
[Bo72]   E. Bodenstedt: *Experimente der Kernphysik und ihre Deutung,* Teil 1 (Bibliographisches Institut, Mannheim 1972)
[Bo75]   F. Borkowski et al.: Nucl. Phys. **B93** (1975) 461
[Bo92]   F. Boehm, P. Vogel: *Physics of Massive Neutrinos,* 2nd ed. (Cambridge University Press, Cambridge 1992)
[Br64]   C. Brunnée, H. Voshage: *Massenspektroskopie* (Karl-Thiemig, Munich 1964)
[Br65]   D. M. Brink: *Nuclear Forces* (Pergamon Press, Oxford 1965)
[Br67]   G. E. Brown: *Unified Theory of Nuclear Models and Forces* (North Holland, Amsterdam 1967)
[Br69]   M. Breidenbach et al.: Phys. Rev. Lett. **23** (1969) 935
[Br92]   D. I. Britton et al.: Phys. Rev. Lett. **68** (1992) 3000
[Bu57]   E. M. Burbidge et al.: Rev. Mod. Phys. **29** (1957) 547
[Bu85]   H. Burkard et al.: Phys. Lett. **B160** (1985) 343
[Bu88]   C. Budtz-Jørgensen, H.-H. Knitter: Nucl. Phys. **A490** (1988) 307
[Bu91]   H. Burkhardt, J. Steinberger: Ann. Rev. Nucl. Part. Sci. **41** (1991) 55
[Ca35]   H. Casimir: Physica **2** (1935) 719
[Ca63]   N. Cabibbo: Phys. Rev. Lett. **10** (1963) 531
[Ca69]   C. G. Callan Jr., D. J. Gross: Phys. Rev. Lett. **22** (1969) 156
[Ch64]   J. H. Christenson, J. W. Cronin, V. L. Fitch, R. Turlay: Phys. Rev. Lett. **13** (1964) 138
[Ch73]   P. R. Christensen et al.: Nucl. Phys. **A207** (1973) 33
[Ch89]   R. E. Chrien, C. B. Dover: Ann. Rev. Nucl. Part. Sci. **39** (1989) 113
[Cl79]   F. Close: *An Introduction to Quarks and Partons* (Academic Press, London, New York, San Francisco 1979)
[Co56a]   C. L. Cowan Jr, F. Reines et al.: Science **124** (1956) 103
[Co56b]   L. N. Cooper: Phys. Rev. **104** (1956) 1186
[Co57]   C. W. Cook et al.: Phys. Rev. **107** (1957) 508
[Co73]   E. D. Commins: *Weak Interactions* (McGraw-Hill, New York 1973)
[Co74]   M. D. Cooper, W. F. Hornyak, P. G. Roos: Nucl. Phys. **A218** (1974) 249

[Co87]   E. R. Cohen, B. N. Taylor: Rev. Mod. Phys. **59** (1987) 1121
[Co88]   G. Costa et al.: Nucl. Phys. **297** (1988) 244
[Da62]   G. Danby et al.: Phys. Rev. Lett. **9** (1962) 36
[Du91]   D. Dubbers: Progr. in Part. and Nucl. Phys., Hrsg. A. Faessler (Pergamon Press) **26** (1991) 173
[Dy73]   A. F. Dylla, J. G. King: Phys. Rev. **A7** (1973) 1224
[El61]   R. P. Ely et al.: Phys. Rev. Lett. **7** (1961) 461
[El82]   J. Ellis et al.: Ann. Rev. Nucl. Part. Sci. **32** (1982) 443
[EL92]   *Energy Levels of Light Nuclei*
          $A =$ 3: D. R. Tilley, H. R. Weller, H. H. Hasan: Nucl. Phys. **A474** (1987) 1;
          $A =$ 4: D. R. Tilley, H. R. Weller, G. M. Hale: Nucl. Phys. **A541** (1992) 1;
          $A =$ 5 − 10: F. Ajzenberg-Selove: Nucl. Phys. **A490** (1988) 1;
          $A = 11 − 12$: F. Ajzenberg-Selove: Nucl. Phys. **A506** (1990) 1;
          $A = 13 − 15$: F. Ajzenberg-Selove: Nucl. Phys. **A523** (1991) 1;
          $A = 16 − 17$: F. Ajzenberg-Selove: Nucl. Phys. **A460** (1986) 1;
          $A = 18 − 20$: F. Ajzenberg-Selove: Nucl. Phys. **A475** (1987) 1;
          $A = 21 − 44$: P. M. Endt: Nucl. Phys. **A521** (1990) 1
[En64]   F. Englert, R. Brout: Phys. Rev. Lett. **13** (1964) 321
[Er66]   T. Ericson, T. Mayer-Kuckuk: Ann. Rev. Nucl. Sci. **16** (1966) 183
[Fa82]   A. Faessler et al.: Nucl. Phys. **A402** (1982) 555
[Fa88]   A. Faessler in: *Progress in Particle and Nuclear Physics,* ed. A. Faessler, **20** (1988) 151
[Fa90]   S. Fanchiotti, A. Sirlin: Phys. Rev. **D41** (1990) 319
[Fe34]   E. Fermi: Z. Phys. **88** (1934) 161
[Fe49]   E. Feenberg, K. C. Hammack: Phys. Rev. **75** (1949) 1877
[Fe53]   E. Fermi: *Nuclear Physics,* 5th ed. (University of Chicago Press, Chicago 1953)
[Fe75]   G. J. Feldman, M. L. Perl: Phys. Rep. **19 C** (1975) 233
[Fe85]   R. P. Feynman: *QED − The Strange Theory of Light and Matter* (Princeton University Press, Princeton 1985)
[Fo58]   L. L. Foldy: Rev. Mod. Phys. **30** (1958) 471
[Fr82]   J. Friedrich, N. Vögler: Nucl. Phys. **A373** (1982) 192
[Fu90]   R. Fulton et al.: Phys. Rev. Lett. **64** (1990) 16
[Fu98]   Y. Fukuda et al.: Phys. Lett. **B436** (1998) 33;
          Phys. Rev. Lett. **81** (1998) 1562
[Ga66]   S. Gasiorowicz: *Elementary Particle Physics* (John Wiley & Sons, New York 1966)
[Ga81]   S. Gasiorowicz, J. L. Rosner: Am. J. Phys. **49** Nr. 10 (1981) 954
[Ge55]   M. Gell-Mann, A. Pais: Phys. Rev. **97** (1955) 1387
[Ge80]   M. Gell-Mann in: *The Nature of Matter,* Wolfson College Lectures 1980, ed. J. H. Mulvey (Clarendon Press, Oxford)
[Ge92]   C. Gerthsen, H. O. Kneser, H. Vogel: *Physik,* 20th ed. (Springer, Berlin, Heidelberg, New York 1999)
[Gi97]   L. K. Gibbons et al.: Phys. Rev. **D55** (1997) 6625
[Gl89]   P. Glässel et al.: Nucl. Phys. **A502** (1989) 315c
[Go51]   M. Goldhaber, A. W. Sunyar: Phys. Rev. **83** (1951) 906
[Go52]   M. Goldhaber, R. D. Hill: Rev. Mod. Phys. **24** (1952) 179
[Gö55]   M. Göppert-Mayer, J. H. D. Jensen: *Elementary Theory of Nuclear Shell Structure* (John Wiley & Sons, New York 1955)
[Go58]   M. Goldhaber et al.: Phys. Rev. **109** (1958) 1015
[Go79]   R. Golub et al.: Scientific American **240** (Juni 1979) 106

[Go84]   K. Gottfried, V. F. Weisskopf: *Concepts of Particle Physics,* Vol. 1, (Clarendon Press, Oxford, New York 1984)

[Go86]   K. Gottfried, V. F. Weisskopf: *Concepts of Particle Physics,* Vol. 2, (Clarendon Press, Oxford, New York 1986)

[Go94a] R. Golub, K. Lamoreaux: Phys. Rep. **237** (1994) 1

[Go94b] SLAC E139, R. G. Arnold et al.: Phys. Rev. Lett. **52** (1984) 727; korrigierte Daten: J. Gomez et al.: Phys. Rev. **D49** (1994) 4348

[Gr72]   V. N. Gribov, L. N. Lipatov: Sov. J. Nucl. Phys. **15** (1972) 438

[Gr87]   D. J. Griffiths: *Introduction to Elementary Particles,* (John Wiley & Sons, New York 1987)

[Gr91]   P. Große-Wiesmann: Cern Courier **31** (April 1991) 15

[Gr93]   C. Grupen: *Teilchendetektoren* (Bibliographisches Institut, Mannheim 1993)

[Ha48]   O. Haxel, J. H. D. Jensen, H. E. Suess: Die Naturwissensch. **35** (1948) 376; O. Haxel, J. H. D. Jensen, H. E. Suess: Z. Phys. **128** (1950) 295

[Ha49]   O. Haxel, J. H. D. Jensen, H. E. Suess: Phys. Rev. **75** (1949) 1766

[Ha73]   F. J. Hasert et al.: Phys. Lett. **B46** (1973) 138

[Ha94]   O. Haxel: Physikalische Blätter **50** (April 1994) 339

[Ha96]   W. Hampel et al.: Phys. Lett. **B388** (1996) 384

[He50]   G. Herzberg: *Spectra of Diatomic Molecules* (Van Nostrand, New York 1950)

[He77]   S. W. Herb et al.: Phys. Rev. Lett. **39** (1977) 252

[Hi64]   P. W. Higgs: Phys. Rev. Lett. **13** (1964) 508

[Ho46]   F. Hoyle: Mon. Not. Roy. Astr. Soc. **106** (1946) 343; F. Hoyle: Astrophys. J. Suppl. **1** (1954) 121

[Ho55]   F. Hoyle, M. Schwarzschild: Astrophys. J. Suppl. **2** (1955) 1

[Ho57]   R. Hofstadter: Ann. Rev. Nucl. Sci. **7** (1957) 231

[Hu57]   L. Hulthén, M. Sugawara: *Encyclopaedia of Physics,* **39** (Springer, Berlin 1957)

[Hu65]   E. B. Hughes et al.: Phys. Rev. **B139** (1965) 458

[In77]   W. R. Innes et al.: Phys. Rev. Lett. **39** (1977) 1240

[Kl52]   P. F. A. Klingenberg: Rev. Mod. Phys. **24** (1952) 63

[Kl92a] K. Kleinknecht: *Detektoren für Teilchenstrahlung,* 3rd ed. (Teubner, Stuttgart 1992)

[Kl92b] K. Kleinknecht: Comm. Nucl. Part. Sci. **20** (1992) 281

[Ko56]   H. Kopfermann: *Kernmomente* (Akademische Verlagsgesellschaft, Frankfurt a. M. 1956)

[Ko73]   M. Kobayashi, T. Maskawa: Prog. Theor. Phys. **49** (1973) 652

[Kö86]   K. Königsmann: Phys. Rep. **139** (1986) 243

[Ko88]   W. Korten: Dissertation, Heidelberg (1988)

[Kö89]   L. Köpke, N. Wermes: Phys. Rep. **174** (1989) 67

[Ko93]   W. Kossel, G. Möllenstedt: Physikalische Blätter **49** (Jan. 1993) 50

[Ko95]   S. Kopecky et al.: Phys. Rev. Lett. **74** (1995) 2427

[Kü88]   J. H. Kühn, P. M. Zerwas: Phys. Rep. **167** (1988) 321

[Ku89]   T. K. Kuo, J. Pantaleone: Rev. Mod. Phys. **61** (1989) 937

[Kw87]   W. Kwong, J. L. Rosner, C. Quigg: Ann. Rev. Part. Nucl. Sci. **37** (1987) 325

[La91]   J. Lach: Fermilab-Conf. (1991) 200; J. Lach: Fermilab-Conf. (1990) 238

[La95]   P. Langacker in: *Precision Tests of the Standard Electroweak Model,* ed. P. Langacker (World Scientific, Singapur 1995)

[Le68]   L. M. Lederman, M. J. Tannenbaum in: *Advances in Particle Physics,* eds. R. L. Cool, R. E. Marshak, Vol. 1 (Interscience, New York 1968)

[Le78]   C. M. Lederer, V. S. Shirley: *Table of Isotopes,* 7th ed. (John Wiley & Sons, New York 1978)

[Le94]   W. R. Leo: *Techniques for Nuclear and Particle Physics Experiments,* 2nd ed. (Springer, Heidelberg 1994)

[Li75]   L. N. Lipatov: Sov. J. Nucl. Phys. **20** (1975) 94

[Li97]   P.Liaud et al.: Nucl. Phys. **A612** (1997) 53

[Lo92]   E. Lohrmann: *Hochenergiephysik,* 4. Aufl. (Teubner, Stuttgart 1992)

[Ma63]   E. Marshalek, L. Person, R. Sheline: Rev. Mod. Phys. **35** (1963) 108

[Ma73]   G. Mairle, G. J. Wagner: Z. Phys. **258** (1973) 321

[Ma76]   R. D. Mattuck: *A guide to Feynman diagrams in the many-body problem,* 2nd ed. (McGraw-Hill, New York 1976)

[Ma89]   W. Mampe et al.: Phys. Rev. Lett. **63** (1989) 593

[Ma91]   W. J. Marciano: Ann. Rev. Nucl. Part. Sci. **41** (1991) 469

[Me33]   W. Meissner, R. Ochsenfeld: Die Naturwissensch. **21** Heft 44 (1933) 787

[Mo71]   E. J. Moniz et al.: Phys. Rev. Lett. **26** (1971) 445

[Na90]   O. Nachtmann: *Elementary Particle Physics: Concepts and Phenomena* (Springer, Berlin 1990)

[Ne91]   B. M. K. Nefkens: *Proc. of the Workshop on Meson Production, Interaction and Decay,* Kraków/Poland (World Scientific, Singapur 1991)

[Ne97]   B. Nemati et al.: Phys. Rev. **D55** (1997) 5273

[Ni55]   S. G. Nilsson: Mat. Fys. Medd. Dan. Vid. Selsk. **29** Nr. 16 (1955)

[No49]   L. W. Nordheim: Phys. Rev. **76** (1949) 1894

[Nu98]   *Proceedings of the XVIII International Conference on Neutrino Physics and Astrophysics,* Takayama, Japan, 4-9 June 1998, edited by Y. Suzuki and Y. Totsuka. To be published in Nucl. Phys. B (Proc. Suppl.)

[Pa89]   A. Paschos, U. Türke: Phys. Rep. **178** (1989) 145

[PD92]   *Review of Particle Properties*
          Particle Data Group, K. Hikasa et al.: Phys. Rev. **D45** (1992) S1

[PD94]   *Review of Particle Properties*
          Particle Data Group, L. Montanet et al.: Phys. Rev. **D50** (1994) 1173

[PD98]   *Review of Particle Properties*
          Particle Data Group, C. Caso et al.: Eur. Phys. J. **C3** (1998) 1

[Pe65]   A. A. Penzias, R. W. Wilson: Astrophys. J. **142** (1965) 419

[Pe75]   M. L. Perl et al.: Phys. Rev. Lett. **35** (1975) 1489

[Pe83]   P. W. Petley: Nature **303** (1983) 373

[Pe87]   D. H. Perkins: *Introduction to High Energy Physics,* 3rd. (Addison-Wesley, Wokingham 1987)

[Po81]   B. Povh: Prog. Part. Nucl. Phys. **5** (1981) 245

[Po95]   J. Pochodzalla et al.: Phys. Rev. Lett. **75** (1995) 1040

[Pr63]   M. A. Preston: *Physics of the Nucleus,* 2nd ed. (Addison-Wesley, Wokingham 1963)

[Ra86]   I. Ragnarsson, S. Åberg: Phys. Lett. **B180** (1986) 191

[Re59]   F. Reines, C. L. Cowan: Phys. Rev. **113** (1959) 273

[Ro50]   M. N. Rosenbluth: Phys. Rev. **79** (1950) 615

[Ro94]   M. Rosina, B. Povh: Nucl. Phys. **A572** (1994) 48

[Sa52]   E. E. Salpeter: Astrophys. J. **115** (1952) 326

[Sc35]   H. Schüler, T. Schmidt: Z. Phys. **94** (1935) 457

[Sc37]   T. Schmidt: Z. Phys. **106** (1937) 358

[Sc95]   F. Schwabl: *Quantuum Mechanics,* (Springer, Heidelber 1995)

[Sc95]   P. Schmüser: *Feynman-Graphen und Eichtheorien für Experimentalphysiker,* 2nd ed. (Springer, Heidelberg 1995)

[Se77]   E. Segrè: *Nuclei and Particles* (Benjamin, New York 1977)

384   References

[Se79]   R. Sexl, H. Sexl: *Weiße Zwerge – Schwarze Löcher*, 2nd ed. (Friedrich
         Vieweg & Sohn, Braunschweig 1979)
[Sh90]   J. Shapey-Schafer: Physics World **3** Nr. 9 (1990) 31
[Si79]   I. Sick et al.: Phys. Lett. **B88** (1979) 245
[Sm89]   J. M. Smith: *Did Darwin Get It Right?* (Chapman & Hall, New York,
         London 1989)
[So64]   A. A. Sokolov, J. M. Ternov: Sov. Phys. Dokl. **8** (1964) 1203
[St75]   SLAC E61, S. Stein et al.: Phys. Rev. **D12** (1975) 1884
[St88]   U. Straub et al.: Nucl. Phys. **A483** (1988) 686
[SY78]   *Symbols, Units and Nomenclature in Physics*
         S. U. N. Commission: Physica **93A** (1978) 1
[Ta67]   R. E. Taylor: *Proc. Int. Symp. on Electron and Photon Interactions at
         High Energies* (Stanford 1967)
[Te62]   V. A. Telegdi: Scientific American **206** (Jan. 1962) 50
[Tr92]   A. Trombini: Dissertation, Univ. Heidelberg (1992)
[Tw86]   P. J. Twin et al.: Phys. Rev. Lett. **57** (1986) 811
[We35]   C. F. von Weizsäcker: Z. Phys **96** (1935) 431
[Wh74]   R. R. Whitney et al.: Phys. Rev. **C9** (1974) 2230
[Wh92]   L. Whitlow et al.: Phys. Lett. **B282** (1992) 475
[Wi38]   G. C. Wick: Nature **142** (1938) 994 (abgedruckt in [Br65])
[Wi74]   D. H. Wilkinson, B. E. F. Macefield: Nucl. Phys. **A232** (1974) 58
[Wi78]   D. H. Wilkinson, A. Gallmann, D. E. Alburger: Phys. Rev. **C18** (1978)
         401
[Wi92]   K. Wille: *Physik der Teilchenbeschleuniger und Synchrotronstrahlungs-
         quellen* (Teubner, Stuttgart 1992)
[Wi93]   H. Wiedemann: *Particle Accelerator Physics I+II* 2nd ed. (Springer, Hei-
         delberg 1999)
[Wu57]   C. S. Wu et al.: Phys. Rev. **105** (1957) 1413.
[Yu35]   H. Yukawa: Proc. Phys. Math. Soc. Japan **17** (1935) 48

Index

Springer
and the
environment

At Springer we firmly believe that an
international science publisher has a
special obligation to the environment,
and our corporate policies consistently
reflect this conviction.
We also expect our business partners –
paper mills, printers, packaging
manufacturers, etc. – to commit
themselves to using materials and
production processes that do not harm
the environment. The paper in this
book is made from low- or no-chlorine
pulp and is acid free, in conformance
with international standards for paper
permanency.

Springer